Formation and Evolution of Low Mass Stars

NATO ASI Series

Advanced Science Institutes Series

A Series presenting the results of activities sponsored by the NATO Science Committee, which aims at the dissemination of advanced scientific and technological knowledge, with a view to strengthening links between scientific communities.

The Series is published by an international board of publishers in conjunction with the NATO Scientific Affairs Division

A	Life Sciences	Plenum Publishing Corporation
B	Physics	London and New York
C	Mathematical and Physical Sciences	Kluwer Academic Publishers Dordrecht, Boston and London
D	Behavioural and Social Sciences	
E	Applied Sciences	
F	Computer and Systems Sciences	Springer-Verlag
G	Ecological Sciences	Berlin, Heidelberg, New York, London,
H	Cell Biology	Paris and Tokyo

Formation and Evolution of Low Mass Stars

edited by

A. K. Dupree

Smithsonian Astrophysical Observatory,
Cambridge, MA, U.S.A.

and

M.T. V. T. Lago

Universidade do Porto,
Porto, Portugal

Kluwer Academic Publishers

Dordrecht / Boston / London

Published in cooperation with NATO Scientific Affairs Division

Proceedings of the NATO Advanced Study Institute on
Formation and Evolution of Low Mass Stars
Viano do Castelo, Portugal
September 21 – October 2, 1987

Library of Congress Cataloging in Publication Data

Formation and evolution of low mass stars : proceedings of the NATO
  advanced study institute held at Viana do Castelo, Portugal,
  September 21-October 2, 1987 / edited by A.K. Dupree and M.T.V.T.
  Lago.
       p.   cm. -- (NATO advanced study institute series. Series C,
  Mathematical and physical sciences ; 241)
     "Proceedings of the NATO Advanced Study Institute on Formation and
  Evolution of Low Mass Stars"--Introd.
     Includes indexes.
     ISBN 9027727821
     1. Cool stars--Congresses. 2. Stars--Formation--Congresses.
  3. Stars--Masses--Congresses.   I. Dupree, Andrea K.   II. Lago, M.
  T. V. T. (M. Teresa V. T.)   III. NATO Advanced Study Institute on
  Formation and Evolution of Low Mass Stars (1987 : Viana do Castelo,
  Portugal)   IV. Series.
  QB843.C6F67 1988
  523.8--dc19   88-12937
                                                              CIP

ISBN 90–277–2782–1

Published by Kluwer Academic Publishers,
P.O. Box 17, 3300 AA Dordrecht, The Netherlands.

Kluwer Academic Publishers incorporates the publishing programmes of
D. Reidel, Martinus Nijhoff, Dr W. Junk, and MTP Press.

Sold and distributed in the U.S.A. and Canada
by Kluwer Academic Publishers,
101 Philip Drive, Norwell, MA 02061, U.S.A.

In all other countries, sold and distributed
by Kluwer Academic Publishers Group,
P.O. Box 322, 3300 AH Dordrecht, The Netherlands.

Near Infrared Image of M17

Displayed in this figure is a near infrared image of one of the most active star formation sites in the Galaxy: the luminous H II region M17, the Omega Nebula. This image is a mosaic of more than 200 individual frames obtained with the near infrared array camera of the National Optical Astronomy Observatories. The individual frames are about 50×50 arc seconds on a side and conatin 58×62 pixels. These observations were obtained in the broad K band at 2.2 μm by Ian Gatley, Charles Lada, and Daren Depoy with the 2.1 meter telescope at Kitt Peak. The image clearly delineates the size and extent of the recently formed cluster of OB stars that energizes the H II region and the adjacent molecular cloud, which itself appears as a dark region devoid of sources. More than 1000 infrared sources have been found in the cluster and the vast majority are probably members. More than 75% of these sources were unknown prior to this observation. Since the K band filter also includes emission from the Brγ infrared recombination line, the image also reveals with unprecedented high angular resolution, the prominent ionization fronts at the interface between the molecular cloud and the H II region.

TABLE OF CONTENTS

IV. MAIN SEQUENCE EVOLUTION

V. FUTURE PROSPECTS FOR OBSERVATIONS

Near Infrared Image of the Reflection Nebula Elias 21

This figure shows a near infrared image in the K band (2.2 μm) of the spectacular reflection nebula Elias 21 (GSS 30). The image was taken by H. Zinnecker with IRCAM at the United Kingdom Infrared Telescope on Mauna Kea, Hawaii. The size of the array (58 $\times$ 62 pixels) corresponds to 35 $\times$ 38 arc seconds, with a scale of 0.6 arcsecond/pixel. The nebula is extended in the NE/SW direction, and a second source is revealed to lie towards the NE, about 15 arc second from from the peak of the main source. It is not known whether the second source is associated with the main nebula. The elongation of the nebula is probably due to scattering of 2μm photons off the dust in the bipolar lobes emanating from a protostellar disk.

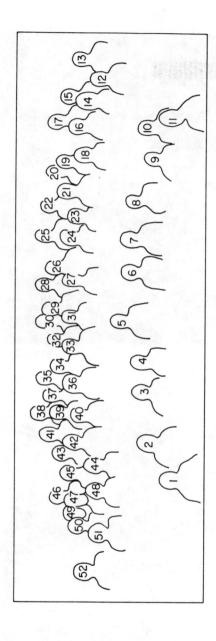

1. J.-P. Caillault 2. J. Emerson 3. H. Zinnecker 4. M. G. O. Fernandes 5. L. Hartmann 6. A. K. Dupree 7. T. Lago 8. J. E. Pringle 9. H. P. Singh 10. A. C. Cameron 11. T. Montmerle 12. D. Gough 13. G. Gahm 14. C. A. Torres 15. C. Chiuderi 16. A. Pascoal 17. M. Skrutskie 18. L. Ribeiro 19. J. Carneiro 20. P. Bodenheimer 21. B. Foing 22. A. Hjalmarsen 23. A. Garcia 24. B. Bach 25. P. Friberg 26 A. Burkert 27. C. E. Walker 28. M. Balluch 29. C. K. Walker 30. R. Liseau 31. S. Catalano 32. S. Kenyon 33. G. Basri 34. C. Lada 35. D. Soderblom 36. N. Calvet 37. J. Buj i Aris 38. W. J. Markiewicz 39. J. Bouvier 40. N. Kiziloglu 41. B. R. Petersen 42. I. Kucuk 43. J. Kuijpers 44. R. Celikel 45. E. Guinan 46. J.-P. Arcoragi 47. V. Keskin 48. S. Yorke 49. S. Baliunas 50. J. Stauffer 51. J. Hughes 52. C. Akan

LIST OF PARTICIPANTS

Mustafa Can Akan, *Ege Üniversitesi Fen Fakültesi, Bornova, Izmir, TURKEY*

Guillem Anglada, *Universidad de Barcelona, Barcelona, SPAIN*

Jean-Pierre Arcoragi, *Max-Planck-Institut für Astrophysik, Garching bei München, F. R. GERMANY*

Brigitte G. Bach, *Institut für Astronomie, Universität Wien, AUSTRIA*

Sallie L. Baliunas, *Harvard-Smithsonian Center for Astrophysics, Cambridge, MA USA*

Martin Balluch, *Institut für Astronomie, Universität Wien, AUSTRIA*

Gibor B. Basri, *University of California, Berkeley, CA USA*

Claude Bertout, *Institut d'Astrophysique, Paris, FRANCE*
CA USA

Peter Bodenheimer, *Max-Planck-Institut für Physik und Astrophysik, Garching bei München, F. R. GERMANY*

Roger Bonnet, *European Space Agency, Paris, FRANCE*

Jérome Bouvier, *Institut d'Astrophysique, Paris, FRANCE*

Dieter Breitschwerdt, *University of Manchester, Manchester, ENGLAND*

Jordi Buj i Arís, *Universidad de Barcelona, Barcelona, SPAIN*

Andreas Burkert, *Institut für Astronomie und Astrophysik der Universität München, München, F. R. GERMANY*

Jean-Pierre Caillault, *University of Georgia, Athens, GA USA*

Nuria Calvet, *Centro de Investigaciones de Astronomia, Merida, VENEZUELA*

Andrew C. Cameron, *University of Sussex, Brighton, ENGLAND*

Ruth Carballo, *Instituto de Astrofisica de Canarias, Tenerife, SPAIN*

Jorge Carneiro, *Observatório Astronómico Prof. Manuel de Barros, Vila Nova de Gaia, PORTUGAL*

John Carr, *University of Texas, Austin, TX USA*

Santo Catalano, *Citta Universitaria di Catania, Catania, ITALY*

F. Rikkat Celikel, *Middle East Technical University, Ankara, TURKEY*

Claudio Chiuderi, *Osservatorio Arcetri, Firenze, ITALY*

Andrea K. Dupree, *Harvard-Smithsonian Center for Astrophysics, Cambridge, MA USA*

James Emerson, *Queen Mary College, London, ENGLAND*

Robert Estalella, *Universidad de Barcelona, Barcelona, SPAIN*

Maria G. O. Fernandes, *Institut für Astronomie, Universität Wien, AUSTRIA*

Mario F. S. Ferreira, *Universidade do Aveiro, Aveiro, PORTUGAL*

Bernard H. Foing, *Laboratoire de Physique Stellaire et Planétaire, Verrieres-le-Buisson, FRANCE*

José Franco, *Instituto de Astronomia, Mexico D.F., MEXICO*

Apostolos Ch. Frangos, *Scientific Group for Space Research, Athens, GREECE*

Per Friberg, *Onsala Space Observatory, Onsala, SWEDEN*

Gösta Gahm, *Stockholm Observatory, Saltsjöbaden, SWEDEN*

Maria A. G. Garcia, *Universidade de Coimbra, Coimbra, PORTUGAL*

Douglas Gough, *Institute of Astronomy, Cambridge, ENGLAND*

Edward F. Guinan, *University of Villanova, Villanova, PA USA*

Patrick Hartigan, *Harvard-Smithsonian Center for Astrophysics, Cambridge, MA USA*

Lee Hartmann, *Harvard-Smithsonian Center for Astrophysics, Cambridge, MA USA*

Åke Hjalmarson, *Onsala Space Observatory, Onsala, SWEDEN*

Joanne Hughes, *Queen Mary College, London, ENGLAND*

Scott Kenyon, *Harvard-Smithsonian Center for Astrophysics, Cambridge, MA USA*

Varol Keskin, *Ege Üniversitesi Fen Fakültesi, Bornova, Izmir, TURKEY*

Nilgün Kiziloglu, *Middle East Technical University, Ankara, TURKEY*

Ibrahim Kücük, *Middle East Technical University, Ankara, TURKEY*

Jan Kuijpers, *Sonnenborgh Observatory, Utrecht, THE NETHERLANDS*

Charles Lada, *Steward Observatory, University of Arizona, Tucson, AZ USA*

M. Teresa V. T. Lago, *Universidade do Porto, Porto, PORTUGAL*

Carlos Lazaro, *Instituto de Astrofisica de Canarias, Tenerife, SPAIN*

René Liseau, *Istituto di Fisica dello Spazio Interplanetario, Frascati, ITALY*

Rosario López, *Universidad de Barcelona, Barcelona, SPAIN*

Wojtek J. Markiewicz, *Max-Planck-Institut für Kernphysik, Heidelberg, F. R. GERMANY*

Thierry Montmerle, *Section d'Astrophysique Centre d'Etudes Nucleaires Saclay, Gif-sur-Yvette, FRANCE*

Reinhard Mundt, *Max-Planck-Institut für Astronomie, Heidelberg, F. R. GERMANY*

Antonella Nota, *Space Telescope Science Institute, Baltimore, MD USA*

Antonio J. Pascoal, *Observatório Astronómico Prof. Manuel de Barros, Vila Nova de Gaia, PORTUGAL*

Bjorn R. Pettersen, *University of Oslo, Oslo, NORWAY*

James E. Pringle, *Institute of Astronomy, Cambridge, ENGLAND*

Victor Reglero, *Universidad de Valencia, Burjassot (Valencia) SPAIN*

Bo Reipurth, *European Southern Observatory, Garching bei München, F. R. GERMANY*

Lígia M. Ribeiro, *Observatório Astronómico Prof. Manuel de Barros, Vila Nova de Gaia, PORTUGAL*

Catarina C. Sá, *Universidade do Porto, Porto, PORTUGAL*

Frank Shu, *University of California, Berkeley, CA USA*

Harindar P. Singh, *University of Delhi, New Delhi, INDIA*

Michael Skrutskie, *University of Massachusetts, Amherst, MA USA*

David R. Soderblom, *Space Telescope Science Institute, Baltimore, MD USA*

John Stauffer, *Ames Research Center, National Aeronautics & Space Administration, Moffett Field, CA USA*

Carlos A. Torres, *Laboratorio Nacional de Astrofisica, Itajuba-(MG), BRASIL*

Christopher K. Walker, *Steward Observatory, University of Arizona, Tucson, AZ USA*

Constance E. Walker, *Steward Observatory, University of Arizona, Tucson, AZ USA*

Hans Zinnecker, *Max-Planck-Institut für Astrophysik, Garching bei München, F. R. GERMANY*

INTRODUCTION

This book represents the Proceedings of the NATO Advanced Study Institute on *Formation and Evolution of Low Mass Stars* held from 21 September to 2 October 1987 at Viana do Castelo, Portugal. Holding the meeting in Portugal recognized both the historical aspects and the bright future of astronomy in Portugal. In the early sixteenth century, the Portugese played an important role in the critical diffusion of classical and medieval knowledge which formed so large a part of scientific activity at that time. Navigation and course setting, brought to a high level by Portugese explorers, relied on mathematics and astronomy to produce precise tables of solar positions. In contemporary Portugal, astronomy is the focus of renewed interest and support at the universities. It is thus particularly appropriate that the NATO Advanced Study Institute was held on the coast of the Atlantic Ocean in the friendly surroundings of the Costa Verde.

The topic of this meeting represents a major challenge of contemporary astrophysics that only recently has become amenable to observation and quantitative anaysis. Sensitive photon-counting detectors now permit detailed spectral studies of young objects; powerful computer codes have been developed to treat the complicated physical conditions encountered in the atmospheres and winds of young stars. Earth-orbiting satellites reveal radiations in the X-ray, ultraviolet, and infrared regions - so crucial to the interpretation of the evolution of these objects. This NATO Advanced Study Institute gathered together both senior and young researchers to focus on the basic physical processes of stellar physics combined with new results from studies in the field of cool stars and star formation.

This book is divided into three major sections: *I. Formation of Low Mass Stars; II. Approach to the Main Sequence; III. Main Sequence Evolution* which mark the early evolutionary sequences of low mass stars. Throughout the volume, fundamental lectures on basic processes are introduced so that students will be prepared for the most advanced and recent material. We are particularly pleased to be able to include results from the recently developed infrared array imaging cameras that promise to contribute substantially towards our understanding of star forming regions. Future plans for both ground and space-based instruments and experiments are also described as envisioned in Europe and in the United States.

The Scientific Organizing Committee for the NATO Advanced Study Institute consisted of Roger Bonnet, Andrea Dupree (Scientific Director), Lee Hartmann, Teresa Lago, Jim Pringle, and Heinz Volk.

The smooth organization of the Institute was due in large part to Sara Yorke and Stephanie Deeley of the Smithsonian Astrophysical Observatory, and Martha Spiegel, and we appreciate their contributions.

The NATO Science Committee was very generous with its support in accepting and facilitating the Advanced Study Institute. We are also grateful to the following Institutions and Companies who supplied additional funds or services to make the meeting a most rewarding one: U. S. National Science Foundation, Smithsonian Astrophysical Observatory, Junta Nacional de Investigação Científica e Tecnológica, Instituto Nacional de Investigação Científica, Direcção Regional de Turismo do Alto Minho, Governo Civil e Câmara Municipal de Viana do Castelo, TAP-AIR Portugal, Instituto do Vinho do Porto, Rank Xerox de Portugal, Câmaras Municipais de Ponte de Lima e de Esposende, and Banco Português do Atlântico. Edward Hewett kindly contributed original drawings to enhance this publication.

The Editors thank also the lecturers in the NATO ASI for their comprehensive manuscripts, photographs, and many original contributions in music and rhyme which complemented the meeting and associated activities as well as this volume.

INTRODUCTORY BACKGROUND LECTURE

Nuria Calvet
Centro de Investigaciones de Astronomía (CIDA)
Ap.P. 264
Mérida 5101-A
Venezuela

ABSTRACT. A general introduction to the topics to be discussed in the NATO ASI "Formation and Evolution of Low Mass Stars" is given. Topics adressed are: differences between low and high mass stars on the main sequence, characteristics and origin of stellar activity, evolution of low mass stars in phases previous to the main sequence, molecular cloud cores, protostars, molecular outflows, properties of pre-main sequence stars.

I. PURPOSE

The purpose of this lecture is to cover in a systematic but necessarily superficial manner the topics to be discussed in this Institute, introducing the most generally used terms as well as giving the observational and theoretical background to the new interpretations to be presented. In this spirit, no explicit citations to authors will be given; rather, I will refer the reader to the chapters in this Volume where the appropiate references can be found.

II. LOW MASS STARS

II.1. Main sequence

The main sequence establishes the range of masses in which stable stars are found. Masses of stars in the main sequence go from roughly $0.01\ M_\odot$ to $100\ M_\odot$. The lower limit of the mass is set by either one of two conditions. This limit may be established by the condition that the contraction of an object of less that the limiting mass can be halted by degeneracy pressure, so that the central temperature never gets to the point of starting nuclear reactions. Alternatively, it may be determined by the lowest mass that can be formed by succesive fragmentation of a

1

A. K. Dupree and M. T. V. T. Lago (eds.), Formation and Evolution of Low Mass Stars, 1–19.
© *1988 by Kluwer Academic Publishers.*

cloud. The upper limit is determined by the fact that in a high mass object, temperatures are high enough for radiation pressure to exceed gravity, so that no stable star can be formed. Among these approximate limits we find the main sequence stars, stars that are obtaining energy from the nuclear fusion of hydrogen into helium in their centers.

The distribution of stars as a function of mass is far from uniform. The *initial mass function* $\xi(M)$ gives the relative number of stars formed, per unit mass interval in a unit volume as a function of mass M. Salpeter found that it can be approximated by

$$\xi(M) = C \ (M/M_\odot)^{-2.35} \tag{1}$$

where C is a constant. More recent work seems to indicate that the exponent is approximately -1.25 for $M < 1 \ M_\odot$, and -3.2 for $M > 3 M_\odot$. In any event, stars with lower masses are formed much more frequently than stars with higher masses.

II.2. Mechanisms of energy transport

Among stars on the main sequence, differences arise between low and high mass stars. One of these differences is in the main mechanism of energy transport in different regions of the star. Possible mechanisms for the transport of the energy generated in the internal regions in of the stars are radiation and convection. In the first, energy is transported by photons, which slowly drift to the surface following a random-walk path as they are absorbed and re-emitted by the atoms of the gas. The equation that describes the transport of photons in inner stellar regions is

$$F_\nu = - \ 4/3 \ (1/X)(dB_\nu/dT) \ (dT/dz) \tag{2}$$

where F_ν is the flux of photons, X the opacity of the material, which measures how much energy transported by photons is absorbed by the medium, T the temperature, and B_ν the Planck function. This equation is referred to as the *diffusion approximation*, in analogy with the heat conduction equation $\Phi = - \ \kappa \ \Delta T$. The term $4/3 \ (1/X)(dB_\nu/dT)$ is referred to as the radiative conductivity. Convection, that is the movement upwards of hot material and downwards of cool material, sets in whenever the temperature gradient required to transport energy by radiation, or radiative gradient, exceeds the adiabatic temperature gradient in the medium. The radiative gradient, as shown in equation (2), is large when either the opacity or the flux is large.

In low mass stars, the surface temperature is low, $T < 7000K$, so hydrogen is not ionized. As the temperature increases inwards hydrogen begins to ionize. Photons traveling outwards from the stellar interior are effectively trapped in the *hydrogen ionization zone*, where the opacity increases by orders of magnitude. In this zone, radiation is not an efficient mechanism to transport energy and convection sets in. The extent of the external convection zone is almost zero for a star of $M \simeq 2\ M_\odot$ and increases as the mass of the star, and therefore the effective temperature, decreases. In a star of $0.8\ M_\odot$, the outer convection zone depth is larger than 30% of the radius, and the lowest mass stars on the main sequence are completely convective.

II.3. Rotational velocity

Another difference between low and high mass stars in the main sequence is their surface rotational velocity. On the average, rotational velocity decreases from 100 to 200 km/s for masses higher than $3\ M_\odot$, to less than 5 km/s for masses lower than $1.5\ M_\odot$ (although see J. Stauffer's and D. Soderblom's chapters, this Volume). This break in rotational velocity has been traditionally related to the disappearance of the outer convection zone for high masses. As we will review later, convection zones are thought to be the site of wave generation. These waves, as they travel up towards layers with decreasing density, dissipate and heat them. At the expense of the energy carried by the waves, the outermost layers expand away from the star, carrying away angular momentum.

These differences suggest that the dividing line between low and high mass stars can be drawn around $3\ M_\odot$. Low mass stars have then spectral types later than AO in the main sequence. The corresponding effective temperatures and luminosities in the main sequence are $T_{ef} \leqslant 9000K$ and $L \leqslant 2\ L_\odot$. For historical reasons, the cooler end of this set is called *late type stars*. In the discussion that follows, we will refer to low mass stars, except stated otherwise.

III. STELLAR ACTIVITY

III.1. Atmosphere

The *atmosphere* of a star is the outer region of the star in which the photon mean free path is comparable to the height of the region. Photons can only escape from the atmosphere. If the atmospheres are in *radiative equilibrium*, that is, is energy is transported by photons only, then $dT/dr < 0$. In the dense inner region of the atmosphere called the *photosphere*, radiative equilibrium prevails. However, there is evidence of the existence of hot gas located above the photosphere. The outer hot region is separated into zones

according to the extent of the zone and the main cooling
mechanism. The *chromosphere* is the region just above the
photosphere, with densities between 10^9 and 10^{11} cm^{-3} and
temperature $\geqslant$ 6000K, where the important coolant is the
emission in the resonance lines of Ca II (H and K) and Mg II
(h and k). The most important lines emitted in this region
are these resonance lines. The extent of the chromosphere is
typically less than 0.1 stellar radii. The *corona* is a region
of extent of several stellar radii and low density, less than
10^9 cm^{-3}, with T $\geqslant$ 10^6K. Cooling in this region is
acomplished by the emission of highly ionized species of
metals, X ray emission, and by conduction to lower
temperature regions. The main emission from this region is X
ray free-free emission. The *transition region* is located
between the chromosphere and the corona. For the temperatures
and densities at the top of the chromosphere, the cooling law
indicates that any quantity of heating cannot be balanced by
any effective cooling, so the temperature increases quickly,
giving rise to the transition region, which ends when dT/dr $\simeq$
0 again as coronal cooling mechanisms become effective. The
main emission of the transition region comes from lines of
ionized metals, like C III, C IV, Si IV, O V, O VI, N V, with
characteristic temperatures of formation between 10^4K and 2 x
10^5K.

III.2. Stellar activity

The observations provided by the ultraviolet and X ray
satellites indicated that late type stars posses
chromospheres and coronae, as our sun does, since their
spectra showed emission lines of ionized metals and emission
in X rays. However, the amount of emission varied from star
to star, from degrees comparable to solar values to as high
as 10 to 100 times the solar. The term *stellar activity*
started to be widely used, as the quality measured by the
degree of emission in spectral indicators of the hot regions
above the photosphere. Hence, the UV and X ray observations
seemed to indicate that a large range of activity existed
among late type stars.

One problem was to find the source of energy for the
heating of the outer regions which could account for the
energy lost in the different emissions from these regions. As
mentioned before, photons could not account for it, so a
source of *non radiative heating* was sought. Early studies on
this problem in the particular case of the sun suggested that
the heating could be provided by *acoustic waves* generated in
the subphotospheric convection zone. This zone is highly
turbulent, and under this condition, acoustic waves are
generated with a power proportional to M^5, where M is the
Mach number. If these waves travel upwards with no energy

dissipation, then the wave power $F_w \simeq 1/2\ \rho(\delta v)^2 v_s \simeq$ const, where ρ is the density, δv the wave amplitude, and v_s the sound velocity. As ρ decreases, δv increases; when it becomes comparable to v_s, the waves become shocks and dissipate energy. Detailed calculations showed that although $\simeq 90\%$ of the energy transported by waves is radiated away, the remaining energy can produce the heating required to produce chromospheres, at least for the least active stars.

III.3. Magnetic fields

Besides of not being able to produce enough energy for the most active stars, there was a more fundamental problem with acoustic wave heating. According to stellar structure theory, stars with the same mass, chemical composition, and age should have similar structures, and in particular, similar convection regions. The output in acoustic waves should then be the same for similar stars, so that the spread in degree of emission could not be explained.

In the sun, there is a tight correlation between the magnetic field strength and the amount of emission in the Ca II K line. This observation suggested that magnetic fields could be acting as heating agents in active stars.

When magnetic fields are present, the turbulent convection zone produces *magnetohydrodynamic (MHD) waves*. The solution of the linearized MHD equations for a perfectly conducting fluid indicates that in a direction of propagation at an angle θ from the magnetic field direction, three types of waves can be found: an Alfvén wave, propagating with phase velocity $v_a \cos\theta$, and two vaves with phase velocity given by

$$v_{f,s} = (1/2[v_a{}^2 + v_s{}^2 \pm$$

$$\{(v_a{}^2 - v_s{}^2)^2 + 4v_a{}^2 v_s{}^2 \sin\theta^2\}^{1/2}])^{1/2} \qquad (3)$$

where v_s is the sound velocity and v_a is the Alfvén velocity, equal to $(B/4\pi\rho)^{1/2}$. The two signs correspond to the *fast* and to the *slow* waves. For propagation along the field direction, the fast and slow waves correspond to an Alfvén wave and to a sound wave, respectively, traveling with velocities v_a and v_s. For propagation perpendicular to the field direction, v_s = 0, and the fast wave corresponds to a *magnetosonic* wave, with phase velocity $(v_a{}^2+v_s{}^2)^{1/2}$, which is similar to a compressional wave along the propagation direction and to an electromagnetic wave in the plane perpendicular to this direction. Approximate calculations showed that in the presence of magnetic fields, MHD waves produced in the turbulent convection zones would carry energy with powers 10 to 100 times higher than acoustic waves.

III.4. Dynamo mechanism

It was also stated that the *dynamo mechanism* should be active in these stars to regenerate and amplify any magnetic field present. In late type stars, convection and rotation can interact to produce *differential rotation*. Differential rotation, in turn, interacts with cyclonic convection to mantain and amplify magnetic fields, according to the dynamo mechanism. In this mechanism, illustrated very schematically in Figure 1, a toroidal field line can be produced from a poloidal field line, as the latter enters the star and differentially rotates "frozen" to the material. The high conductivity of the material warrants that the field and material move together. On the other hand, a loop in a toroidal field line would be transported upwards by an ascending convective cell. These convective cells move to the surface in a cyclonic way, spiraling up in different sense depending on the hemisphere where they are located, because of Coriolis forces. The loops emerge on the surface oriented in the same direction; the fusion of many of these loops would create a poloidal field from the toroidal field.

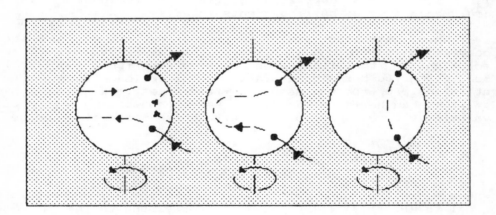

Figure 1a. Effect of distorsion on a poloidal field line introduced by differential rotation. In the Figure, rotational velocity increases inwards.

These effects can be calculated by the set of equations of the dynamo. From the equation

$$\partial\vec{B}/\partial t = \nabla \times (\vec{v} \times \vec{B}) + \eta \, \nabla^2\vec{B} \qquad (4)$$

where $\eta = c^2/4\pi\sigma$ is the magnetic diffusivity, using spherical

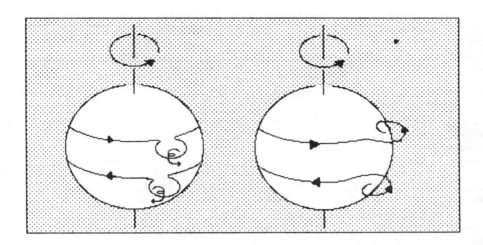

Figure 1b. Effect of cyclonic convection and Coriolis force on "loops" in toroidal field lines.

coordinates, the generation of the toroidal field $B_\varphi(r,\theta)$ from the rotational velocity field $v_\varphi = \omega(r,\theta)r\sin\theta$ and from the poloidal field described by B_θ and B_r, will be given by

$$[\partial/\partial t - \eta(\nabla^2 - 1/z^2)] B_\varphi = (B_\theta \partial\omega/\partial\theta + B_r r\partial\omega/\partial r)\sin\theta \quad (5)$$

where $z = r\sin\theta$. On the other hand, if the poloidal field is described in terms of a toroidal vector potential, $\vec{B}_p = \nabla \times \vec{e}_\varphi A$, then we can write from equation (4)

$$[\partial/\partial t - \eta(\nabla^2 - 1/z^2)]A = (\vec{v} \times \vec{B})_\varphi \quad (6)$$

The LHS term gives the mean rate of generation of the vector potential A, while the RHS gives the relation between the cyclonic convection and the field. The calculation of this term is difficult, among other reasons for the lack of an appropriate theory of turbulent convection. Often, this term is replaced by the term $\Gamma(r,\theta)B_\varphi(r,\theta)$, where the function $\Gamma(r,\theta)$ depends on the details of the convection and its calculation is highly approximate. In any event, these expressions show that the energy in the field comes from the energy in the velocity fields (rotational and convective). They also show that the "ingredients" for the dynamo mechanism to operate are differential rotation and convection.

According to the last considerations, magnetic fields could be present in late type stars, and enough energy could be transported to outer layers to produce the observed heating. The details of the transport and deposition of energy are still under discussion, however.

III.5. Observations of magnetic activity

Observations have confirmed these expectations. Direct measurements of magnetic fields in late type stars have been made, by measuring the distorsion in the magnetically sensitive absorption lines produced by Zeeman splitting. Although the accuracy of these determinations is not very well established, magnetic field strengths of the order of KG have been estimated. Indirect evidence of the action of magnetic field has also been found. Binary star systems of low mass stars show a variability in the energy emitted in broad wavelength bands which anticorrelates with the the strength of UV emission lines. This observation has been interpreted as indicative of the presence of *spots* and associated bright regions on the surface of the star. In analogy with the solar case, spots on the surface are supposed to be regions with high magnetic field strength, of the order of KG, where the magnetic field inhibits the energy transport by convection, resulting in a region cooler than the surroundings, and therefore less bright. As in the sun, spots are expected to be associated with bright regions, with enhanced emission in the resonance lines of Ca II and Mg II and in UV lines. A minimum of light in the broad band measurements would correspond to a measurement of the flux coming from the stellar surface when it is covered by cooler regions, namely spots. An enhancement of emission lines would then be expected, as observed. This type of correlation has been found for a large variety of late type stars (see S. Baliunas' chapter, this Volume).

In analogy to the solar case, the magnetic field is expected to be concentrated in tubes of roughly constant flux. The observed range in emssion properties has then been explained in terms of a variable surface coverage of magnetic field tubes. Observations of the presence of spots in stellar surfaces, and determinations of magnetic fields from the profiles of absorption lines seem to agree on the inhomogeneous nature of the distribution of magnetic fields.

The observed *rotation-activity relation* provides evidence in favor of the dynamo mechanism. It has been found that the X ray luminosity, and the emission in lines such as Ca II K and Mg II k correlates with the rotational velocity or the period of the stars. In simple terms, one would expect that larger rotation would induce a larger amplification

effect on magnetic fields, and then a larger energy
deposition in outer stellar regions.

The outer layers of late type stars are losing material
in the form of winds at a range of rates, from about 10^{-14}
$M_\odot$/yr in main sequence stars, to as high as 10^{-7} $M_\odot$/yr in
giants. In the solar-type stars, the wind is thermal, that
is, because of the high temperature of the corona the thermal
velocity becomes larger than the escape velocity. In some
cases, however, thermal winds are not enough to provide
observed mass loss rates. It has been proposed, specially for
low gravity giant late type stars, that Alfvén waves created
at the surface by distorsion of field lines, as they travel
upward, can deposit energy and momentum, inducing an
expansion of the outer layers.

IV. FORMATION AND EVOLUTION TO THE MAIN SEQUENCE

IV.1. Characteristic times

Stars, in general, form from the gravitational collapse
of fragments of molecular clouds in the interstellar medium.
In very schematic terms, the phases of evolution can be
divided according to the characteristic times in each phase.
When gravity overcomes all the cloud support mechanism, the
characteristic time is the *free-fall time*, given by

$$t_{ff} \simeq 4.7 \times 10^7/n_H^{1/2} \text{ years} \qquad (7)$$

where n_H is the hydrogen density. Collapsing objects in this
phase correspond to *protostars*. When the protostar collapses,
part of the gravitational energy goes into heating the
material and into producing the dissociation of the H
molecule and the H atom. When a temperature and a density
gradients are established, and when the density is high
enough for photons to be trapped by the object, then a
pressure gradient comparable to the gravity force is
established and the collase slows down. The star is in quasistatic equilibrium. The contraction, however, continues as it
is the main energy source for the star. In this phase, the
characteristic time is the *Kelvin time*

$$t_K \simeq 2 \times 10^7 \ M^2/RL \text{ years} \qquad (8)$$

where M, L, and R are the mass, luminosity, and radius of the
star in solar units. Contracting objects spend a much longer
time in this phase, so it is more likely to find them. As a
pre-main sequence star, the star evolves first keeping its
effective temperature approximately constant and decreasing
its luminosity, along the so called *Hayashi track*. The star
is completely convective in this phase. As the luminosity

decreases, a point is reached in which the radiative temperature gradient in inner regions, where the opacity is lower because H in the interior is completely ionized, becomes less than the adiabatic gradient. The core of the star, then, becomes radiative. The star evolves afterwards increasing its effective temperature with almost constant luminosity along the *radiative track*. The size of the radiative core increases as the star evolves, until only the outer convection zone is left when the star arrives in the main sequence. A star of 1 $M_\odot$ spends 9×10^6 years in the Hayashi track and 4×10^7 years in the radiative track until it finally settles on the main sequence. However, during the last 2.5×10^7 years, it would be observationally difficult to distinguish its luminosity and effective temperature from those of a main sequence star.

The contraction, with consecuent central temperature and density increase, continues until conditions are adecuate for nuclear fusion of hydrogen into helium. When the reactions are fully in operation the collapse stops, and the star enters the main sequence phase, where the loss of energy by radiation (and neutrinos) is balanced by the nuclear energy generation mechanism. The characteristic time in this phase of hydrogen burning is the *nuclear time*

$$t_n \simeq 2 \times 10^{10} \text{ M/L years} \qquad (9)$$

The star remains on the main sequence as long as it has nuclear hydrogen fuel. As the central hydrogen gets exhausted, the star evolves by increasing slightly its radius and luminosity and decreasing its surface temperature.

IV.2. Molecular clouds

Stars form from giant molecular clouds (GMC), which are complexes of interstellar molecular gas, with sizes of 50 to 100 pc and masses of 10^4 to 10^5 $M_\odot$, located in the plane of the galaxy. GMC have densities of approximately 15 $M_\odot/pc^3$ and are gravitationally confined.

Most stars form in associations. The OB associations contain massive stars, the O and B stars from which the name is taken, and also low mass stars. On the other hand, the T associations contain only low mass stars. Observed associations have space densities less than the critical density for a cluster to be stable against disruption due to galactic tides in the solar neighborhood, so they must be young, less than ten millions years old. Associations are gravitationally unbound, although they form from GMC, which are bound systems.

IV.3. Jeans Mass

If only gravity and thermal pressure are acting in a cloud, then there is a minimum mass, the *Jeans mass* M_J, given by

$$M_J = (\pi kT/mG)^{3/2} \, 1/\rho^{1/2}$$

$$\simeq 0.8 \; (T/10K)^{3/2} (n_H/10^6 cm^{-3})^{-1/2} \; M_\odot \qquad (10)$$

such that if the mass of the cloud exceeds it, it will collapse under its own gravitation. For conditions typical of molecular clouds, M_J is low, so that it is expected that they contain many subunits that can collapse independently. Once a cloud, under these conditions, starts to collapse, and if T decreases or stays constant, then as the density increases, M_J decreases. The collapsing cloud can break up into fragments of smaller mass, which can themselves collapse independently. On the other hand, as the cloud collapses, part of its potential energy is converted into heat. If there is no effective cooling mechanism, then the temperature will increase and the fragmentation process will stop at a certain minimum mass.

IV.4. Mechanisms of support

Thermal pressure is not the only means of balancing gravity in real clouds. Other mechanisms that have been claimed for supporting GMC against its self gravity are turbulent and magnetic pressures. Profiles of molecular lines, especially of CO, show a broadening which is much larger than that expected from thermal motions alone. The excess broadening has been atributed to turbulence, that is, to motions on scales comparable to the photon mean-free-path in the cloud. One of the explanations that has been given for this turbulence is the impact of winds of young objects on the molecular material surrounding them.

Another important mechanism that may provide support of the cloud are magnetic forces. Interstellar clouds are permeated by fields of order 1 – 100 μG. If a homogeneous field of strength B_0 exist in a cloud, and if the material is assumed frozen with the field, gravitational forces can overcome magnetic forces and initiate a gravitational collapse only if $M > M_c$, where

$$M_c = 0.15/G^{1/2} \; \Phi = 3.2 \times 10^4 \; B_0 (\mu G)^3 n_H^{-2} \; M_\odot \qquad (11)$$

where Φ is the magnetic flux through the cloud.

A process that can reduce the magnetic flux in a cloud

is *ambipolar diffusion*. The magnetic field acts only in the plasma of ions and electrons, and affects the neutral gas only indirectly through friction between the plasma and the neutral gas. As the neutral gas contracts, the ionized plasma and the field frozen with it drift outwards relative to it, with a drift velocity given by the balance between the magnetic force and the frictional force provided by collisions between the plasma and the neutral material. The diffusion time, a measure of the time required for the field to drift out of the cloud, is given by

$$t_D = 5 \times 10^{13} \, n_i/n_H \text{ years} \tag{12}$$

In the dense cores of molecular clouds, shielding of ionizing UV radiation can be effective, and the ionization fraction n_i/n_H can be very low, $\simeq 10^{-7}$ or less, so that ambipolar diffusion can be an effective mechanism for separating the bulk of the matter, which is in the neutral stage, from its means of support, the magnetic field through the ions. This mechanism could be important in the formation of low mass stars (see F. Shu's chapter, this Volume).

Observed molecular clouds are far from uniform. Dense cores are found in them which have been identified with the sites of star formation. These cores have densities of 10^4 to 10^5 particles per cm^3, temperatures of $\simeq 10K$, sizes of 0.05 to 0.2 pc and masses of 0.3 to 10 M_θ. Their characteristic free-fall time is of 1 to 4×10^5 yr. They are in close association, forming apparent clusters of diameters of $\simeq 2$ pc, with well known optical young stars, as well as with embedded infrared sources, also identified with the early stages of stellar evolution (see F. Shu's chapter, this Volume).

IV.5. Protostars

Once gravitational collapse starts, the contraction proceeds in a nonhomologous fashion, that is, the inner parts contract faster than the outer parts. From expression (7) one can see that the inner, denser parts will contract first than the outer parts. The inner parts will get to a condition close to hydrostatic equilibrium, while the outer envelope will still be accreting material into it. At the surface of the slowly contracting core a shock wave is developed, which produces enough UV radiation to heat the inner parts of the infalling envelope. However, this inner part is hidden from view by dust in the outer material. Only in the outermost parts, where the dust density is low enough, photons are allowed to escape. One can see only the *dust photosphere*, corresponding to the radius where the optical depth of the dust is around 1. Photons emitted from the dust photosphere would reflect the local low temperature, so the main emission of these objects would be in the infrared (Wien's law, $\lambda(\mu m)$ = 2.89 x 10^3/T). The object would then appear as an IR source (see C. Lada's and J. Emerson's chapters, this Volume).

The core is initially radiative. However, at some point, the central temperature becomes high enough to start the nuclear fusion of deuterium. The energy released by this reaction is not enough to alter significantly the contraction, but the radiative temperature gradient becomes high enough for convection to set in. The core then becomes convective.

It is to be expected that the collapsing cloud has an initial angular momentum. In fact, observed estimates of cloud angular momentum indicate that if the star contracted without losing angular momentum, then centrifugal forces would stop the contraction at some point; hence, the fact that stars exist implies that angular momentum must be lost in the collapse process.

The collapse of a rotating cloud would proceed faster in the direction parallel to the axis of rotation than in the perpendicular direction, where centrifugal forces would be active. The object would then collapse to a form a disk.

IV.6. Accretion disks

If the configuration being formed in the collapse process is such that the central concentration of mass is much higher than the mass distributed in the disk, then the subsequent evolution could be described by the theory of accretion disks around a central potential (see J. Pringle's chapter, this Volume).

If a mass of gas is rotating in a central potential, and if the time scale for any viscous processes (which redistribute angular momentum among gas elements) to occur is much longer than both the radiative and the orbital time scales, then the mass of gas can get rid of as much energy as possible, by collisions, shock heating, radiative cooling, while retaining angular momentum. Since the orbit of least energy for a given angular momentum is a circular one, then the gas settles down to moving on circular orbits in the form of a thin disk around the potential. In these circumstances, the central gravitational force is balanced by the centrifugal force at each radius r.

In general, the fluid rotates differentially. If there is any viscosity present in the gas, the energy of the shearing motions is dissipated as heat and radiated away. The energy lost by viscous processes can only be compensated by gravitational energy, so that the gas moves deeper into the potential well. However, because of conservation of angular momentum, part of the mass must move out. Then, dissipative processes tend to spread the disk out (see J. Pringle's chapter, this Volume).

Most of the solutions that have been carried out involve *steady disks*. In this case, we find that a constant rate of mass is accreting through the disk, given by

$$\dot{M} = 2\pi r \; \Sigma \; v_r \tag{13}$$

where v_r is the radial velocity, and Σ the surface density.

The circular velocity remains Keplerian $\Omega^2 = GM/r^3$, where M is the mass of the central object, all the way down to a narrow *boundary layer*, where the angular velocity varies rapidly from $(GM/R^3)^{1/2}$, where R is the radius of the central object, to Ω_0, the angular velocity of the stellar surface. The width of the boundary layer is much smaller than the stellar radius.

For a steady disk, the energy dissipation at radius r is given by

$$D(r) = 3GM\dot{M}/4\pi r^3 \; [1 - (R/r)^{1/2}] \tag{14}$$

independent of the viscosity law. The total disk luminosity is then

$$L_{disk} = 1/2 \; GM\dot{M}/R \tag{15}$$

This is only half of the total available accretion energy, since the total potential drop from infinite is $GM\dot{M}/R$. The other half of the available energy is emitted in the boundary layer.

The structure of a thin disk can be calculated assuming hydrostatic equilibrium in the vertical direction, that is, that the pressure gradient balances the component of gravity parallel to the axis of the disk. From this expression, to order of magnitude in thin disks, the disk thickness is given by $H \simeq r(v_s/v_\phi)$. The surface temperature at each radius r of the disk can be estimated by equating σT_{eff}^4 to $1/2 \; D(r)$. For typical accretion rates, effective temperatures are low, so the main disk emission will be in the infrared. The emission from accretion disks seems to account for the infrared emission of young objects (see chapters by L. Hartmann, S. Kenyon, C. Lada, F. Shu, this Volume). In the calculations done so far, the disk has been

assumed either flat or "flaring", that is, with a thickness
that increases with radius. Also, the energy emitted by the
disk has been considered either as coming from the
dissipation processes associated with accretion, or from
radiation energy emitted at the stellar surface which hits
the disk surface where it is *reprocessed* and emitted at the
local temperature.

Since one half of the potential energy of the
accretion disk is expected to dissipate in the narrow
boundary layer, this region must be hot. The boundary layer
temperature can be estimated by equating the emission of a
black body with temperature and area as those attributed to
the boundary layer to $1/2$ $GM\dot{M}/R$, where it has been assumed
that the boundary layer is optically thick. The boundary
layer temperature, then, depends on the layer radial
thickness, which in turn is an unknown parameter, so its
determination is uncertain, and so is its spectral emission
(see chapters by G. Basri and by S. Kenyon, this Volume).

IV.7. Outflows

Even if most of the accretion occurs in a disk, matter
is still infalling from all directions, occulting the stars.
However, objects as young as 10^4 years old are seen
optically, which means that the accretion must stop at an
early stage. The suggestion has been that, as the central
core becomes convective and since it is rotating
differentially (it is difficult to image how infalling matter
would accrete so that the core rotated as a solid body), then
seed magnetic fields would be amplified by the dynamo
mechanism. These magnetic fields, in turn, would induce an
expansion of the outer layers, at least at some latitudes
(see F. Shu's chapter, this Volume), which could eventually
stop accretion.

Observations indicate the presence of this expansion
of matter. Molecular line profiles, especially CO, observed
at millimeter and submillimeter waves in regions of active
star formation show an excess of flux in the wings of the
line, relative to the gaussian type profile generally
observed in clouds. This excess of flux has been interpreted
as an indication of the presence of material moving at
velocities larger than that of the average of the cloud.

The regions where this peculiar profiles are observed
are always located around infrared sources or other young
stellar objects, so that the high velocity molecular material
is interpreted as material in interaction with the wind of a
young object, from which it has obtained energy and momentum.
The material would be moving away from the central object,

that is it would be an *outflow*. In most cases, the excess flux is localized in confined regions with a common symmetry axis and such that the excess occurs in the blue wing of the line in one of the regions and in the red wing in the other region. In these cases, we talk about *bipolar outflows* (see C. Bertout's chapter, this Volume). In the case of bipolar outflows, the young stellar object is always near the center of symmetry of the system.

The observed motions could not be an infall. The binding mass for a test particle at radius r and velocity v would be given by $M_B = v^2 r/2G$. For an object such as L1551, where the extent of the perturbed region is about 0.5 pc and the maximum velocity is about 15 km/s, M_B should be 1.2×10^4 $M_\odot$, which is much higher than the mass of any embedded object or than the mass of the cloud interior to the flow.

Most observed outflows are associated with high mass stars. Among the low mass stars, outflows have been found among stars either with masses near 3 $M_\odot$ (L 1551 IRS 5, 2.7 $M_\odot$; T Tau, 2.7 $M_\odot$) or with lower masses but extremely young (HL Tau, 1 $M_\odot$, age 1×10^4 yr). The wind mass loss rate can be estimated from the energy and momentum transferred to the molecular material. For the low mass associated with outflows, $M \simeq 9 \times 10^{-9}$ to 4×10^{-7} $M_\odot$/yr.

In most of the observed outflow sources, the central object is surrounded by a region of high density which has been interpreted as a disk. This region is observed in molecules such as NH_3 and CS, rather than in CO. In the interstellar medium, to see an emission transition above the background radiation field, which is basically the isotropic background radiation of 2.7K, a mechanism must exist which populates effectively the upper level of the transition. One of the most important mechanisms for accomplishing this is collisional excitation. For each molecule, there exist a minimum density n_c required to excite a line significantly above the background, which is calculated by setting $n_{H2}\gamma = A$, where n_{H2} is the density of molecular hydrogen, with which collisions are more likely, γ the collisional cross section, and A the Einstein coefficient of the transition. For CO, $n_c = 4 \times 10^4$ cm^{-3}, for NH_3, $n_c = 1 \times 10^5$ cm^{-3}, and for CS, $n_c = 1 \times 10^6$ cm^{-3}. The regions observed in NH_3 and in CS are then denser than those observed in CO. As stated, emission from high density indicators such as NH_3 and CS is observed around outflow central objects. In some cases, the molecular profiles show a systematic velocity shift across the disk, interpreted as due to rotation of the disk.

Highly collimated regions with optical emission have been found associated with the molecular outflows; these regions have been interpreted as *jets* ejected from the stars. The spectrum of these regions is characteristic of a shock, where the strength of the forbidden lines of [O I], [O II], [O III], [S II], [N II] is high relative to the Balmer lines (see R. Mundt's chapter, this Volume).

In regions of active star formation, small, concentrated nebulae called *Herbig-Haro objects* are found. Originally, it was thought that they had a forming star inside. However, when their spectrum showed to be similar to that of a low velocity (50-140 km/s) shock, it was realized that they formed by the effect on the medium of the wind of young objects always found near them.

Surveys of the properties of the exciting stars to HH objects indicate that most of them are associated with objects with luminosities between 0.7 to 60 L_θ, which implies that they are associated with stars with the highest masses among the low mass range, or with very young low mass stars. When HH objects are found in the vecinity of a molecular outflow, they are physically associated with it, since the direction of their proper motions is similar to the direction of the outflow.

IV.8. Pre-main sequence stars

Theoretical calculations indicate that once the accretion process stops, the core settles down rapidly at the tip of the Hayashi track, and continues evolving as a pre-main sequence star. Although some of the details of this picture may change, observations show that stars exist in this part of the HR diagram, with peculiarities in agreement to what would be expected from the evolutionary picture described above.

T Tauri stars are stars in this phase of evolution, with ages between a few times 10^4 to 10^7 years. They are characterized by emission lines of H, metals once and twice ionized, UV line emission 10^4 times stronger than in the sun, and excess continuum flux above that expected for a star of the same spectral type in the blue, ultraviolet, and infrared. They also show in most cases an absorption spectrum characteristic of cool photospheres, $T_{eff} < 5500K$. In some cases, the absorption lines appear *veiled*, that is, the depth of the lines are less than those in the spectrum of a standard star with the same lines. In extreme cases, as in the *continuum stars*, no absorption lines appear. The observed veiling is such that the weak absorption lines are less affected than the strong ones. In the case of the continuum stars, broad emission lines appear superimposed on the stellar continuum (see chapters by G. Basri and by T. Lago, this Volume).

From their position in the HR diagram it is inferred that their mass is between 0.2 to 3 $M_\odot$, so they are progenitors of the main sequence low mass stars.

P Cygni profiles in the emission lines are taken as evidence of outflow of matter. Mass loss rates of the order of 10^{-9} to 10^{-8} $M_\odot$/yr have been inferred from theorical models for the expanding envelopes, as well as from studies of the molecular surroundings of the stars.

T Tauri stars show variability both in broad wavelength bands and in line strengths and profiles. In some cases, the light varies periodically, and such that the star becomes redder when fainter, indicative of the presence of spots in the surface (see G. Gahm's chapter, this Volume). Their X ray variability could be explained in terms of *flares*, which in analogy with the solar case, are dense regions on the surface where the release of magnetic energy occurs rapidly giving rise to enhanced emission (see chapter by T. Montmerle, this Volume). These evidences seem to point to the presence of strong magnetic fields with high coverage on the stellar surface. With these magnetic fields, Alfvén waves could originate at the surface and as they travel upwards they could deposite energy and angular momentum to initiate winds of the strength expected for T Tauri stars.

Models to explain the emission characteristics involve a chromosphere much denser than that in the sun, as well as an extended wind region where most of the emission in Hα arises. These models have been successful in explaining the order of magnitude of the emission. However, they cannot account for the observed IR excess; the emission of an accretion disk, either by reprocessing star light or by emitting accretion energy can, on the other hand, account for this excess. It seems then that stars that show IR excess, which in turn are the most active, have accretion disks around the central object (see chapters by G. Basri and S. Kenyon, this Volume).

Accretion disks could also explain the *FU Ori* phenomenon. This phenomenon involves the brightening of a former T Tauri star by two orders of magnitude in time scales of months to years. After the rise, the luminosity declines slowly. The spectrum of the object after the outburst is reminiscent of that of a low gravity star, although the spectral type becomes later with wavelength, that is, the lines correspond to a cooler emitting gas as the wavelength increases. These features can be explained if instead of a star, we are observing an accretion disk with a high M and therefore with a high accretion luminosity (see L. Hartmann's chapter, this Volume).

If accretion disks are present around T Tauri stars, then the observed blue and UV emission excesses could come either from the chromosphere-wind region or from the boundary layer. The question is not settled yet (see chapters by G. Basri and by S. Kenyon).

Not all stars in the pre-main sequence region of the HR diagram would be classified as T Tauri stars. Many stars in the same region, and then presumely with the same age, do not show signs of great activity in their atmospheres. Most of the very weak-line stars show emission in the Ca II K line (and probably in the Mg II k line), but little in other lines. They do not show IR excess either. This suggests that the difference between stars in the pre-main sequence phase comes from the presence of the accretion disk.

The indication is then that, schematically, two types of activity co-exist in different degree in pre-main sequence stars: a non-solar type, related to the presence of an accretion disk, an a solar type, with a stellar energy source and similar to that found in other type of late type stars. The detailed characteristics and the evolution of each type of activity remain to be determined.

INFRARED EMISSION PROCESSES

J. P. Emerson
Department of Physics
Queen Mary College
Mile End Road
London E1 4NS, England

ABSTRACT. This introductory tutorial contribution covers infrared emission processes involving dust grains in star forming regions. These grains produce emission, absorption, reddening, scattering, and polarisation effects which are described. The optics of dust grains are outlined together with current ideas about how these are determined by the composition, dielectric constants, size, structure and shape of the dust grains. Sources of suitable optical constants for grain materials are suggested. Fundamental constraints on the frequency variation of the far infrared absorption efficiency of grains are explained and used to deduce a simple expression for the equilibrium temperature of grains surrounding a star. The interpretation of optically thin infrared emission spectra to deduce dust masses and temperature is discussed, and the usefulness of sub-millimetre observations emphasised. Convenient methods of displaying dust emission spectra are discussed. Ways in which power law spectra may arise from thermal emission by dust are illustrated, including a centrally heated cloud with a density gradient, and an optically thick disk.

1. INTRODUCTION

With the availabilty of many "common user" instruments operating throughout the electromagnetic spectrum on various telescopes and satellites wavelength based astronomy is becoming increasingly rarer, as many problems are best approached by observations at many frequencies. Instrument technology of course distinguishes observations in various spectral regions, but physical processes are not so specific to wavelength regions. The infrared (henceforth IR) region, which I shall take to run from 1 to 1000 μm, is important in studies of star formation for two main reasons. Firstly because cool bodies, such as protostars, tend to produce most of their emission in the IR, (for example the flux per decade of frequency $\nu S(\nu)$ of a 367 K blackbody peaks at 10 μm and of a 37 K blackbody at 100 μm). Secondly because extinction by the dust grains, which are usually associated with regions of star formation, becomes much smaller in the IR than in the visible and ultraviolet where dust opacity is often $>> 1$, so that it is only in the IR that we can actually detect some radiation from these regions. But what processes are specifically IR ones?

Although line and continuum emission from neutral and ionised gas may be very usefully studied in the IR region of the spectrum the basic processes occuring

A. K. Dupree and M. T. V. T. Lago (eds.), Formation and Evolution of Low Mass Stars, 21–44.

are much the same as those producing line and continuum emission from gas at other wavelengths. What is almost unique to the IR region of the spectrum is the importance of dust in producing both continuum and band radiation. This is so even if the ultimate source of the energy in a region is a hot object, for example a star whose luminosity is predominately in the optical-UV region of the spectrum. If such a hot object is embedded in a dust cloud the high frequency photons will be absorbed and scattered by the dust grains, causing extinction of the light, the grains will be heated by the absorbed photons and will reach a temperature that will balance the absorbed energy by reradiating it as thermal emission at IR wavelengths. Thus the dust grains serve to reprocess the emitted photons from optical/UV to IR frequencies. It is to be expected that stars forming or recently formed will be embedded in such dust clouds and so, until these dust clouds dissipate, they will strongly affect the emitted radiation from these regions and so must be understood.

Therefore this Introductory contribution will concentrate on describing some of the basic physical ideas that must be taken into account to understand the role played by emission, absorption and scattering by dust grains associated with star formation regions, particularly in the far IR where observations are now becoming increasingly available. I will try to outline the important processes and indicate where more details of them may be found, so that readers inexperienced in dealing with IR data may learn how to do so. The material discussed is more fully treated in some of the monographs, reviews and conference proceedings on dust (e.g. Wickramsinghe(1967), Martin (1978), Bohren and Huffman (1983), Andriesse (1977), Huffmann (1977), Greenberg (1978), Savage and Mathis (1979), Stein and Soifer (1983), Wolstencroft and Greenberg (1984), Nuth and Stencel (1986), Tielens and Allamandola (1987)), as well as the other references given in the text.

Section 2 discusses the fundamental parameters which control the interaction between grains and electromagnetic radiation, and Section 3 treats observations of interstellar extinction and polarisation. Section 4 discusses evidence for the composition of dust that may be gleaned from various emission and absorption bands. Section 5 explains some fundamental constraints on the absorption efficiency of grains in the far IR, and its variation with frequency, and 6 points out some of the sources of detailed values of absorption efficiency. Section 7 shows how the equilibrium temperature of a dust grain depends on its distance from the source of radiative heating, and 8 discusses stochastic heating. Section 9 explains how observations of flux density at frequencies which are optically thin may be interpreted in terms of temperature and mass of dust, and of the luminosity of the embedded heating object. Section 10 discusses useful units for representing the energy distribution of IR emission and the location of the spectral peak. Section 11 shows how a power law spectrum may arise from a dust cloud with a range of temperature, either in an optically thin dust cloud with density gradient heated by a central star, or from an optically thick disc, and these results are tabulated, in various notations, in Section 12.

2. OPTICS OF DUST GRAINS

2.1. Basic ideas

When a beam of electromagnetic radiation is incident on a particle of geometrical cross sectional area A the radiation may be either absorbed or scattered. The absorbing effect is quantified in terms of the effective cross sectional area that the particle presents to the incident beam for absorption C_{abs}. Similarly the scattering effect is quantified by C_{sca}. These effective cross sectional areas can be larger than the geometrical cross sectional areas due to diffraction and interference effects, and are dependent on the frequency of the incident radiation. The ratio of the effective to geometrical cross sectional areas are known as absorption and scattering efficiencies Q where

$$Q_{abs} = \frac{C_{abs}}{A} \quad \text{and} \quad Q_{sca} = \frac{C_{sca}}{A}.$$

Note that, unlike most efficiencies, these can be > 1. The scattering is direction dependent and is not completely described by Q_{sca}, but we shall ignore this effect here.

To an observer looking along the beam absorption and scattering both reduce the intensity of the beam and the combined effect is called extinction. The extinction efficiency Q_{ext} is defined as

$$Q_{ext} = Q_{abs} + Q_{sca}.$$

An optically thin dust cloud of thickness l, containing spherical grains of radius a, with extinction efficiency Q_{ext}, and grain number density N will have an optical depth for extinction of

$$\tau = \pi a^2 Q_{ext} N l$$

and the intensity of the emergent radiation I will be related to that of the incident radiation I_0 by

$$I = I_0 e^{-\tau}.$$

Dust may emit thermal radiation as well as absorb radiation, but Kirchoff's Laws tell us that $Q_{em} = Q_{abs}$. The power, P_ν, radiated per unit frequency interval around frequency ν over all directions by a single dust grain of radius a, surface area $4\pi a^2$, absorption efficiency Q_{abs}, and temperature T is

$$P_\nu = 4\pi a^2 Q_{abs}(\nu)\pi B(\nu, T)$$

where $B(\nu, T)$ is the Planck function.

Both the above situations of extinction and emission, as well as scattering, are important in understanding IR radiation, and hence knowledge of the efficiencies Q and their frequency dependence is fundamental to understanding the role of dust.

2.2. Calculation of Q

The principle by which Q is calculated involves solving Maxwell's equations for the incident, scattered and absorbed beams with appropriate boundary conditions at the grain surface. The boundary conditions involve the frequency dependent complex dielectric constant of the grain material $\epsilon = \epsilon_1 + i\epsilon_2$ because the incident electro-magnetic wave induces polarisation in the grain material and this contributes to the resulting scattering and absorption. The efficiencies also depend on the particle shape and orientation with respect to the incident beam, as well as on the ratio of the particle size to the wavelength, λ, of the incident beam in the form of the "size parameter" $x = 2\pi a/\lambda$. For a particle of arbitrary shape this is a difficult problem but analytic solutions are available for the cases of particles that are spheres and for normally illuminated infinitely long cylinders. These solutions are in the form of infinite series in increasing powers of $2\pi a/\lambda$ (Bohren and Huffman 1983).

In the IR for $\lambda > 4\mu$m and typical interstellar grains sizes of 0.3μm, $2\pi a/\lambda \ll 1$ and so we may neglect all but the first term in these expansions to find

$$Q_{abs} = -4\frac{2\pi a}{\lambda} Im\left(\frac{\epsilon - 1}{\epsilon + 2}\right) \quad \text{and} \quad Q_{sca} = \frac{8}{3}\left(\frac{2\pi a}{\lambda}\right)^4 Re\left(\frac{(\epsilon - 1)^2}{(\epsilon + 2)^2}\right)$$

where Im and Re indicate the imaginary and real parts. Calculating Q is therefore relatively simple in the IR, compared to the UV and optical.

Note:
1) because $2\pi a/\lambda \ll 1$ in the IR, absorption is usually more important than scattering.
2) Q_{abs}/a is independent of a. The form in which many of the later equations will be writen will reflect this important fact.
3) Q_{abs}/a depends on λ both through the $1/\lambda$ term and the fact that the dielectric constant is a function of λ.

Because the complex dielectric constant of a material is related to the complex refractive index m by $\epsilon_1 + i\epsilon_2 = \epsilon = m^2 = (n + ik)^2$ knowledge of the dielectric constants or refractive indices are equivalent. Thus knowing either $\epsilon(\lambda)$ or $m(\lambda)$ for the grain material Q_{abs} and Q_{sca} can be calculated. A much more thorough discussion and listings of FORTRAN programs to calculate Q_{abs} and Q_{sca} for homogeneous spheres, coated spheres, and normally illuminated infinite cylinders, are contained in the monograph by Bohren and Huffmann (1983).

To decide on appropriate values to adopt for $\epsilon(\lambda)$ for interstellar grains one must decide what material the dust grains are composed of. I now discuss some observations that give clues about the dust size and composition, and which are also important in interpreting observations in the IR.

3. INFRARED EXTINCTION AND POLARISATION

Absorption and scattering will remove from the beam light that was heading for the observer and so cause extinction. As the amount of extinction is stronger at shorter wavelengths the spectrum of the beam is reddened. The extinction produced in magnitudes, A_λ, is related to the optical depth τ_λ by:

$$A_\lambda = 2.5 \, log_{10}e \, \tau_\lambda = 1.086 \, \tau_\lambda = 1.086 N \pi a^2 Q_{ext} l$$

To determine the intrinsic luminosity and energy distributions of objects this extinction must be accounted for, even though in the IR it is generally much less than at shorter wavelengths. This is usually done by determing from observations the differential extinction $A_{\lambda_1} - A_{\lambda_2}$ between two wavelengths λ_1 and λ_2 from the difference in the observed and intrinsic colours of an object. Although any pairs of λs will do the B (0.44 μm) and V (0.55 μm) photometric bands are often used where the colour excess E_{B-V} in magnitudes is defined as

$$E_{B-V} = A_{0.44} - A_{0.55} = (B - V)_{obs} - (B - V)_{intrinsic}$$

and A_λ is determined from the ratio A_λ/E_{B-V} given by an adopted extinction law. A value of $R = A_V/E_{B-V} = 3.1$ appears to be characteristic of the diffuse interstellar medium and is known as the ratio of total to selective extinction, R. Table I shows an average extinction law for the diffuse interstellar medium tabulated at wavelengths of frequently used photometric bands as given by Savage and Mathis (1979) with the addition of M and N from Rieke and Lebofsky (1985). Landini et al (1984) show that the IR extinction can be fitted by the approximate relation $A_\lambda \simeq 0.37 A_V \lambda^{-1.85}$, where λ is in μm, which agrees with the theoretical curve, known as van de Hulst curve number 15, for extinction by small ice grains (van de Hulst 1947) which is tabulated by Rieke and Lebofsky (1985).

Table I

Infrared Extinction Law

Band	λ/μm	A_λ/E_{B-V}	A_λ/A_V
	∞	0.0	0.0
N	10	0.16	0.052
M	5	0.07	0.023
L	3.4	0.16	0.058
K	2.2	0.38	0.112
J	1.25	0.87	0.281
I	0.9	1.50	0.482
R	0.7	2.32	0.748
V	0.55	3.10	1.000
B	0.44	4.10	1.324

If the shape of this reddening law is the same in all directions we would expect a constant value of R. However it is found that R can attain higher values than 3.1, particularly in the direction of some of the dark clouds in which low mass star formation is occurring, for example reaching ~ 4.2 in the ρ Ophiuchus dark cloud (Chini 1981). The increased value of R in these regions is thought to indicate the presence of larger dust grains than in the diffuse interstellar medium, perhaps due to accretion of volatile mantles onto the usual grain material or to coagulation of grains.

One should therefore beware of blind application of the diffuse ISM extinction law with $R = 3.1$ when trying to remove the effects of extinction in studying regions of low mass star formation, because if the true value of R is really higher than this one can underestimate the visual luminosities of the objects by large factors, causing one to assume that one is studying low mass stars rather than what are really high mass ones. Luckily observations in the IR, where extinction is relatively low, suffer less from this problem than visible wavelengths, but one must still be careful to consider what value of R is appropriate when deducing a luminosity from IR data.

Ultraviolet to optical extinction, and polarisation, in the diffuse interstellar medium have been successfully modelled by a power law size distribution of un-coated crystalline silicate and graphite grains, where the number of grains, $n(a)$, of radius a, $n(a) \propto a^{-3.5}$ for a from 0.005 to 1 μm for crystalline graphite and 0.025 to 0.25 μm for crystalline silicate (Mathis, Rumpl and Nordsieck 1977). This "MRN" model is also consistent with the available abundances of the heavier elements as any grain model must be. Other models, notably that of Hong and Greenberg

(1980), assume a trimodal distribution of grains size, and that some of the grains have volatile mantles, and can also fit the optical extinction curve but are less commonly adopted in the literature (Mathis 1985). Biermann and Harwit (1980) have argued that $n(a) \propto a^{-3.5}$ is expected from grain-grain collisions and Mathis and Wallenhorst (1981) have discussed the effect on the extinction law of variations of the MRN model grain radius cut-offs.

Observations of both linear and circular polarisation from light passing through interstellar material provide constraints on the size, shape and composition of the grain material. The observed correlation of polarisation with extinction for stars seen through the diffuse interstellar medium indicates that elongated grains have the direction of the spin axes aligned by the local magnetic field, and the direction of polarisation can be used to indicate the direction of the magnetic field, although the details of the alignment mechanism are not yet well enough understood to determine accurate field strengths.

In the IR the reduced extinction allows polarisation measurements of background stars seen through dark clouds and so makes possible the mapping of magnetic field direction across the dark cloud complexes that are the sites of low mass star formation. It is frequently found that the magnetic field direction is perpendicular to the major axis of the dark clouds, which are often elongated. This suggests that magnetic fields have supported the cloud material in one direction, and allowed collapse along the field lines in another (Moneti et al 1984).

Scattering, even off spherical dust grains, can also produce polarisation because of the transverse nature of electromagnetic radiation. In this case the normals to the polarisation vectors point back to the location of the scattered light source. For example the object L1551 NE discovered with IRAS by Emerson et al (1984) in the North East CO lobe of the outflow driven by L1551 IRS 5 can also be located by its scattered light (Draper et al 1985). Polarisation seen in the direction of bipolar outflows is thought to arise from scattering off anisotropic dust clouds surrounding the IR sources. The polarisation tends to be orthogonal to the local field direction, as determined from polarisation of background stars, suggesting that the magnetic fields are perpendicular to the disks that are thought to lie orthogonal to the outflow directions (Sato et al 1985).

4. DUST ABSORPTION AND EMISSION BANDS

The presence of absorption and emission bands in the IR give us direct spectroscopic clues to the identity of some of the grain material. Reviews by Aitken (1981), Willner (1984) and Tielens and Alamandola (1987) have covered this topic in detail. These features have, until relatively recently been mostly studied towards objects of high luminosity such as compact HII regions (eg Willner et al 1982) including the bright but highly extinguished ($A_V \sim 100$) W33A (Soifer et al 1979, Geballe et al 1985).

4.1. Silicate bands

A broad unstructured feature of full width at half maximum (FHWM) $\sim 2 - 3\mu m$ and peaking around 9.7 μm is attributed to Si-O stretching in silicate materials and is referred to as the 10 μm silicate feature. It is seen in absorption towards highly extinguished objects, in emission in the Trapezium region of the Orion Nebula (Gillett et al 1975), and also in emission or absorption around late type oxygen rich stars with circumstellar dust shells. The extinction across the silicate absorption band arising from the interstellar medium (ISM) is tabulated by Rieke and Lebofsky (1985) and the ratio of visual extinction A_V to the silicate optical depth is $A_V/\tau_{sil} \sim 16.6 \pm 2.1$. The shape of the astronomical feature is different in different environments. For instance the ISM silicate absorption feature width is broader than that in emission in O rich late-type stars, and the wavelength at which τ_{sil} peaks differs between the emission features found in O rich late-type stars and absorption in the ISM (Pegourie and Papoular 1985). The 10 μm feature has been studied in T Tauri stars by Cohen (1980) and Cohen and Witteborn (1985), and for eleven low mass protostars and pre main sequence objects by Myers et al (1987). Cohen and Witteborn (1985) point out that in T Taus the feature is sometimes in emission, sometimes in absorption, and that it is sometimes narrower than the Trapezium feature. The strength of the silicate feature is an important parameter in the procedures by which Adams and Shu (1986) fit their accreting, rotating, protostellar disk and envelope models that seem successful in reproducing the overall spectral shape of these type of objects. There is also a less frequently studied silicate feature at 18 μm attributed to Si-O-Si bending.

Laboratory studies of candidate silicate materials are hard to carry out in conditions of temperature that approximate those in the interstellar medium, but they show that crystalline materials have considerable structure across the 10 μm band in contrast to the rather smooth observed shape in astronomical sources. It has therefore been suggested that the silicate particles are amorphous, and laboratory studies show that amorphous silicates have much less structure (see Donn 1985 for a summary), although there is still no one material studied in the laboratory that can completely reproduce the observed astronomical feature.

There is also another broad band in the 10 μm region at 11.8 μm attributed to SiC but this is seen only around late-type Carbon stars and will not be further discussed here.

4.2. Ice bands

An absorption feature at 3.07 μm and with a FWHM $\sim 0.3\mu$m is found in the spectra of objects embedded in, or seen through, dark molecular clouds and is mainly due to water ice. This feature is asymmetrical often having a long wavelength wing extending to $\sim 3.5\mu$m, attributed to hydrocarbon ices. A short wavelength wing has also been observed and attributed to ammonia ice, but as these observations require an airborne observatory the frequency of this phenomenon is not well determined. At one time it was thought that ice was only found in the immediate vicinity of deeply embedded young IR sources but more extensive studies have shown that ice absorption may be found in general lines of sight to background stars seen through the Taurus dark molecular cloud (Whittet et al 1983), indicating that ice is not confined to protostellar regions but can be found in ambient dark molecular cloud material, although there seems to be an extinction threshold ($A_V \sim 3$ in Taurus but apparently much higher in Ophiuchus and Orion) before ice becomes detectable. $A_V/\tau_{ice} \sim 10$ seems to hold above the threshold for the stars seen through the Taurus cloud. As ice is a volatile material it is not surprising that the ice feature has never been seen in emission.

The literature does not contain many positive detections of ice features towards low mass stars, with the deep feature seen towards HL Tau (Cohen 1975) being a notable exception, but ice absorption has recently been detected (Emerson et al 1988) towards many of the deeply embedded low mass protostellar objects studied by Beichman et al (1986) and Myers et al (1987). These data can be used to determine the column density of dust near these objects that is too cool to lose its ice mantles (T $< \sim$ 100 K), and hence to test some aspects of models (Adams and Shu 1986) of their structure.

Detailed modelling of the shape of the ice feature has proved complex (van de Bult et al 1985) but it seems that amorphous, rather than crystalline, ice is required to explain the smoothness of the feature (Leger et al 1983), and that the long wavelength wing is related to the mixture of hydrocarbon or NH_3 ices with H_2O ice, and to the range of grain sizes present.

Absorption features due to ice are also found at 6.0 and 42 μm, but are not acessible to ground based telescopes. The predicted 12.5μm ice feature has only been seen in two objects, neither of which are low mass pre main sequence objects. This is partly due to the difficulty of disentangling this feature from the much broader silicate absorption bands found in the same objects, and partly because the feature probably becomes broadened and suppressed when volatile materials condense onto the silicate cores in molecular clouds (Aitken and Roche 1984).

4.3. Other features

Absorption features due to various volatile materials are seen towards deeply embedded objects (W33A being a favourite in which to search) which also show ice absorption. At present these have told us more about grain temperature and composition than about the environment of low mass stars, but there is clearly potential to use these absorption features as probes of the cold outer envelopes of young accreting protostars. The features seen include NH_3 ice at 2.95 μm (Knacke et al 1982), H_2S at 3.9 μm (Geballe et al 1985), CN ice at 4.62 μm and CO ice at 4.67 μm (Lacy et al 1984), S bearing ice at 4.9 μm (Geballe et al 1985), as well as others not yet clearly identified.

Other emission features have been found in environments where there is lots of ultraviolet radiation present, but not yet in low mass star formation regions. These features at 3.28, 3.40, 3.43, 3.53, 6.20, 7.7, 8.6 and 11.3 μm were known as the "unidentified features", but have been recently proposed to mostly arise from fluorescence of PAHs (Polycyclic Aromatic Hydrocarbons) which will be mentioned again in section 8 below.

In the ultraviolet region the 220 nm extinction feature is attributed to vibrations in the basal plane of crystalline graphite grains of size $\sim 0.03 \mu$m. This feature cannot be produced by amorphous carbon and arises in the diffuse ISM.

Thus there is direct spectroscopic evidence for the presence of graphite and silicate grains in the diffuse interstellar medium, and that in dark cloud regions there are also grains of water ice and other volatile materials. It seems likely that these volatiles condense out in these regions onto the more ubiquitous graphite and silicate cores, leading to a core-mantle structure. This picture also explains why the ratio of total to selective extinction R tends to be greater in some dark clouds. The silicate and ice have amorphous, rather than crystalline, structure.

5. GRAINS IN THE FAR INFRARED.

It turns out that in the far IR region of the spectrum one can make some firm predictions of how the dust absorption efficiency for grain materials should depend on frequency. These predictions use only basic physics and are therefore independent of the exact details of the grain's composition and must be obeyed by any grain material.

The first such relationship involves the Kramers Kronig dispersion relationships which relate the real and imaginary parts of the dielectric constant of a material through an integral over all frequencies. They are derived in standard electromagnetism texts (eg Panofsky and Phillips 1969) and by Andriesse (1977) and are *based only on the fundamental assumption of causality*. These Kramers Kronig relations relationships (first derived in 1927) are completely independent of any specific model of the grain material, or of its shape or of whether it is crystalline or amorphous. They *must* be obeyed by any plausible grain material. The Kramers Kronig relation that is used in this derivation is

$$\epsilon_1(\omega) = 1 + \frac{2}{\pi} \int_0^\infty \frac{x \epsilon_2(x)}{x^2 + \omega^2} dx$$

where ϵ_1 is the real and ϵ_2 the imaginary part of the dielectric constant and ω is $2\pi\nu$, where ν is the frequency.

This relationship can be used (Purcell 1969) to show that

$$\int_0^\infty \frac{Q_{abs}}{a} \frac{d\nu}{\nu^2} \leq \frac{4\pi^2}{c}$$

where a is the grain radius and c the velocity of light. Now if we assume a power law dependence of Q_{abs}/a on ν with exponent n, $Q_{abs}/a = Q_0\nu^n$, where Q_0 is independent of a, and integrate between the limits ν' (some low IR frequency) and 0 we have

$$\frac{Q_0\nu'^{n-1}}{n-1} < \frac{4\pi^2}{c}$$

As $\nu' \to 0$ the left hand side of the inequality will only be finite, and hence consistent with the inequality, if n > 1. Whatever the behaviour of Q_{abs} at high frequencies a behaviour $Q \propto \nu^n$ in the far IR requires n > 1 to be consistent with causality!

Thus for any grain material Q_{abs}/a must decrease faster than ν^1 as $\nu \to 0$ in the far IR to be consistent with causality.

The second relationship also concerns how Q_{abs}/a varies with frequency as $\nu \to 0$ and applies to crystalline grains only. Although amorphous grains fit the observed IR 3.1 μm ice and 10 μm silicate features much better than do crystalline grains it has been common in the literature to consider crystalline grains when modelling dust. This proves to be quite successful in modelling the visual extinction and polarisation, and of course the evidence from the 220 nm graphite feature is that the graphite grains are crystalline. Additionally it is easier to physically understand crystalline grains as they have a regular structure the consequences of which those astronomers who were brought up as physicists studied in solid state physics (see eg Kittel 1968). We recall that in the IR

$$Q_{abs} = -4\frac{2\pi a}{\lambda} Im \frac{(m^2-1)}{(m^2+2)}$$

as $\epsilon = m^2$. In a crystalline material the atoms behave as Lorentz ocillators in their lattice and electromagnetism (eg Panofsky and Phillips, 1969) and optics (eg

Longhurst, 1967) texts use this fact to derive the Clausius-Mosotti equation for the frequency dependence of the complex index of refraction

$$\frac{(m^2 - 1)}{(m^2 + 2)} = A + \frac{B}{(\nu_0^2 - \nu^2) + i\nu\gamma}$$

where ν_0 is the oscillator frequency A and B are constants and γ is the damping constant. Taking the imaginary part of the Clausius-Mosotti equation and substituting into the equation for Q_{abs} we find

$$Q_{abs} = \frac{-8\pi a}{\lambda} \frac{-\nu\gamma B}{(\nu_0^2 - \nu^2)^2 + (\nu\gamma)^2}$$

$\nu_0^2 >> \nu^2$ as $\nu \to 0$ so

$$\frac{Q_{abs}}{a} \to \frac{8\pi\gamma B}{c\nu_0^4}\nu^2$$

and we expect a *quadratic* dependence of Q_{abs}/a on frequency for *crystalline* materials at frequencies well below that of the last resonance, at frequency ν_0, in the grain material.

Tielens and Allamandola's (1987) interesting discussion of the far IR absorption of amorphous material, is briefly summarised here. Amorphous material is expected to show the same quadratic frequency dependence as crystalline, although for a completely different physical reason, namely that it is determined by the density of vibrational states. An exception to this law occurs for amorphous *layered* materials, such as amorphous carbon and layer-lattice silicates where the structure limits the phonons to two dimensions which changes the far IR frequency dependence of Q_{abs} from ν^2 to ν^1. If there are still some cross linking bonds between the layers in the material an intermediate frequency dependence will arise. Also for very small grains the material will behave as if 2 dimensional, again leading to a $Q_{abs} \propto \nu^1$.

Thus in the far infared, as $\nu \to 0$, $Q_{abs} \propto \nu^n$ is expected on theoretical grounds with the exponent, n, of the frequency dependence lieing between 1 and 2, its precise value depending on the structure of the grain material.

6. SOURCES OF Q_{abs}

In interpretation of IR emission, absorption and scattering by dust it is necessary to adopt values for the complex dielectric constant ϵ from which absolute values of Q may be calculated. Because the composition and structure of grain material is not well known many different assumptions have been made by many different authors, which can lead to confusion when comparing the results of these interpretations. There are two approaches to determing ϵ, and hence Q, for grain materials in the IR, either to deduce it from the astronomical spectra in a modelling process, or to measure it directly for candidate grain materials in the laboratory. The first method ensures that we are dealing with the right material but with the difficulty that one does not really know the state and physical conditions in the regions modelled, and therefore has to make assumptions. The difficulty with the laboratory method is that it is difficult to do the measurements under conditions appropriate to those encountered astrophysically, and even more difficult to know that one is measuring the right material.

Many of the laboratory measurements are referenced and discussed by Donn (1985), Hasegawa and Koike (1984) and Bohren and Huffmann (1983). These confirm the theoretical expectations about frequency dependence of Q_{abs} discussed above.

In addition to the direct spectroscopic astronomical evidence suggesting that silicate and ice grains are amorphous theoretical considerations also suggest that grains forming under circum- and interstellar conditions should be amorphous (Leger et al 1979, Seki and Hasegawa 1981) rather than crystalline.

Historically the greatest effort has gone into modelling the UV and optical extinction, and the model for this that has been most used in recent years is the MRN model of Mathis et al (1977) which uses a power law size distribution of crystalline graphite and silicate grains (see above). Building on this model Draine and Lee (1984) reviewed physical constraints and experimental results on crystalline graphite and silicate grains. They then calculated values of $\epsilon_1(\nu)$ from adopted values of $\epsilon_2(\nu)$ to ensure consistency with the Kramers Kronig relations. For silicate they synthesised ϵ to produce the observed Trapezium and interstellar medium "astronomical silicate" feature. They then use these optical constants to calculate $Q(\nu)$ in a self consistent manner. This work probably remains the most thorough and self consistent source of values of $Q(\nu)$ and reproduces many observations well. The resulting complex dielectric functions and values of Q_{abs} for graphite and "astronomical silicate" for $a = 0.01\mu$m and 0.1μm are tabulated between 20nm and 2000 μm by Draine (1985).

These are probably the best values to adopt but there is still a difficulty in the far IR where Draine and Lee's grains have $Q_{abs} \propto \nu^2$ for both graphite and silicate, whereas in many astronomical sources a flatter frequency dependence is found, indicating that layered lattice grains are present. Radiative transfer calculations (Rowan-Robinson 1986) confirm that the astronomical spectra require a shallower far IR opacity law.

The Draine and Lee (1984) model applies to the diffuse interstellar medium, but there is evidence (see above) that in the denser regions of molecular clouds icy mantles condense out onto the silicate and graphite cores. Lee and Draine (1985) describe the effect of adding ice mantles to the core grains of Draine and Lee (1984) to model the ice feature and polarisation in the BN object between 2 and 15 μm. Again they use a Kramers Kronig analysis to synthesise dielectric constants that reproduce the observed 3.1 μm ice band (and the lack of 12 μm band). They do not tabulate their values of ϵ or Q_{abs}, but show them graphicaly between 2 and 15 μm. For $\lambda > 15\mu$m Q_{abs} for ice should be computed using the refractive indices, m, tabulated by Leger et al (1983) out to 80 μm, and the core-mantle grain FORTRAN program listed in Bohren and Huffmann (1983).

Further improvements in our knowledge of Q_{abs} for interstellar grain materials can be expected as laboratory, theoretical and observational work continues.

7. EQUILIBRIUM TEMPERATURE OF DUST AROUND A STAR

Now that we know that Q_{abs}/a is likely to have a power law frequency dependence in the far IR we can calculate grain emission in the IR. We consider the temperature that grains around a star are likely to reach. This is important both in modelling IR emission from circumstellar shells and in determining if grains will vaporise, retain ice mantles etc. The dust destruction radius is, for example, an important parameter in models of collapsing protostars (Stahler et al 1981).

Consider the energy balance of a single dust grain of radius a which is in equilibrium, with grain temperature T, with the radiation field produced at distance R from a star of specific luminosity $L(\nu)$. The rate of heating by absorption of high frequency photons (UV and optical from starlight) absorbed by the grains cross sectional area will be balanced by the rate of cooling by blackbody thermal emission with efficiency Q_{abs} from the grain's surface area. ($Q_{abs} = 1$ for a true blackbody).

$$\int_0^\infty \pi a^2 Q_{abs}(\nu) \frac{L(\nu)}{4\pi R^2} d\nu = \int_0^\infty 4\pi a^2 Q_{abs}(\nu) \pi B(\nu, T) d\nu$$

The frequencies contributing most to the left hand side of this equation are at optical and UV wavelengths where most stellar energy is radiated. Q_{abs} is roughly constant in this region (we shall call its value $Q_{abs}(opt)$) and so it may be taken outside the integral. For typical circumstances grains reach temperatures of a few 10s or 100s of degrees Kelvin and thus the frequencies contributing most to the right hand side of the equation are in the far IR where $Q_{abs}(\nu)/a = Q_0 \nu^n$ with Q_0 a constant depending on the grain material, *but not its radius*. Substituting in the expression for the Planck function we have

$$\pi a^2 Q_{abs}(opt) \frac{L_*}{4\pi R^2} = \int_0^\infty 4\pi a^3 Q_0 \nu^n \frac{2\pi h\nu^3}{c^2(e^{h\nu/kT} - 1)} d\nu$$

where $L_* = \int_0^\infty L(\nu) d\nu$ is the stars luminosity. Note that the dependence of the right hand side on grain radius is through the grain volume ($\propto a^3$) as Q_0 is independent of a. To do this, and any integral involving the Planck function, it is neccesary to make the change of variable $x = h\nu/kT$ so that $\nu = (kT/h)x$ and $d\nu = (kT/h)dx$. With these changes of variable we have

$$\pi a^2 Q_{abs}(opt) \frac{L_*}{4\pi R^2} = \int_0^\infty 4\pi a^3 Q_0 (kT/h)^n x^n \frac{2\pi h(kT/h)^3 x^3}{c^2(e^x - 1)} (kT/h) dx$$

which simplifies to

$$\frac{Q_{abs}(opt)L_*}{4\pi R^2} = \frac{8\pi a h Q_0}{c^2}(kT/h)^{n+4} \int_0^\infty \frac{x^{n+3}}{(e^x - 1)} dx.$$

Now $\int_0^\infty \frac{x^{n+3}}{(e^x-1)} dx$ is a standard integral with value $(3+n)! \sum_{p=1}^\infty p^{-(4+n)}$ where $\sum_{p=1}^\infty p^{-(4+n)}$ lies within 8 % of 1 for all $n \geq 0$. Thus

$$\frac{L_*}{R^2 a} = \frac{32\pi^2 h Q_0 (3+n)!}{Q_{abs}(opt)c^2}(kT/h)^{4+n} \sum_{p=1}^\infty p^{-(4+n)}$$

so that grain temperature, T falls off with distance from the star as $T \propto R^{\frac{-2}{4+n}}$ and is also different for different grain sizes a. Thus for $Q_{abs} \propto \nu^1, (n = 1)$, we have $T \propto R^{-0.4}$ or for $Q_{abs} \propto \nu^2, (n = 2)$, we have $T \propto R^{-0.33}$.

We shall use this result several times in what follows. If the above approximations about Q_{abs} at optical frequencies seem over simplistic one can use more exact values of Q_{abs} and the Planck mean absorption coefficient

$$< Q > (T) = \int_0^\infty Q_{abs}(\nu)\pi B(\nu, T) d\nu / \int_0^\infty \pi B(\nu, T) d\nu$$

with T the stellar effective temperature, and similarly for far IR frequencies with T the grain temperature. Draine and Lee (1984) show $< Q > /a$ for their grain models as a function of T. Note that we have neglected any effects of the grains attenuating the star light, of scattering, and of the grains heating each other by IR radiation. These assumptions are justified when the optical depth of the dust is $<< 1$ at optical/UV and IR wavelengths, but in general a radiative transfer calculation needs to be performed (eg Scoville and Kwan 1976, Rowan-Robinson 1980, Wolfire and Cassinelli 1986, Crawford and Rowan-Robinson 1986, Adams and Shu 1986). In what follows we shall continue to assume, for simplicity, optical thinness.

8. NONEQUILIBRIUM DUST TEMPERATURES, SMALL GRAINS AND PAHS

In the preceding section we did not explicitly consider the energy quantisation of the photons absorbed and emitted by the dust, but rather assumed steady heating and cooling. This is a good approximation if the heat capacity of a grain is $>>$ the energy of individual arriving photons, so that the grain temperature remains constant at its equilibrium temperature. If, however, the heat capacity of a grain is $\simeq$ or $<$ the energy of individual arriving photons, the grain temperature will fluctuate wildly. A very small grain will have a very small heat capacity, and will become very hot on absorption of an optical/UV photon, gradually cooling as IR photons are emitted. Thus for most of the time such small grains will be cool, but sometimes they will be very hot. This stochastic effect will lead to a non black body spectrum, with a colour temperature that increases towards shorter IR wavelengths.

This process has been recognised by many authors over the past years (eg Purcell 1976) but it was not until 1984 that observations indicated that it was actually taking place when Sellgren (1984) interpreted near IR continuum emission in reflection nebulae in this way, and IRAS discovered extensive warm cirrus emission in L255 with 12 to 25 μm colour temperatures much higher than the 60 to 100 μm colour temperatures of the same region, and much too high to be explained by equilibrium heating from the nearby starlight (Beichman et al 1984). Grains of radius $a = 0.003\mu m$ are required to explain the colour temperatures found for cirrus, and this process has been modelled in detail by Draine and Anderson (1985) and Weiland et al (1986).

But when grains get this small they do not have very many atoms in them so do they perhaps behave more like molecules ? This leads us to consider PAHs (PolyAromatic Hydrocarbons) which have been reviewed by Allamandola, Tielens and Barker (1987) and Duley (1986), and discussed in many papers in the PAH workshop proceedings (Leger et al 1987).

Duley and Williams (1981) pointed out that various groups on carbon rings would tend to produce IR emission bands and Leger and Puget (1984) expanded on this idea to give plausible identifications of the "unidentified" IR emission features (eg 3.3 and 3.4 μm C-H stretch, 6.2 μm C-C stretch, 8.6 μm C-H in plane bend, 11.3 μm out of plane C-H bend). These PAH molecules become excited by absorption of individual photons, but produce band emission, rather than a continuum as with dust grains. Puget et al (1985) predict the emitted spectra of PAHs in various radiation fields and Barker et al (1987) have succeeded in using PAHs to explain quantitatively as well as qualitatively the detailed emission band structure found in the 3μm region. PAHs look increasingly promising as an explanation for the "unidentified" IR emission features mentioned in section 4. These PAHs might also explain the high IRAS 12-25 μm colour temperatures mentioned above. IR spectroscopy of the cirrus is required to check this possibility, but this is observationally extremely difficult as cirrus is extended emission, which is discriminated against by the beamswitching techniques use on all IR telescopes, except the now extinct IRAS!

Emisson features associated with PAHs have not yet been detected in low mass pre main sequence stars, but high colour temperature extended 12 and 25 μm emission around such objects in ρ Ophiuchus suggests that small grains or PAHs may be present (Young et al 1986).

9. DUST COLOUR TEMPERATURE, LUMINOSITY AND MASS

In general dust colour temperature is best calculated by interpolation in tables of flux density ratio against temperature. In the Rayleigh-Jeans region the spectral slope is independent of temperature. In the Wien region (eg at near IR wavelengths with cold dust) we have $\frac{S(\nu_1)}{S(\nu_2)} = (\frac{\nu_1}{\nu_2})^{n+3} e^{h(\nu_2 - \nu_1)/kT}$ for $Q_{abs}/a = Q_0 \nu^n$ so that

$$T = \frac{h(\nu_2 - \nu_1)/k}{ln\frac{S(\nu_1)}{S(\nu_2)} - (n+3)ln\frac{\nu_1}{\nu_2}} \quad \text{or} \quad T = \frac{hc(1/\lambda_2 - 1/\lambda_1)/k}{ln\frac{S(\nu_1)}{S(\nu_2)} + (n+3)ln\frac{\lambda_1}{\lambda_2}}$$

in terms of frequency and wavelength respectively. Note that to determine a temperature we need to assume a value for n, the exponent of the frequency variation in $Q_{abs}/a \propto \nu^n$, or to know the form of Q_{abs}/a more precisely.

If the dust cloud is optically thin the power per unit frequency interval radiated over all directions by N_d grains is

$$L(\nu) = N_d 4\pi a^2 Q_{abs}(\nu)\pi B(\nu, T)$$

so the observed flux density a distance D away will be

$$S(\nu) = N_d \frac{4\pi a^2}{4\pi D^2} Q_{abs}(\nu)\pi B(\nu, T)$$

or multiplying both sides by $\frac{4}{3}a\rho$ where ρ is the solid density of the grain material

$$\frac{4}{3}a\rho S(\nu) = N_d \frac{4\pi a^3}{3}\rho Q_{abs}(\nu)\frac{B(\nu, T)}{D^2}$$

these N_d grains will have a mass $M_d = N_d\frac{4}{3}\pi a^3 \rho$ so

$$M_{dust} = \frac{4a\rho}{3Q_{abs}(\nu)}\left(\frac{S(\nu)D^2}{B(\nu, T)}\right)$$

Note that since Q_{abs}/a is independent of a in the regime $2\pi a/\lambda << 1$, even though a appears in the above equation *no assumption about the dust grain size is needed in estimating the dust mass.*

Thus the procedure to determine a dust mass is to measure $S(\nu)$, determine grain temperature from colour temperature, assume a distance D and then, if

$\frac{4a\rho}{3Q_{abs}(\nu)}$ is known, the dust mass may be calculated. As the gas to dust ratio by mass is probably fairly constant (≈ 200) this procedure allows us to estimate the mass of all the material, both gas and dust.

Hildebrand (1983) discusses the empirical determination of $\frac{4a\rho}{3Q_{abs}(\nu)}$ from observational data and recommends a value of 0.1 gm cm^{-2} at 250 μm, corresponding to $Q_{abs}/a = 40$ cm^{-1}, $\rho = 3$ gm cm^{-3} and $n = 2$ beyond 250μm.

At submillimetre wavelengths (350, 450, 750, 850 and 1000 μm) $2\pi a/\lambda << 1$ and the emission is surely optically thin by these wavelengths so that submillimeter observations are the best way to determine the dust mass and dust distribution in dust clouds, and presumably this will also trace the gas distribution in the same clouds. Also, unless the dust is very cold, we are in the Rayleigh Jeans region at sub-mm wavelengths so that $M_d \propto \frac{S(\nu)}{T}$ and so is relatively insensitive to the uncertainty with which T is known. Submillimetre flux densities are also useful in defining the turnover frequency of spectra and in extrapolating to infinite wavelength to determine bolometric luminosities. Dust masses determined from sub-mm observations in this way will be one of the important products of the new sub-mm telescopes coming into operation on dry sites (eg the 15m James Clerk Maxwell Telescope, JCMT, on Mauna Kea).

10. SPECTRA OF SINGLE TEMPERATURE DUST EMISSION.

10.1. Format of display of IR spectra

Observational results are usually quoted in terms of a flux density (flux per unit frequency interval) $S(\nu)$ in Janskies (1 Jy = 10^{-26} Watts m^{-2} Hz^{-1}) or in terms of flux per unit wavelength interval $S(\lambda)$ in units of Watts cm^{-2} μm^{-1}, however these two forms look different when plotted and neither indicate immediately where most energy is actually radiated. Also the shapes of blackbody curves in such plots depend on the temperatures of the blackbodies. It has therefore become popular amongst many authors to use $\nu S(\nu)$ as a unit in which to represent spectra, and we adopt this convention in what follows. This quantity has units of Watts m^{-2} decade^{-1} where decade means a range of a factor of ten in frequency (= a factor of 1 in logν) centred on the observed frequency. If we had tall enough pieces of graph paper it would be most useful to plot $\nu S(\nu)$ against $log\nu$ since that would give an obvious indication of where most of the energy was since

$$\int \nu S(\nu)dlog\nu = \int \nu S(\nu)log_{10}e \, dlog_e\nu = \int \nu S(\nu)log_{10}e\frac{d\nu}{\nu} = 0.43\int S(\nu)d\nu$$

which is just $log_{10}e\times$ the flux over the range of integration, and so such plots would give a direct visual indication of where energy is radiated.

However the use of a logarithmic vertical axis is necessary in practise and is extremely convenient as, for example, in plots of $log\nu S(\nu)$ against $log\nu$ blackbody curves at any temperature have the same shape. The virtues of such plots have been discussed by Disney and Sparks (1980), although their use in IR work seems to have originated with E. P. Ney and colleagues at Minnesota.

For those that like to see wavelength rather than frequency as the horizontal scale plots of $log(\lambda S(\lambda))$ against $log\lambda$ may be preferred but the vertical scales will, happily, be the same as $\lambda S(\lambda) = \nu S(\nu)$ and these plot have similar properties to those just discussed.

It is straightforward to calculate equivalents to the Wien displacement law (eg $\lambda T = 3669.7 \mu$m K and $\nu/T = 8.17 \times 10^{10}$ Hz K^{-1}) for such representations of flux and to use them in conjunction with such plots to deduce temperatures etc. Theoreticians tend to prefer plots against $log\nu$ and observers against $log\lambda$ but by the simple expedient of putting both horizontal axes on the graph (perhaps at the upper and lower edges) everyone can be satisfied and a physically useful representation of spectral distributions be produced!

10.2. Single dust temperature spectrum

Consider the spectrum produced by isothermal dust grains of temperature T emitting as modified blackbodies with $Q_{abs}/a = Q_0\nu^n$ then the spectral shape will be given by

$$I(\nu) = Q_0\nu^n B(\nu, T) = \frac{2Q_0 h}{c^2} \frac{\nu^{3+n}}{e^{\frac{h\nu}{kT}} - 1}$$

to find where such a spectrum peaks we make the change of variable $x = \frac{h\nu}{kT}$ so that

$$I(\nu) = \frac{2Q_0 h}{c^2} \left(\frac{kT}{h}\right)^{3+n} \frac{x^{3+n}}{e^x - 1}$$

and differentiate both sides with respect to x and set $\frac{dI(\nu)}{dx} = 0$

$$0 = \frac{2Q_0 h}{c^2} \left(\frac{kT}{h}\right)^{3+n} \left(\frac{(3+n)x^{2+n}}{e^x - 1} - \frac{x^{3+n}e^x}{(e^x - 1)^2}\right)$$

which simplifies to
$$x = (3+n)(1 - e^{-x}).$$

For $n = 0$, a greybody, $x = 2.82$ which leads to the well known relationships for the peak in $I(\nu)$ $\lambda T = 5099.6 \mu$m K or $\nu/T = 5.88 \times 10^{10}$ Hz K^{-1}. As $B(\lambda) = \frac{\nu^2}{c}B(\nu)$ one can deduce the position of the peak of $B(\lambda)$ by putting $n = 2, Q_0 = 1/a$ into the above equation whence we find $x = 4.965$ giving $\lambda T = 2897.9 \mu$m K or $\nu/T = 10.34 \times 10^{10}$ Hz K^{-1} the first of these being Wien's displacement law. Similarly to find the peak in $\nu B(\nu)$ we put $n = 1, Q_0 = 1$, and find the peak at $x = 3.93$ giving $\lambda T = 3669.7 \mu$m K or $\nu/T = 8.17 \times 10^{10}$ Hz K^{-1}. Note that for the values of interest to us $x > 2.8$ and so $1 - e^{-x}$ does not differ much from 1. This is borne out by the fact that the exact values of x are within 6 % of those that we would deduce from the approximate relation $x = 3 + n$ for $n \geq 0$. Thus we give the approximate (to better than 6%) position of the peak of the spectrum, in the preferred form $\nu S(\nu)$, for blackbody emission modified by grains with $Q_{abs}/a = Q_0\nu^n$ as

$$\lambda T = 3669.7 \left(\frac{4}{4+n}\right) \mu\text{m K} \quad\text{or}\quad \frac{\nu}{T} = 8.17 \times 10^{10} \left(\frac{4+n}{4}\right) \text{Hz K}^{-1}$$

which are quite useful in interpreting spectra, although they are approximate and $x = (3+n)(1 - e^{-x})$ should really be solved (numerically via trial values of x) for greater accuracy at a given n. Thus the grain temperature may be estimated from the peak of the $\nu S(\nu)$ spectrum if the opacity power law is known. These useful relationships, and other similar ones, are tabulated in Table II.

Table II

Spectral peak for thermal emission from dust at temperature T

peak of	blackbody ×	wavelength	frequency
		λ T (μm K)	ν/T (Hz K^{-1})
$S(\nu)$	$Q_{abs} = $ const	5099.6	5.88×10^{10}
$S(\nu)$	$Q_{abs}/a \propto \nu^n$	$\simeq 5099.6 \left(\frac{3}{3+n}\right)$	$\simeq 5.88 \times 10^{10} \left(\frac{3+n}{3}\right)$
$S(\lambda)$	$Q_{abs} = $ const	2897.9	10.34×10^{10}
$S(\lambda)$	$Q_{abs}/a \propto \nu^n$	$\simeq 2897.9 \left(\frac{5}{5+n}\right)$	$\simeq 10.34 \times 10^{10} \left(\frac{5+n}{5}\right)$
$\nu S(\nu)$ and $\lambda S(\lambda)$	$Q_{abs} = $ const	3669.7	8.17×10^{10}
$\nu S(\nu)$ and $\lambda S(\lambda)$	$Q_{abs}/a \propto \nu$	2897.9	10.34×10^{10}
$\nu S(\nu)$ and $\lambda S(\lambda)$	$Q_{abs}/a \propto \nu^{1.5}$	2632.7	11.39×10^{10}
$\nu S(\nu)$ and $\lambda S(\lambda)$	$Q_{abs}/a \propto \nu^2$	2409.7	12.44×10^{10}
$\nu S(\nu)$ and $\lambda S(\lambda)$	$Q_{abs}/a \propto \nu^n$	$\simeq 3669.7 \left(\frac{4}{4+n}\right)$	$\simeq 8.17 \times 10^{10} \left(\frac{4+n}{4}\right)$

11. SPECTRUM FROM DUST WITH A RANGE OF TEMPERATURE

In the above we have referred to the temperature of individual dust grains. In most real situations all dust grains are not isothermal and so the emitted spectrum is not a blackbody modified by Q_{abs} but rather a superposition of such emission for a range of temperatures. (We will still assume a single grain type, although more realistically we would expect a range of dust types).

In section 7 it was shown that the temperature of a dust grain in equilibrium with a radiation field depended on the grain size as well as the distance from the heating star. Mezger et al (1982) have discussed how such a size dependence will affect the observed spectra (qualitatively the range of grain temperatures will broaden the spectrum) and conclude that, for the MRN power law grain size distribution, the resulting spectral shape is close to that of an isothermal cloud with a temperature corresponding to that of the average sized grains. We therefore do not treat this effect further here.

We consider the emitted spectrum in two situations where there is a range of temperature, firstly a spherical, optically thin dust cloud, with a density gradient, heated by a central star, and secondly an optically thick dust disc with specified temperature variation between its inner and outer boundaries. In both these cases the resulting spectrum is a power law within a certain frequency range, the power law exponent depending on the temperature variation and, for the optically thin case, on the density variation and the dust emissivity law.

11.1. Optically thin dust cloud with density gradient, heated by a central star

We saw above that for N_d grains at temperature T

$$S(\nu) = N_d \frac{\pi a^2}{D^2} Q_{abs}(\nu) B(\nu, T)$$

We represent the volume density of grains at distance R from the central star as $n(R)$ and their temperature as $T(R)$. The number of grains in a spherical shell between radius R and $R + dR$ is $n(R)4\pi R^2 dR$ so that the total flux density at frequency ν from grains at all distances R is

$$S(\nu) = \frac{4\pi^2 a^3}{D^2} \frac{Q_{abs}(\nu)}{a} \int_{R_1}^{R_2} n(R)R^2 \frac{2h\nu^3}{c^2(e^{h\nu/kT(R)} - 1)} dR$$

where R_1 and R_2 are the inner and outer radii of the dust shell. We assume a power law variation of grain density $n(R) = n_1(R/R_1)^{-q}$ (for example $q = 3/2$ in free fall collapse Adams and Shu (1985)) and a power law variation of grain temperature $T(R) = T_1(R/R_1)^{-p}$ where n_1 and T_1 are the grain density and temperature at the inner shell radius R_1. To carry out this integral over R we need to change the variable to $x = h\nu/kT(R) = (h\nu/kT_1)(R/R_1)^p$ so that

$$R = R_1 \left(\frac{kT_1 x}{h\nu}\right)^{\frac{1}{p}} \quad \text{and} \quad dR = R_1 \left(\frac{kT_1 x}{h\nu}\right)^{\frac{1}{p}} \left(\frac{1}{p}\right) x^{-1} dx$$

substituting for n(R), R and dR in terms of x we have

$$S(\nu) = \frac{4\pi^2 a^3}{D^2} \frac{Q_{abs}(\nu)}{a} \int_{x_1}^{x_2} n_1 \left(\frac{kT_1 x}{h\nu}\right)^{\frac{-q}{p}} R_1^2 \left(\frac{kT_1 x}{h\nu}\right)^{\frac{2}{p}} \frac{2h\nu^3 R_1}{c^2(e^x - 1)} \left(\frac{kT_1 x}{h\nu}\right)^{\frac{1}{p}} \left(\frac{1}{p}\right) x^{-1} dx$$

where $x_1 = h\nu/kT(R_1)$ and $x_2 = h\nu/kT(R_2)$ are the values of x at the inner and outer radii of the shell R_1 and R_2. Collecting like terms

$$S(\nu) = \frac{4\pi^2 a^3}{D^2} \frac{Q_{abs}(\nu)}{a} \int_{x_1}^{x_2} n_1 \left(\frac{kT_1}{h\nu}\right)^{\frac{-q+2+1}{p}} x^{\frac{-q+2+1-p}{p}} R_1^3 \frac{2h\nu^3}{c^2(e^x - 1)} \left(\frac{1}{p}\right) dx$$

which simplifies to

$$S(\nu) = \frac{8\pi^2 h n_1 a^3 R_1^3}{D^2 c^2 p} \left(\frac{kT_1}{h}\right)^{\frac{-q+3}{p}} \frac{Q_{abs}(\nu)}{a} \nu^{\frac{q-3}{p}+3} \int_{x_1}^{x_2} \frac{x^{\frac{-q+3-p}{p}}}{e^x - 1} dx$$

The integral is independent of frequency and most of the emission will be in the far IR for which $Q_{abs}/a = Q_0\nu^n$ so that

$$\nu S(\nu) \propto \nu^{4+n+\frac{q-3}{p}}$$

Note that $S(\nu)$ depends on $n_1 a^3 R_1^3$ and Q_{abs}/a the former depending on the volume of the grains and the latter being independent of their size. Thus, for example, for the case n = 1 ($Q_{abs}/a = Q_0\nu$) and p=0.4 ($T = T_1(R/R_1)^{-0.4}$), which was modelled in a first attempt to understand the flat far IR spectra of T Tauri stars and Ae-Be stars in nebulosity, we have $\nu S(\nu) \propto \nu^{2.5(q-1)}$ in agreement with Harvey et al (1979).

It has already been shown that for a centrally heated optically thin cloud $T(R) \propto R^{\frac{-2}{4+n}}$ so that $p = \frac{2}{4+n}$ and so

$$\nu S(\nu) \propto \nu^{\frac{(4+n)(q-1)}{2}}$$

although we might prefer the previous form with p included explicitly in case other things than photons from the central star control the dust grain heating which might lead to a value of p different from $\frac{-2}{4+n}$!

We define the spectral index of a power law spectrum $\nu S(\nu) \propto \nu^\alpha$ as $\alpha = \frac{dlog\,\nu S(\nu)}{dlog\,\nu}$ and for centrally heated optically thin emission

$$\frac{d\alpha}{dq} = 2 + \frac{n}{2} \quad \text{and} \quad \frac{d\alpha}{dn} = \frac{q-1}{2}$$

Typically $n \sim 1$ so $\frac{d\alpha}{dq} = 2.5$, and $q = 1.5$ for free fall collapse so $\frac{d\alpha}{dn} = 0.25$ and we see that the spectral index in this situation is controlled 10 times mores strongly by the density gradient than it is by the variation of the dust emissivity.

The main thing to note about this is that we can get a power law spectrum by the superposition of many blackbody curves at different temperatures. The power law form does not extend over all frequencies but is valid between the frequencies ν_1 and ν_2 at which the Planck function (modified by Q_{abs}) peaks for the maximum and minimum dust temperatures found at radii R_1 and R_2. Shortward of ν_1 the spectrum falls off according to the Rayleigh Jeans law and longward of ν_2 according to the Wien Law. Diagrams illustrating this behaviour for a superposition of blackbody spectra are shown by Beall et al (1984).

11.2. Optically thick dust disc

In this case, as the emission is optically thick, neither the dust density nor the grain emissivity are involved and each annulus of disc at radius R radiates like a blackbody and each annulus of radius R to $R + dR$ corresponds to an element of solid angle $d\Omega = 2\pi \frac{R}{D} \frac{dR}{D}$, where D is the source distance. We assume we are seeing the disk face on, and so ignore projection effects which will effect the absolute value of the flux density, but not its spectral shape.

$$S(\nu) = \int B(\nu, T(R)) d\Omega = \int_{R_1}^{R_2} \frac{2h\nu^3}{c^2(e^{h\nu/kT(R)} - 1)} 2\pi \frac{R}{D} \frac{dR}{D}$$

where R_1 and R_2 are the inner and outer radii of the disc.

Now making the change of variable $x = h\nu/kT(R)$, and taking a power law variation of grain temperature $T(R) = T_1(R/R_1)^{-p}$ where T_1 is the temperature at the inner disk radius R_1, as in the example of an optically thin centrally heated dust cloud, we find

$$S(\nu) = \int_{x_1}^{x_2} \frac{4\pi h\nu^3}{c^2 D^2} \frac{1}{e^x - 1} R_1 \left(\frac{kT_1 x}{h\nu}\right)^{1/p} R_1 \left(\frac{kT_1 x}{h\nu}\right)^{1/p} \left(\frac{1}{p}\right) x^{-1} dx$$

where x_1 and x_2 are the values of $x = (h\nu/kT(R))$ at the innner and outer disc radii R_1 and R_2, and so rearranging

$$S(\nu) = \frac{4\pi h R_1^2}{pc^2 D^2} \left(\frac{kT_1}{h}\right)^{2/p} \nu^{3-2/p} \int_{x_1}^{x_2} \frac{x^{\frac{2-p}{p}}}{e^x - 1} dx$$

again the integral is independent of frequency and we have

$$\nu S(\nu) \propto \nu^{4-\frac{2}{p}}.$$

Thus, for example, for a canonical accretion disk p=3/4 and so $\nu S(\nu) \propto \nu^{\frac{4}{3}}$ as expected (Lynden Bell and Pringle 1974, Beall 1987). To produce a "flat spectrum source" ($\alpha = 0$) therefore requires $p = \frac{1}{2}$ as has been suggested to be the case for some T Tauri stars by Adams Lada and Shu (1987 & 1988).

The same discussion about the range of the power law spectrum applies here as to the previous example, as both result from the superposition of blackbody spectra.

When interpreting flux densities arising from disks in terms of disk luminosities it must always be remembered that, becuase the disks have planar rather than spherical surfaces, the viewing direction will affect what is observed, and that if this is not take into account incorrect disk luminosities can be deduced.

12. NOTATIONS FOR SPECTRAL INDEX

Because of the different units in which different authors choose to express measured or predicted fluxes various different definitions of spectral index have come into usage in the IR, unlike the usage in, for example, the radio region where the spectral index is taken as $\beta = d\,logS(\nu)/dlog(\nu)$ for $S(\nu) \propto \nu^\beta$ where $\beta = \alpha - 1$ as defined above. (We ignore here the use of magnitudes which is common in the IR regions accessible from the ground below 20μm). It is therefore useful to point out the relationship between the indices because, although trivial, it can be a source of some confusion to reader's of the literature, especially from another field of astrophysics.

For these various ways of plotting or quoting spectra we have power law spectral indices of the form

$$\nu S(\nu) \propto \nu^\alpha \quad \text{and} \quad \lambda S(\lambda) \propto \lambda^{-\alpha} \quad \text{and} \quad S(\nu) \propto \nu^{\alpha-1} \quad \text{and} \quad S(\lambda) \propto \lambda^{-(\alpha+1)}$$

Three recent papers on low mass star formation choose different indices to represent observed spectra, which can be confusing. Adams, Lada and Shu (1987) use an index α as defined here. Lada (1987) used $dlog(\lambda S(\lambda))/dlog\lambda = -\alpha$ as his "spectral index", whilst Myers et al (1987) used $dlogS(\nu)/dlog\lambda = -\alpha + 1$ as their "spectral slope", and the radio astronomer's definition of spectral index corresponds to $\alpha - 1$. We therefore give for reference, in Table III, the spectral indices in each of these 4 representations, resulting from various density distributions and dust emissivity laws for centrally heated optically thin dust shells, and for an optically thick disk, with specified temperature distribution. Recall that the range of frequencies over which these power laws are expected to hold will be finite and are determined by the maximum and minimum dust temperatures.

Table III

Spectral index for dust emission with a temperature range.

optically thin shell		optically thick disc	resulting spectral index			
$Q/a \propto \nu^n$						
$n=1$	$n=2$		$\alpha - 1$	$-\alpha + 1$	$-\alpha$	α
$\rho \propto R^{-q}$		$T_d \propto R^{-p}$				
q	q	p	$\frac{d\log S(\nu)}{d\log\nu}$	$\frac{d\log S(\nu)}{d\log\lambda}$	$\frac{d\log\lambda S(\lambda)}{d\log\lambda}$	$\frac{d\log\nu S(\nu)}{d\log\nu}$
-0.2	0	2/7	-4	4	3	-3
0.2	0.33	1/3	-3	3	2	-2
0.6	0.66	2/5	-2	2	1	-1
1.0	1.00	1/2	-1	1	0	0
1.4	1.33	2/3	0	0	-1	1
1.53	1.44	3/4	1/3	-1/3	-4/3	4/3
1.8	1.66	1	1	-1	-2 3	2
2.2	2.0	2	2	-2	-3	3

13. CONCLUSION

Various facets of the physics of dust emission have been described in the hope that this may provide a useful source of material for newcomers to the interpretation of IR observations of dust. In the short space available I have tried to get across the most important processes and to give some of the most important references where more detailed explanations can be found. It should be remembered that radiative transfer calculations should really be made where the simplifying assumptions of optical thinness, made here, are invalid. Although not discussed here, it should not be forgotten that gaseous emission in the IR contains much useful information on star forming regions.

With the ready availability of data from IRAS from 12 to 100 μm, the opening of large telescopes at dry sites capable of working at 350, 450, 750, 850 and 1200 μm, where dust emission is usually optically thin, and where angular resolution of tens of arcsec is obtainable, and with the European Space Agencies forthcoming Infrared Space Observatory (ISO) one can expect to see further advances in observational far IR astronomy that will help us to better understand the formation and early evolution of low mass stars.

14. ACKNOWLEDGEMENTS

I thank the scientific organizing committee for inviting me to lecture at the ASI, and for their financial support.

15. REFERENCES

Adams F.C. and Shu F.H., 1985, *Astrophys. J.*, **296**, 655.

Adams F.C. and Shu F.H., 1986, *Astrophys. J.*, **308**, 836.

Adams F.C. Lada C.J. and Shu F.H., 1987, *Astrophys. J.*, **312**, 788.

Adams F.C. Lada C.J. and Shu F.H., 1988, *Astrophys. J.*, in press.

Aitken D.K., 1981, in *Infrared Astronomy, IAU Symposium 96*, editors Wynn-Williams C.G. and Cruikshank D.P., D. Reidel, 207.

Aitken D.K. and Roche P.F., 1984, *Mon. Not. R. Astr. Soc.*, **209**, 338.

Allamandola L.J. Tielens A.G.G.M. and Barker J.R., 1987, in *Interstellar Processes*, editors Hollenbach D.J. and Thronson H.A., D. Reidel, 471.

Andriesse C.D., 1977, *Vistas in Astronomy, Pergamon Press*, **21**, 107.

Barker J.R. Allamandola L.J. and Tielens A.G.C.M., 1987, *Astrophys. J.*, **315**, L61.

Beall J.J., 1987, *Astrophys. J.*, **316**, 227.

Beall J.H. Knight K.K. Smith H.A. Woods K.S. Lebofsky M. and Rieke G., 1984, *Astrophys. J.*, **284**, 745.

Beichman C.A. Leene A. Emerson J. Harris S. Jennings R.E. Young E. Baud B. Rice W. Hacking P. Gillett F.C. Low F.J. and Neugebauer G., 1984, *Bull AAS*, **16**, 522.

Beichman C.A. Myers P.C. Emerson J.P. Harris S. Mathieu R. Benson P.J. and Jennings R.E., 1986, *Astrophys. J.*, **307**, 337.

Biermann P. and Harwit M., 1980, *Astrophys. J.*, **241**, L105.

Bohren C.E. and Huffman D.R., 1983, *Absorption and Scattering of Light by Small Particles*, J. Wiley.

Chini R., 1981, *Astron. Astrophys.*, **99**, 346.

Cohen M., 1975, *Mon. Not. R. Astr. Soc.*, **173**, 279.

Cohen M., 1980, *Mon. Not. R. Astr. Soc.*, **191**, 499.

Cohen M. and Witteborn F.C., 1985, *Astrophys. J.*, **294**, 345.

Crawford J. and Rowan-Robinson M., 1986, *Mon. Not. R. Astr. Soc.*, **221**, 923.

Disney M.J. and Sparks W.B., 1982, *Observatory*, **102**, 234.

Donn B., 1985, in *Interrelationships Among Circumstellar, Interstellar and Interplanetary Dust*, editors Nuth J.A. and Stencel R.E., NASA CP-2403, 109.

Draine B.T. and Anderson N., 1985, *Astrophys. J.*, **292**, 494.

Draine B.T. and Lee H.M., 1984, *Astrophys. J.*, **285**, 89.

Draine B.T., 1985, *Astrophys. J. Supp. Ser.*, **57**, 587.

Draper P.W. Warren-Smith R.F. and Scarrott S.M., 1985, *Mon. Not. R. Astr. Soc.*, **216**, 7P.

Duley W.W., 1986, *Quart. J. R. Astr. Soc.*, **27**, 403.

Duley W.W. and Williams D.A., 1981, *Mon. Not. R. Astr. Soc.*, **196**, 269.

Emerson J.P. Harris S. Jennings R.E. Beichman C.A. Baud B. Beintema D.A. Marsden P.L. and Wesselius P.R., 1984, *Astrophys. J.*, **278**, L49.

Emerson J.P. et al, 1988, in preparation

Geballe T.R. Baas F. Greenberg J.M. and Schutter W., 1985, *Astron. Astrophys.*, **146**, L6.

Gillett F.C., Forrest W.J., Merrill K.M., Capps R.W. and Soifer B.T., 1975, *Astrophys. J.*, **200**, 609.

Greenberg J.M., 1978, in *Infrared Astronomy*, editors Setti G. and Fazio G.G., D. Reidel, 51.

Harvey P.M. Thronson H.A. and Gatley I., 1979, *Astrophys. J.*, **231**, 115.

Hasegawa H. and Koike C., 1984, in *Obsevational Infrared Spectra of Interstellar Dust*, editors Wolstencroft R.D. and Greenberg J.M., Occassional Reports of the Royal Observatory Edinburgh No 12, 137.

Hildebrand R.H., 1983, *Quart. J. R. Astr. Soc.*, **14**, 267.

Hong S.S. and Greenberg J.M., 1980, *Astron. Astrophys.*, **88**, 194.

Huffmann D.R., 1977, *Advances in Physics*, **26**, 129.

Kittel C., 1968, *Introduction to Solid State Physics, 3rd Edn*, Wiley.

Knacke R.F. McCorkle S. Puetter R.C. Erickson E.F. and Kratschmer W., 1982, *Astrophys. J.*, **260**, 141.

Lacy J.H. Baas F. Allamandola L.J. Persson S.E. McGregor P.J. Lonsdale C.J. Geballe T.R. and van de Bult C.E.P.M., 1984, *Astrophys. J.*, **276**, 533.

Lada C.J., 1987, in *IAU Symposium 115, Star Formation*, editors Peimbert M. and Jugaku J., D. Reidel, 1.

Landini M. Natta A. Oliva E. Salinari P. and Moorwood A.F.M., 1984, *Astron. Astrophys.*, **134**, 284.

Lee H.M. and Draine B.T., 1985, *Astrophys. J.*, **290**, 211.

Leger A. d'Hendecourt L. and Boccara N. editors, 1987, *Polycyclic Aromatic Hydrocarbons in Astrophysics*, D. Reidel.

Leger A. Gauthier S. Defourneau D. and Rouan D., 1983, *Astron. Astrophys.*, **117**, 164.

Leger A. Klein J. de Cheveigne S. Guinet C. Defourneau D. and Belin M., 1979, *Astron. Astrophys.*, **79**, 256.

Leger A. and Puget J.L., 1984, *Astron. Astrophys.*, **137**, L5.

Longhurst R.S., 1967, *Geometrical and Physical Optics, 2nd Edn, Chap 20*, Longmans.

Lynden-Bell D. and Pringle J., 1974, *Mon. Not. R. Astr. Soc.*, **168**, 603.

Martin P.G., 1978, *Cosmic Dust*, Oxford Univ Press.

Mathis J.S., 1985, in *Interrelationships Among Circumstellar, Interstellar and Interplanetary Dust*, editors Nuth J.A. and Stencel R.E., NASA CP-2403, 29.

Mathis J.S. Rumpl W. and Nordsieck K.H., 1977, *Astrophys. J.*, **217**, 425.

Mathis J.S. and Wallenhorst S.G., 1981, *Astrophys. J.*, **244**, 483.

Mezger P.G. Mathis J.S. and Panagia N., 1982, *Astron. Astrophys.*, **104**, 372.

Moneti A. Pipher J.L. Helfer H.L. McMillan R.S. and Perry M.L., 1984, *Astrophys. J.*, **282**, 508.

Myers P.C. Fuller G.A. Mathieu R.D. Beichman C.A. Benson P.J. Schild R.E. and Emerson J.P., 1987, *Astrophys. J.*, **319**, 340.

Nuth J.A. and Stencel R.E. editors, 1986, *Interrelationships Among Circumstellar, Interstellar and Interplanetary Dust*, NASA CP-2403 .

Panofsky W.K.H. and Phillips M., 1969, *Classical Electricity and Magnetism, 2nd Edn, Chap 22*, Addison Wesley.

Pegourie B. and Papoular B., 1985, *Astron. Astrophys.*, **142**, 451.

Puget J.L. Leger A. and Boulanger F., 1985, *Astron. Astrophys.*, **142**, L19.

Purcell E.M., 1969, *Astrophys. J.*, **158**, 433.

Purcell E.M., 1976, *Astrophys. J.*, **206**, 685.

Rieke G.H. and Lebofsky , 1985, *Astrophys. J.*, **288**, 618.

Rowan-Robinson M., 1980, *Astrophys. J. Supp. Ser.*, **44**, 403.

Rowan-Robinson M., 1986, *Mon. Not. R. Astr. Soc.*, **219**, 737.

Sato S. Nagata T. Nakajima T. Nishida M. Tanaka M. and Yamashita T., 1985, *Astrophys. J.*, **291**, 708.

Savage B.D. and Mathis J.S., 1979, *Ann. Rev. Astron. Astrophys.*, **17**, 73.

Scoville N.Z. and Kwan J., 1976, *Astrophys. J.*, **206**, 718.

Seki J. and Hasegawa H., 1981, *Progr. Theor. Phys.*, **66**, 903.

Sellgren K., 1984, *Astrophys. J.*, **277**, 623.

Soifer B.T. Puetter R.C. Russell R.W. Willner S.P. Harvey P.M. and Gillett F.C., 1979, *Astrophys. J.*, **232**, L53.

Stahler S.W. Shu F.H. and Taam R.E., 1981, *Astrophys. J.*, **248**, 727.

Stein W.A. and Soifer B.T., 1983, *Ann. Rev. Astron. Astrophys.*, **21**, 177.

Tielens A.G.G.M. and Allamandola L.J., 1987, in *Interstallar Processes*, editors Hollenbach D.J. and Thronson H.A., D. Reidel, 397.

van de Bult C.E.P.M. Greenberg J.M. and Whittet D.C.B., 1984, *Mon. Not. R. Astr. Soc.*, **214**, 289.

van de Hulst H.C., 1949, *Rech.Astr.Obs.Utrecht*, **11**, Part2.

Weiland J.L. Blitz L. Dwek E. Hauser M.G. Magnani L. and Rickard L.J., 1986, *Astrophys. J.*, **306**, 401.

Whittet D.C.B. Bode M.F. Longmore A.J. Baines D.W.T. and Evans A., 1983, *Nature*, **303**, 218.

Wickramsinghe N.C., 1967, *Interstellar Grains*, Chapman Hall.

Willner S.P., 1984, in *in Galactic and Extragalactic Infrared Spectroscopy*, editors Kessler M.F. and Phillips J.P., D. Reidel, 13.

Willner S.P. Gillett F.C. Jones B. Krassner J. Merrill K.M. Pipher J.L. Puetter R.C. Rudy R.J. Russell R.W. and Soifer B.T., 1982, *Astrophys. J.*, **253**, 174.

Wolfire M.G. and Cassinelli J.P., 1986, *Astrophys. J.*, **310**, 207.

Wolstencroft R.D. and Greenberg J.M., editors, 1984, *Laboratory and Obsevational Infrared Spectra of Interstellar Dust*, Occassional Reports of the Royal Observatory Edinburgh No 12.

Young E.T. Lada C.J. and Wilking B.A., 1986, *Astrophys. J.*, **304**, L45.

OBSERVATIONS OF YOUNG STELLAR OBJECTS

Claude Bertout*
Astronomy Department
University of California
Berkeley, California 94720

Protostars are the Holy Grail of infrared astronomy
C.G. Wynn-Williams (1982)

Abstract

After a brief description of the main properties of embedded protostellar sources, I investigate in some detail the hypothesis that CO bipolar flows are pushed by winds emanating from the infrared central sources associated with the molecular outflows. I reach the conclusions that radiation-pressure driven winds from high-luminosity embedded sources (with $L_{bol} \gtrsim 10^3 L_\odot$) can drive their associated CO flows, and that the molecular flow is more likely to be driven by the thermal pressure of the wind than by its momentum. The ionized winds of these high-luminosity objects are massive enough to drive the CO flows.

The situation is not nearly as clear for low-luminosity embedded sources. There is some indication that their CO flows might also be energy-driven by hot winds, but in most cases the winds' momentum rate appears larger than the stellar radiative momentum output rate unless the circumstellar medium is very optically thick. Differences found between high-luminosity and low-luminosity sources suggest that the physical mechanisms driving their winds are quite different. Radiation pressure is probably ubiquitous in driving winds from high-luminosity objects. Winds from low-luminosity objects remain much more mysterious, although it is becoming increasingly evident that a neutral wind component is present in addition to the observed ionized component. This investigation tentatively suggests that the neutral component's momentum rate could be comparable to that of the ionized wind.

Finally, I discuss the role of mass-accretion in the T Tauri phase and find that about 50% of the late-type T Tauri stars in the Taurus-Auriga association show evidence for boundary-layer emission characteristic of mass-accretion rates in the range $1 - 5 \cdot 10^{-7}$ $M_\odot$/yr.

1. Introduction

Five years after Wynn-Williams's review and twenty years after the discovery

---

* on leave from the Institut d'Astrophysique de Paris

A. K. Dupree and M. T. V. T. Lago (eds.), Formation and Evolution of Low Mass Stars, 45–64.

of the BN object, clear, direct evidence for a classical solar-type protostar, i.e., a cloud fragment undergoing gravitational collapse, is still missing. While IRAS gives us a much better understanding of the contents of dense molecular cores (cf. Beichmann *et al.* 1986), it didn't help in identifying protostars because direct evidence for gravitational collapse requires high-resolution spectroscopic observations like the one reported by Walker *et al.* (1986) for IRAS 16293-2422, the only ρ-Ophiuchi molecular outflow source.

The $C^{32}S(J=5-4)$ line (at 244.93 GHz) observed in the direction of this source is asymmetric, and Walker *et al.* suggest that it is formed in an infalling envelope. Although this interpretation based on the analysis of a single line profile is model-dependent, derived properties of the infalling region agree remarkably well with theoretical expectations. In particular, the derived infall velocity of about 1 km/s at 10^{16}cm from the source center is typical of current low-mass protostellar models (Shu *et al.* 1987). While this result is encouraging, clear-cut evidence for a low-mass protostar would ideally require the detection of an inverse P Cygni profile. One such profile was observed in HCO^+ toward a bright HII region in W49; there, it unambiguously indicates that an OB association is currently being formed by gravitational collapse (Welch *et al.* 1987).

Today's best protostar candidate IRAS 16293-2422 is associated with a molecular outflow. In this respect, it is not unlike most other young stellar objects, which show no evidence for mass infall but instead display extensive outflows, as a review of the observed properties of young, low-mass objects demonstrates. Section 2 offers a partial guided tour through the current literature. The energetics of molecular and ionized outflows associated with young stellar objects are reviewed in Section 3, while in Section 4 we discuss the role of accretion during the T Tauri phase.

2. Properties of Young Stellar Objects

Considerable progress has been made over the last few years in our knowledge of the observational properties of sources deeply embedded in dense molecular cores (see the reviews by J. Emerson and C. Lada in this volume). A systematic survey using IRAS is reported by Beichmann *et al.* (1986) and Myers *et al.* (1987). Well-studied young low-luminosity objects include: IRS5 in L1551; SVS13, which apparently powers Herbig-Haro objects 7 to 11; the central source (VLA1) powering Herbig-Haro objects 1 and 2; and T Tauri South. The presence of energetic, often bipolar outflow phenomena around these embedded sources is suggested by molecular line emission, radio continuum emission, hydrogen infrared line emission, optical jets and Herbig-Haro objects.

The best-studied CO high-velocity outflow is that of IRS5 in L1551 (cf. Stocke *et al.* 1987; Campbell *et al.* 1987). Snell *et al.* (1980) suggested that the CO gas is contained in a shell pushed by the hot wind emanating from IRS5. Further evidence for this model is presented by Snell and Schloerb (1985) and Moriarty-Schieven *et al.* (1987). Optical reflection nebulosity traces the boundaries of the expanding shell, which suggests the presence of a cavity within the CO flow. This cavity is filled up by hot gas that excites various emission nebulosities, including Herbig-Haro objects No. 28 and 29. Spectroscopy of the emission knots indicate wind velocities of about 200 km/s in the cavity. Spectra of scattered light from IRS5 (Mundt *et al.* 1985) indicate maximum wind velocities in the 500 km/s range. Implications of these observations for the energetics of the outflow driving source will be discussed in Section 3.

Lizano *et al.* (1987) recently reported the discovery of high-velocity neutral hydrogen gas in the vicinity of SVS 13, which is also associated with a bipolar CO outflow (Edwards and Snell 1983) as well as with extended H_2 emission tracing the outflow axis and surrounding the Herbig-Haro objects (Lightfoot and Glencross 1986). H_2-emission, usually thought to originate from moderate velocity shocks, is also observed in the vicinity of T Tauri (Beckwith *et al.* 1978). The CO flow associated with T Tauri is almost monopolar in nature, in the sense that the blue-displaced component is much more extended than the red-displaced one (Edwards and Snell 1982). No CO flow was detected in the region of Herbig-Haro objects 1-2, probably because the outflow is in the plane of the sky.

Embedded sources driving molecular outflows are often radio continuum sources (Snell and Bally 1986). Data are consistent with the hypothesis that radio emission is thermal and originates from an ionized wind. However, detailed models indicate that typical spectra of low-luminosity sources cannot arise in a constant-velocity wind (Bertout 1987; André 1987). The interpretation which we favor currently is that the ionized wind is decelerated in the radio-emitting region (Bertout *et al.* 1988). This contrasts with the radio emission from high-luminosity sources, which is often interpreted in terms of constant-velocity winds. Another difference between high-luminosity and low-luminosity protostellar sources appears when comparing mass-loss rates derived from the ionized winds with those derived from the CO flows (see Section 3).

Brackett and Pfundt lines have been detected in an increasingly large sample of embedded and optically visible young stellar objects (cf. J. Emerson's contribution). The flux of these lines is usually greatly in excess of what is expected from the Lyman continuum photons emitted from the star (Thompson 1985). Mass-loss rates derived from the infrared line fluxes do not agree with mass-loss rates derived either from the radio continuum observations or from the CO molecular flows (Evans *et al.* 1987). While uncertainties associated with each method for determining the mass-loss rates are large enough that these discrepancies may not be as serious as they appear, several possible solutions have been proposed, and the most attractive one assumes that the wind is largely neutral (Natta *et al.* 1987).

Extremely well-collimated jets of ionized gas originate from a number of embedded objects (e.g. L1551/IRS5, HH1-2/VLA1), as well as from T Tauri stars. Both their properties and their relationship to Herbig-Haro objects are discussed in detail by R. Mundt in this volume.

It is expected that circumstellar disks play a key role in collimating and, possibly, in powering the outflows. Flattened large-scale molecular structures perpendicular to the outflows have been reported for several sources (Torrelles *et al.* 1983), including the radio source VLA1 associated with Herbig-Haro objects 1 and 2 (Torrelles *et al.* 1985). Millimeter-wave interferometry permits the detection of smaller structures, and a recent paper reports the detection of a rotating disk with diameter about 4000 AU surrounding HL Tauri (Sargent and Beckwith 1987). Infrared imaging allowed Strom *et al.* (1985) to infer the presence of an elongated structure with diameter 750 AU around L1551/IRS5. Recently, Bastien and Ménard (1987) showed that polarization maps observed around a number of young stellar objects could be best interpreted by scattering in a dusty circumstellar disk.

Indirect evidence for the presence of disks around optically visible T Tauri stars is provided by the shape of their forbidden lines (Appenzeller *et al.* 1984; Edwards *et al.* 1987), and by their spectral energy distributions (cf. Bertout *et al.* 1987 and references therein), as will be discussed in Section 4.

3. Energetics of Molecular and Ionized Outflows

Although they were discovered almost a decade ago, high-velocity molecular outflows remain as puzzling as ever. There is no consensus at this point about the physical mechanism(s) responsible for driving and collimating them. The challenges that CO bipolar flows pose to theoretical models have been summarized by Lada (1985), and Figure 1 shows slightly modified versions of two key diagrams from his paper. The mechanical luminosity in the CO flows, L_{CO}, is plotted against the bolometric luminosity of the embedded infrared objects associated with the flows in Fig. 1a. There, symbols for the data points differ according to the flow view angle, estimated by comparing the flow morphology to synthetic maps computed by Cabrit and Bertout (1986; 1988). Open squares denote flows viewed at less than $20°$ (i.e., seen almost pole-on); open circles are for view angles in the range $20°$ - $40°$; filled squares indicate view angles in the range $40°$ - $60°$; and filled triangles are for view angles larger than $60°$. Only those flows in Lada's list whose view angle could be reasonably well estimated have been included in Fig. 1. L_{CO} is defined by

$$L_{CO} = M_{CO}V_{CO}^3/2R_{CO},$$ (1)

where M_{CO}, V_{CO}, and R_{CO} are, respectively, mass, velocity, and scalelength of the moving CO gas; these quantities are derived directly from the millimetric observations of CO outflows. Fig. 1a shows that a correlation exists between the flow mechanical energy and the bolometric luminosity of the central source. This suggests that the energy contained in the radiation field of the central source is sufficient to drive the flows provided they have been steady over their dynamical lifetimes.

Figure 1b displays a plot of the CO momentum rate F_{CO} versus the stellar bolometric luminosity L_*, with F_{CO} defined by

$$F_{CO} = M_{CO}V_{CO}^2/R_{CO}.$$ (2)

Again, a correlation between the CO gas momentum rate and the overall radiative momentum rate of the central object is found. But the momentum rate in the molecular gas is usually much larger than L_*/c, which apparently rules out a radiatively driven wind as the powering mechanism. There are, however, several reasons to look more closely at this interpretation of Figure 1.

(a) Dyson (1984) and Kwok and Volk (1985) adapted the theory of wind bubble expansion to physical conditions found in interstellar clouds. They showed that

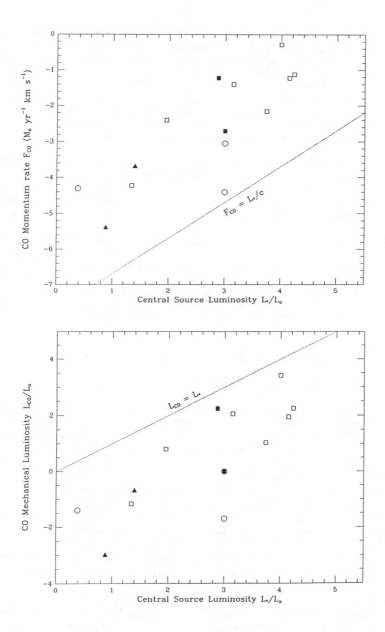

Figure 1: (a) Momentum rate of the CO molecular outflows plotted against the bolometric luminosity of the embedded central objects. (b) Mechanical luminosity of the molecular outflows plotted against the bolometric luminosity of the embedded central objects. The meaning of the various symbols is given in the text.

the interaction of the wind with the CO gas can be either energy-conserving or momentum-conserving; and in the former case, the CO momentum rate can be much larger than the wind momentum rate. Their refined theoretical approach must be taken into consideration.

(b) Cabrit and Bertout (1986, 1988) have shown that uncertainties in the CO flow parameters derived from observations can be more than one order of magnitude. Their estimates of the uncertainty can be used to bracket the true value of the flow's mechanical luminosity and and momentum rate.

(c) The bolometric luminosities of the embedded sources associated with CO flows are now more accurately known from the detailed study of IRAS data by Mozurkewich *et al.* (1986). Furthermore, a large number of radio continuum data are available for these sources (e.g. Snell and Bally 1986) which makes it possible to discuss the possible role of the ionized wind component in driving the CO flow.

The picture of a CO outflow envisioned in this paper was first proposed by Snell *et al.* (1980). It assumes that the central source emits a stellar wind which interacts with the ambient, cold circumstellar material, thereby creating an expanding shell of molecular gas that is identified with the CO flow. Questions to be addressed include (i) What is the nature of the interaction between wind and molecular gas? (ii) Can the wind from the central source be driven radiatively? (iii) Can the ionized wind component (seen in the radio continuum) drive the CO flow?

3.1. The Energetics of Molecular Flows Revisited

3.1.1. *Energy vs. Momentum driven winds*

Weaver *et al.* (1977) developed the theory of wind bubble expansion. They showed that the shell of swept-up circumstellar material is driven by the pressure of the shocked wind as long as the radiative losses from the shocked wind region are small. This phase is referred to as the energy-driven regime. But when radiative losses become important, the shocked wind region collapses and the stellar wind collides directly with the shell; this is the momentum-driven regime. [1] As Dyson (1984) emphasizes, one must find out which type of flow is originally set up and also whether the driving mechanism will change in time. The wind velocity determines whether the wind is initially driven by energy or momentum. Assuming that thermal conduction is suppressed by the magnetic field present in the molecular swept-up material, the critical velocity (below which the flow is momentum-driven; cf. Eq. 8 in Dyson 1984) is in the range 200-300 km/s for mass-loss rates up to about 10^{-6} $M_\odot$/yr and for $n_H = 10^4$ cm^{-3}, which is a density typical of molecular cores.

Little is known about the wind velocity of embedded protostellar sources. If high-luminosity infrared sources are similar to optical OB stars – a likely possibility since they have already reached the main-sequence – we expect their winds to have

---

[1] In both cases, the thin shell assumption for the swept-up material implies efficient radiative cooling within the shell. If the circumstellar material is primarily H_2, far-infrared emission from the shell is expected.

velocities about 1000 km/s or larger, and their CO flows to be energy-driven. IRS5 in Lynds 1551 is representative of low-luminosity sources driving molecular outflows, and its H_α profile has been observed by Mundt *et al.* (1985). It is a typical P Cygni profile, and its blue absorption wing indicates that the maximum velocity in the H_α flow is about 500-600 km/s. If this wind actually pushes the CO bipolar gas in L1551, then the interaction is likely to be pressure-driven rather than momentum-driven. From his study of the flow evolution, Dyson (1984) concludes that all the molecular outflows he investigates are likely to be energy-driven during most of their lifetime. We shall nevertheless consider both types of flows in the following discussion.

An important point stressed by Dyson (1984) and Kwok and Volk (1985) is that the momentum rate and mechanical luminosity of the molecular flow cannot be used directly to find both the wind momentum rate and mechanical luminosity because the conserved quantity depends on the flow type. In a momentum-driven flow, most of the wind kinetic energy is radiated away, while in an energy-driven flow the momentum of the swept-up gas depends both on the pressure inside the bubble and on the wind momentum. Assuming that the stellar wind expands isotropically with a velocity V_w in a medium with density gradient given by

$$\rho(r) = \rho_0 r^\beta, \ \beta \leq 0, \tag{3}$$

Dyson (1984) and Kwok and Volk (1985) find that the wind mechanical luminosity L_w and momentum rate F_w are given, respectively, by

$$L_w^m = \frac{4+\beta}{2} \frac{V_w}{V_{CO}} L_{CO} \tag{4a}$$

and

$$F_w^m = \frac{4+\beta}{2} F_{CO} \tag{5a}$$

if the wind is momentum-driven, and by

$$L_w^e = \frac{77 + 29\beta + 2\beta^2}{9} L_{CO} \tag{4b}$$

and

$$F_w^e = \frac{77 + 29\beta + 2\beta^2}{9} \frac{V_{CO}}{V_w} F_{CO} \tag{5b}$$

if the wind is energy-driven. In these equations, V_{CO} is the velocity of the molecular gas, while L_{CO} and F_{CO} are given by Eqs. 1 and 2. Since Eqs. 4 and 5 assume spherical symmetry, we need to assess their usefulness for bipolar, rather than for spherical, winds.

The evolution of a stellar wind bubble in a plane-parallel, stratified cloud was investigated by Sakashita and Hanami (1986), who find that the bubble becomes bipolar and that the flow collimation depends on the type of flow established as well as on the details of the interstellar density distribution. Their work demonstrates that conservation of momentum (viz. energy) occurs globally in the two lobes, so that the spherical case mentioned above is a useful first approximation of the bipolar situation, provided the flow collimation is caused by density gradients in the interstellar medium (see Königl 1982). While other collimation mechanisms have been proposed for which the above picture would certainly fail – in particular, mechanisms in which collimation is intrinsic to the star/disk system – observational evidence (e.g. Bertout 1987) for strong interstellar density gradients in the immediate vicinity of many young stellar objects that show collimated flows justifies our interest in this model.

3.1.2. Uncertainties in the Derivation of Flow Parameters

Another area of concern is the large uncertainties associated with the physical parameters of molecular flows, i.e., the quantities which enter Eqs. 1 and 2 above. Cabrit and Bertout (1986; 1988) estimated these uncertainties quantitatively by assuming various kinematic models for the CO flows and then solving the CO line formation problem in some detail. Specifically, power-law velocity fields $v(r) \propto r^\alpha$ with $-1 \leq \alpha \leq 1$ were investigated. Synthetic maps obtained for known flow parameters were analyzed in the same way as observed CO maps, and from there the main sources of error in the derived flow parameters were identified.

One disturbing conclusion of these investigations is that the flows most easily detected are those whose physical parameters are most likely to be in error. About one half of the CO flows appearing in Fig. 1 are what the study calls Case I flows, i.e., flows seen at small view angles. This is because the flow radial velocities are largest, and detection of high velocity wings easiest, when the angle between observer and the flow's symmetry axis is small. But the CO flow dynamical time, found from its extent and velocity, is particularly uncertain in this configuration because of projection effects, so that dynamical parameters are not well constrained. More specifically, the momentum rate appears to be overestimated in that case.

Using this study of several CO velocity distributions, we determine a typical uncertainty range on a given physical parameter as a function of the view angle (this is done for each of the methods used to determine that physical parameter). We then define an "error bar" from the minimum and maximum errors found in computations for the different velocity fields. While this "error bar" is in principle valid only for the few simple velocity fields we studied, it does provide a first estimate of the magnitude of possible systematic effects on the determination of the flow's parameters; and it allows us to discuss whether the momentum problem is really as critical as it was assumed to be so far.

3.1.3. Results and Discussion

We use 10 molecular outflows in Lada's (1985) list, chosen according to two criteria. Their integrated intensity maps present some degree of symmetry, so that we can tentatively derive a view angle based on the flow morphology; and the CO mass in the flow can be determined accurately, i.e., by integrating over all lines of sight and by correcting for optical depth effects when the $^{13}CO/^{12}CO$ line ratio

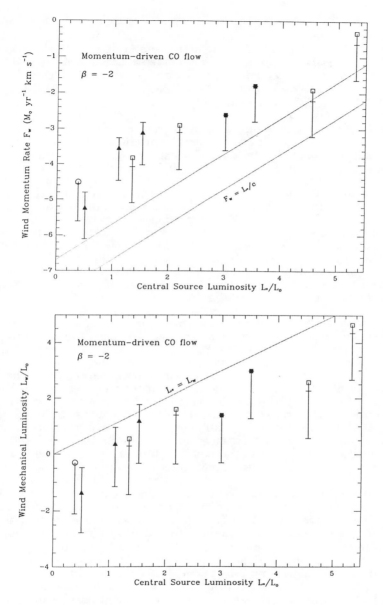

Figure 2: (a) Momentum rate of the stellar winds plotted against the bolometric luminosity of the embedded central objects. It is assumed here that the CO flow is driven by the momentum of the stellar wind. The bars indicate uncertainties on the momentum rate. (see text for details). (b) The mechanical luminosity of the stellar wind is plotted against the bolometric luminosity of the embedded central objets, with the same assumption as (a).

indicates that the $^{12}CO(J = 1 - 0)$ line is optically thick. CO masses were also renormalized to the same CO abundance, using $N(H_2)/N(CO) = 5.6 \ 10^3$. We used CO velocities from Lada (1985) and bolometric luminosities of the central sources from Mozurkewich *et al.* (1986). Since a more detailed account of this investigation will be published elsewhere, we need only give the list of sources here. In order of increasing luminosity, they are: L723, B335[2], RNO43, T Tau, L1551/IRS5, SVS13, LkH$_\alpha$ 234, NGC 2071, S140, and OriA.

Figure 2a displays the wind momentum rate computed from Eq. 5a and plotted as a function of the central source bolometric luminosity for the 10 CO flows. We assume here that the stellar wind is expanding in a medium with $\beta = -2$, a value expected in the outer, isothermal regions of a collapsed molecular core. If the wind were expanding in a constant density medium, all data points would be displaced upwards by 0.3 dex. Error bars associated with data points indicate the uncertainty range on the CO momentum rate, as discussed in the preceding section. The two dotted lines which cross the diagram indicate the maximum stellar radiative momentum output rate $\tau L_*/c$ for $\tau = 1$ and $\tau = 10$. This plot shows that the wind momentum rate is almost always greater than the stellar radiative momentum rate, even when the wind optical depth is as large as 10, which indicates that a radiatively-driven wind does not have enough momentum to drive the CO flow if the interaction is momentum-conserving. The two most luminous sources might represent an exception to this statement.

Figure 2b shows the wind's mechanical luminosity (Eq. 4a) as a function of the central source bolometric luminosity. Symbols have the same meaning as before, and the dotted line represents $L_* = L_w$. In order to draw this diagram, we must first deal with the ratio of wind to CO velocities, so we adopt a single, representative value of 20 for this insufficiently known parameter. The ratio of L_w to L_* is typically 1 to 10 % for the low-luminosity sources, but it diminishes to about 0.1 % for the high-luminosity sources, which is roughly similar to that inferred for optical OB stars. One might, therefore, conclude with Mozurkewich *et al.* (1986) that CO flows around high-luminosity objects can in principle be driven by the momentum of radiatively driven winds. But if the high-luminosity sources have wind velocities comparable to those of OB stars, their associated CO flows are likely to be energy-driven rather than momentum-driven.

Figure 3a displays the wind mechanical luminosity as a function of the bolometric luminosity of the central source for an energy-driven interaction. Here, the mechanical luminosity is found from Eq. 4b, and we assume again that the stellar winds expand in a circumstellar medium with a density distribution proportional to r^{-2}. The high-luminosity sources driving CO flows and optically visible OB stars share the same general region in this diagram. As far as the energy budget is concerned, winds from high-luminosity embedded sources could therefore be driven by radiation-pressure as are winds from OB stars. The force as defined by Eq. 5b is plotted against the central source radiative momentum rate in Fig. 3b. We assumed as before that the ratio of wind to CO gas velocity is 20 for all sources.

---

[2] Both L723 and B335 were recently identified as highly collimated ouflows nearly in the plane of the sky by Cabrit, Goldsmith, and Snell (1988). The revised physical parameters of these flows were kindly communicated by S. Cabrit

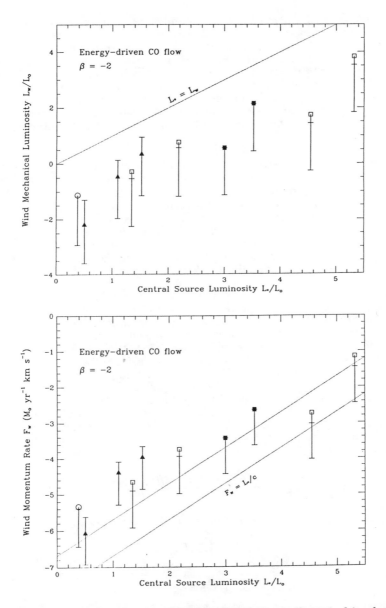

Figure 3: (a) The mechanical luminosity of the stellar wind is plotted against the bolometric luminosity of the embedded central objets. It is now assumed that the CO flow is driven by the thermal pressure of the shocked wind. The bars indicate uncertainties on the mechanical luminosity (see text for details). (b) Momentum rate of the stellar winds plotted against the bolometric luminosity of the embedded central objects, with the same assumption as (a).

Although this last assumption introduces some uncertainty, it does not seem unreasonable to conclude that an optical depth in the range 1-10 would be sufficient for the star's maximum radiative momentum rate $\tau L_*/c$ to account for the wind's momentum rate. When the wind expands in a constant density medium, all points in Fig. 3 go up by 0.46 dex, so that the momentum budget becomes more critical. The wind momentum rate of optical OB stars usually represents a smaller fraction of the total stellar radiative momentum output, even for the favorable case of outwardly decreasing circumstellar density considered above. Protostellar sources appear to be able to transfer their radiative momentum to the wind as efficiently as do Wolf-Rayet stars.

We conclude from this re-examination of the energetics of CO flows that a wind driven by radiation pressure can actually push molecular flows associated with high-luminosity protostellar sources (i.e., $L_{bol} > 10^3 L_o$), provided the density of the interstellar medium decreases away from the star. However, the flow must be driven by the pressure of the shocked wind rather than directly by the stellar wind momentum. If the wind velocities of embedded protostellar sources are comparable to those of OB stars, this is likely. Furthermore, the wind momentum rate represents a significant fraction of the overall stellar radiative momentum output rate, suggesting that the wind is optically thick close to the central source.

CO flows associated with low-luminosity sources might also be energy-driven by stellar winds, although it is unlikely in view of the more stringent momentum rate requirement that these winds are of the same nature as those of high-luminosity sources. The main argument in favor of energy-driven CO flows from low-luminosity sources is the observed high velocity of the L1551/IRS5 wind. Yet, the finding by Edwards et al. (1986) and Clark and Laureijis (1986) that extended infrared emission is associated with the L1551/IRS5 outflow might lead to the dismissal of this model. Edwards et al. (1986) find that the infrared diffuse luminosity enclosing IRS5 is equal to 7 $L_\odot$, a value ten times greater than the mechanical luminosity in the CO flow. They attribute the extended emission to cooling radiation from the shock between ambient and swept-up material. Using the results of Weaver et al. (1977) and Dyson (1984) to derive the luminosity L_{rad} radiated away by the swept-up gas, we find

$$L_{rad} = \left[\frac{6}{11}\left(\frac{77 + 29\beta + 2\beta^2}{9}\right) - 1\right] L_{CO}. \tag{6}$$

Bracketing L_{rad} by using our "error bars" on L_{CO}, we get

$$0.01 \leq L_{rad} \leq 1.53 L_\odot \text{ for } \beta = -2,$$

and

$$0.07 \leq L_{rad} \leq 8.8 L_\odot \text{ for } \beta = 0.$$

Thus, predictions based on the theory of energy-driven flows yield smaller values of L_{rad} than observed for the case $\beta = -2$, but comparable values for the case $\beta = 0$. Whether the extended infrared emission allows one to rule out an energy-driven interaction for L1551/IRS5 is therefore unclear at this point. Further information on far-infrared emission associated with other CO flows would obviously be valuable for understanding the driving mechanism of CO flows.

The physical mechanism responsible for the massive winds from low-luminosity protostellar objects remains elusive. The ratios of both wind mechanical luminosity to stellar luminosity and wind force to stellar radiative momentum output are, on the average, about ten times higher in low-luminosity sources than in sources with $L_{bol} > 10^3$ L_o, which indicates an extremely efficient wind-driving mechanism. The problem of finding such a mechanism for T Tauri winds is well-documented (cf. DeCampli 1981; Hartmann et al. 1982; Hartmann 1986), and obviously we are confronted with a similar problem in the case of low-luminosity embedded sources driving CO flows. Compared to T Tauri stars, protostellar low-luminosity sources have the main advantage of being potentially fast rotators, and a newly developed wind model takes advantage of this possibility (Shu et al. 1987).

Another difference between low-luminosity protostellar sources and optical T Tauri stars lies in their radio properties. While optical pre-main-sequence stars are infrequent radio-emitters (cf. Montmerle and André in this volume), many embedded sources associated with molecular outflows are radio sources and their emission is apparently thermal. The mass-loss in the ionized component can therefore be derived and compared to the wind mass-loss rate derived from the CO observations.

3.2. The Ionized Wind Component

Snell and Bally (1986) have shown that most high-luminosity infrared sources, as well as several low-luminosity sources, associated with molecular outflows are also radio sources. While this suggests that CO flows and ionized winds are somehow connected, the nature of this connection is unclear because mass-loss rates derived by these authors from the radio data are usually lower than those derived from the CO data. It is therefore believed that the ionized wind component cannot drive the molecular gas, and a neutral wind component has been suggested (cf. Shu et al. 1987).

Since we have seen above that CO dynamical parameters are often overestimated, a re-examination of the relationship between CO mass-loss and ionized gas mass-loss is in order. Among the 10 flows investigated in the preceding section, 6 are associated with radio-emitting central objects (T Tau, L1551/IRS5, SVS13, LkH$_\alpha$ 234, NGC 2071, and S140). We derived "radio mass-loss rates" for these objects by assuming that their 5 GHz fluxes originate from a constant-velocity ionized wind. Although this assumption may be incorrect for some low-luminosity objects with relatively flat radio spectra, we ignored the error introduced by possible deviations from a constant velocity wind. An overestimation of our derived radio mass-loss rates is likely to result from this assumption and from the hypothesis of spherical symmetry for the wind. An additional source of uncertainty is the often unknown terminal velocity of the wind. Terminal velocities used here are the same as the wind velocities used in the derivation of wind momentum rates above. In optically visible objects, these velocities are usually lower than those estimated from the H$_\alpha$ line, which underestimates the mass-loss rate. We estimate the overall uncertainty to be a factor 2-3.

We can now compare the wind mass-loss rates derived directly from the wind momentum rates (as found from the assumption of an energy-driven CO flow) with the mass-loss rates in the ionized wind components. The result is shown in Figure 4, where wind mass-loss rates are plotted (with their associated error-bars as defined above) against radio mass-loss rates. The dotted line indicates equal wind and radio mass-loss rates. Again, low-luminosity and high-luminosity sources exhibit qualitatively different behavior.

For high-luminosity sources, the mass-loss in the ionized wind component is higher than the wind mass-loss derived under the assumption that the CO is pressure-driven by the wind. We, therefore, conclude that the ionized wind component is in fact driving the CO flows in these objects.

In low-luminosity sources, the ionized component's mass-loss appears comparable to the wind's mass-loss when the error bars are taken into account. Because of the uncertainties mentioned above, it is difficult at this point to assess the need for an additional, neutral wind. There is however mounting observational evidence for such neutral wind components (cf. Lizano *et al.* 1987). If taken at face value, our data suggest that the momentum rate in the neutral wind component might not need to be larger than that in the ionized component.

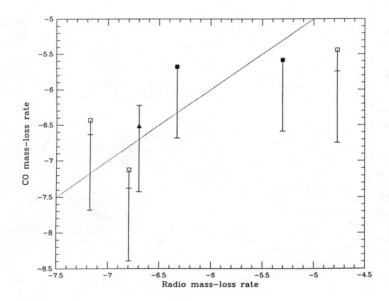

Figure 4: The mass-loss rate derived from radio continuum data is plotted against the CO mass-loss rate. Error bars indicate uncertainties on the CO mass-loss rates.

4. The Role of Accretion in the Pre-Main-Sequence Phase

4.1. Background

The concept of mass-accretion is central to the theory of protostellar evolution. I emphasized in Section 1 that although there is little evidence for infall in the earliest phase of stellar evolution, the problem probably lies in the difficulty of observing direct accretion processes at that stage. In this section, I review observational evidence for accretion processes in young but optically visible low-mass stars.

Before the youth of T Tauri stars was recognized, the T Tauri phenomenon was tentatively interpreted as caused by accretion on field stars passing through nebulae. Herbig (1962) gives several references to this early work. The evolutionary status of T Tauri stars became clear in the 1950's, and this "interaction" hypothesis was soon dismissed.

The accretion hypothesis was revived in a different form in 1972, when Walker found a subclass of T Tauri stars, called YY Orionis stars after their prototype. They have at times inverse P Cygni absorption components at their Balmer lines, which unambiguously indicate that mass accretion is taking place in the Balmer line formation region. At about the same time, Larson (1969, 1972) published the first detailed numerical computations of protostellar collapse, which predicted a long phase (about 10^7 years) of diffuse matter accretion as the optically revealed star moved down its Hayashi track toward the main-sequence. Walker (1972) and Appenzeller and Wolf (1977) identified the YY Ori phenomenon with this phase. A few bright YY Ori stars were discovered and monitored extensively (e.g. Bertout et al. 1982) and computations of line profiles in infalling envelopes were made (cf. Wagenblast et al. 1983 and references therein).

While early models were quite successful in interpreting the data in the framework of *spherical* accretion onto the star, difficulties appeared when high-resolution data became available. It then became obvious that evidence for outflow was also present in the spectra of YY Ori stars. In S CrA, for example, both H_α and the CaII K lines have blue-displaced absorption components while the higher Balmer lines display complex profiles which include a red-displaced absorption component. The contemporaneous detection of high-velocity molecular outflows around many young stellar objects, including a few T Tauri stars, and the discovery of optical jets emanating from a number of young stellar objects clearly showed that large-scale outflow, rather than infall, was characteristic of the observable protostellar phase. Consequently, the picture of spherical collapse began to fade, as did the models of YY Ori and T Tauri stars based on this picture.

Good ideas never die; the accretion hypothesis was revived during the last two years within the new picture of protostellar evolution which emerged from observational advances mentioned in Section 2. Current theoretical thinking about the protostellar phase is summarized in F. Shu's contribution, while L. Hartmann, S. Kenyon, and G. Basri all discuss various aspects of the disk hypothesis for young stellar objects in this volume. My own remarks will therefore be limited to the role of *disk accretion* in the T Tauri phase.

4.2. Disk Accretion in the T Tauri Phase

It was recognized early that T Tauri stars have a rich emission spectrum in the optical range. More recently, it was discovered that line emission extends to the UV and IR ranges, and that they have large continuum excesses when compared to main-sequence stars of similar spectral types. They also possess relatively strong ionized and neutral winds. All of these properties require a large energy input in order to be established and maintained. Two main classes of models have been proposed which differ mainly in their assumptions about the origin of the energy source responsible for the T Tauri characteristics. The first one assumes that the energy generation is intrinsic to the star, while the second one favors an extrinsic energy source.

"Chromospheric" models belong to the first class because they *assume* that the star can somehow heat-up a dense chromosphere (e.g. Herbig 1970; Calvet *et al.* 1984). By analogy with the solar case, magnetic waves are believed to be responsible for chromospheric heating. Calvet (1983) estimated the radiative energy losses due to emission lines, X-ray emission, and UV continuum excess emission and found them to be a sizable fraction of the stellar luminosity in typical T Tauri stars. This requirement is difficult to reconcile with the finding that dynamo-driven magnetic activity is similar in T Tauri stars and RS CVn systems, two classes of objects characterized by similar rotation rates (Bouvier 1987). The fraction of magnetic wave energy flux to radiative luminosity is indeed much smaller in RS CVn stars.

Models involving mass-accretion belong to the second class. In this picture, the star is quite normal, and the T Tauri properties result from the interaction of the star with its surroundings. If this interaction dies out on a time-scale shorter than the pre-main-sequence time-scale, a smooth transition between T Tauri stars and main-sequence solar-type stars is insured. While any plausible extrinsic energy source with this property would do the job, current scenarios indicate that a disk is formed at the same time as the star during the collapse of a molecular core; if so, mass accretion from the disk onto the star is the *unavoidable* result of energy dissipation in the disk.

Interaction between disk and star can take different directions, depending on the mass-accretion rate and the respective specific angular momenta of star and disk material. During early evolutionary phases the transition between star and disk is probably fuzzy, and the stellar equator is probably rotating as fast as the innermost parts of the disk. Mass accretion during that phase will increase the angular momentum of the central object, which will therefore spin faster. Shu *et al.* (1987) have shown that a centrifugally-driven wind can result and indeed, the correlation between wind mechanical luminosity and bolometric luminosity of low-mass protostellar objects discussed in Section 3 could indicate a relationship between accretion and wind luminosities if most of the bolometric luminosity originates in the disk (see also Strom *et al.* 1987a).

The protostellar wind is probably instrumental in braking the stellar rotation until the equatorial velocity reaches values observed in T Tauri stars (vsini = 10-30 km/s). A boundary layer between the fast rotating inner parts of the disk and the photosphere thus develops (Lynden-Bell and Pringle 1974). While the physics of the boundary layer are difficult, as emphasized by J. Pringle in his lecture, it

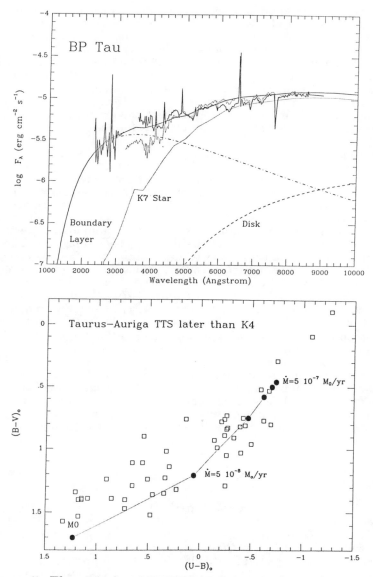

Figure 5: The optical and UV dereddened spectra of BP Tauri. Both models and observations are shown. The total calculated systemic spectrum (star + disk + boundary layer) is the solid smooth curve, while the photospheric + chromospheric stellar continuum is the dotted curve. The boundary layer (dot- dashed) and disk (dashed) contributions are also shown.

Figure 6: The (U-B) – (B-V) plane. Both models and observations are shown. The open squares display the dereddened colors of the Taurus-Auriga T Tauri stars later than K4. Filled dots show the colors of disk models computed for a M0IV central star with 3 $R_\odot$ radius, for five mass-accretion rates between $5 \cdot 10^{-8}$ and $5 \cdot 10^{-7}$ $M_\odot$/yr, and for a view angle of 45°.

is possible to estimate its approximate emitted continuous spectrum from simple assumptions (cf. Bertout *et al.* 1987). To illustrate the effects of emission from the boundary layer on the overall spectrum of T Tauri stars, a comparison of predicted and observed spectral energy distributions for the T Tauri star BP Tau is shown in Fig. 5. It is evident from that figure that about half the total luminosity of BP Tau is caused by accretion, and that the boundary layer emits mostly in the UV range. Since the disk itself emits mostly in the IR, this simple model attributes both IR and UV excesses observed in T Tauri stars to a single physical mechanism. The energy powering the T Tauri activity is ultimately the gravitational energy of the molecular core from which the star/disk system has formed.

There is some debate about how many of the T Tauri stars are actively accreting stars. Kenyon and Hartmann (1987) estimate that only the most extreme members of the T Tauri class (about 10-20%) are surrounded by accretion disks, while Bertout *et al.* (1987) show that at least some of the moderate T Tauri stars are also actively accreting. Strom *et al.* (1987b) try to estimate the accretion luminosity in a large sample of stars, but find that their criterium is insensitive to accretion luminosities less than about twice the stellar luminosity. Bertout *et al.* (1987) mentioned the usefulness of the $(U-B)_o$ vs. $(B-V)_o$ diagram in testing for accretion. I compiled these colors for the T Tauri stars of the Taurus-Auriga association, using data from Rydgren *et al.* (1983) as well as our own database (Bertout and Bouvier, unpublished). Extinction corrections were made by using a standard extinction curve together with Cohen and Kuhi's (1979) visual extinctions. Fig. 6 show $(U-B)_o$ vs. $(B-V)_o$ for all stars later than K4, as well as the computed colors for an M0 star surrounded by an accretion disk seen at 45° (see Bertout *et al.* 1987). The dots correspond to various mass-accretion rates in the range $5 \cdot 10^{-8} - 5 \cdot 10^{-7}$ $M\odot$/yr. Because of the contrast between cool photosphere and hot boundary layer, there is a sharp distinction between late-type stars with and without accretion in this diagram. Fig. 6 demonstrates that about 50% of the Taurus-Auriga T Tauri stars later than K4 (including all stars showing at times spectroscopic signs of accretion: the YY Ori stars) show colors characteristic of boundary layer emission. The second 50% has little radiative excesses; either these stars have accreted their disks, or planet formation has taken place in the disk, thus stopping the accretion process.

Acknowledgments

This work was done while a guest at the Berkeley Astronomy Department. I am indebted to Gibor Basri for making my stay there possible. Partial support was provided by NSF Grant AST 86-16863. I thank Suzan Edwards and Frank Shu for useful discussions, and Sylvie Cabrit's contribution to Section 3 is gratefully acknowledged. I am also grateful to Joli Adams for improving the style of this paper.

References

André, P., 1987, in *Protostars and Molecular Clouds*, Eds. T. Montmerle and C. Bertout (Saclay: CEA/Doc)

Appenzeller, I., Wolf, B., 1977, *Astr. Ap.*, **54**, 713.

Appenzeller, I., Jankovics, I., Oestreicher, R., 1984, *Astr. Ap.*, **141**, 108.

Bastien, P., Ménard, F, 1988, *Ap. J.*, in press

Beckwith, S., Gatley, I., Matthews, K, 1978, *Ap. J.*, **223**, L41.

Beichman, C., Myers, P., Emerson, J., Harris, S., *et al.*, 1986, *Ap. J.*, **307**, 337.

Bertout, C., 1987, in *IAU Symp. 122: Circumstellar Matter*, Eds. I. Appenzeller and C. Jordan (Dordrecht: Reidel)

Bertout, C., Carrasco, L., Mundt, R., Wolf, B., 1982, *Astron. Ap. Suppl.*, **47**, 419.

Bertout, C., Basri, G., Bouvier, J., 1987 *Ap. J.*, in press

Bertout, C., Cabrit, S., Roland, J., 1988, in preparation

Bouvier, J., 1987, *Thèse de Doctorat* (Paris:Université Paris VII)

Cabrit, S., Bertout, C., 1986, *Ap. J.*, **307**, 313.

Cabrit, S., Bertout, C., 1988, in preparation

Cabrit, S., Goldsmith, P.F., Snell, R.L., 1988, *Ap. J.*, submitted

Calvet, N., 1983, *Rev. Mex. Astron. Ap.*, **7**, 169.

Calvet, N., Basri, G., Kuhi, L.V., 1984, *Ap. J.*, **277**, 725.

Campbell, B., Persson, S.E., Strom, S.E., Strom, K.M., Grasdalen, G.L., 1988, *AJ*, in press

Clark, F.O., Laureijis, R.J., 1986, *Astron. Ap.*, **154**, L26.

Cohen, M., Kuhi, L.V., 1979, *Astrophys. J. Suppl.*, **41**, 743.

DeCampli, W.M., 1981, *Ap. J.*, **244**, 128.

Dyson, J.E., 1984, *Ap. Space Science*, **106**, 181.

Edwards, S., Cabrit, S., Strom, S.E., Heyer, I., Strom, K.M., Anderson, E., 1987, *Ap. J.*, **321**, 473.

Edwards, S., Snell, R., 1982, *Ap. J.*, **261**, 151.

Edwards, S., Snell, R., 1983, *Ap. J.*, **270**, 605.

Edwards, S., Strom, S.E., Snell, R.L., Jarrett, T.H., *et al.*, 1986, *Ap. J.*, **307**, L65.

Evans, N.J., Levreault, R.M., Beckwith, S., Skrutskie, M, 1987, *Ap. J.*, **320**, 364.

Hartmann, L., 1986, *Fund. Cosmic Phys.*, **11**, 279.

Hartmann, L., Edwards, S., Avrett, A., 1982, *Ap.J.*, **261**, 279.

Herbig, G.H., 1962, *Adv. Astron. Ap*, **1**, 47.

Herbig, G.H., 1970, *Mem. Soc. R. Sci. Liège*, **19**, 13.

Kenyon, S.J., Hartmann, L., 1987, *Ap. J.*, **323**, 714.

Königl, A, 1982, *Ap. J.*, **261**, 115.

Kwok, S., Volk, K, 1985, *Ap. J.*, **299**, 191.

Lada, C.J., 1985, *Ann. Rev. Astron. Ap.*, **23**, 267.

Larson, R.B., 1969, *Mon. Not. R. Astron. Soc*, **145**, 271.

Larson, R.B., 1972, *Mon. Not. R. Astron. Soc*, **157**, 121.

Lightfoot, J.L., Glencross, W.M., 1986, *Mon. Not. R. Astr. Soc.*, **221**, 993.

Lizano, S., Heiles, C., Rodriguez, L.F., Koo, B.-C., *et al.*, 1987, *Ap. J.*, in press

Lynden-Bell, D., Pringle, J.E., 1974, *Mon. Not. Roy. Astr. Soc.*, **168**, 603.

Moriarty-Schieven, G.H., Snell, R.L., Strom, S.E., Schloerb, F.P., *et al.*, 1987, *Ap. J.*, in press

Mozurkewich, D., Schwartz, P.R., Smith, H.A., 1986, *Ap. J.*, **311**, 371.

Mundt, R., Stocke, J., Strom, S.E., Strom, K.M., 1985, *Ap. J.*, **297**, L41.

Myers, P., Heyer, M., Snell, R., Goldsmith, P., 198,; *Ap. J.*, in press

Natta, A., Giovanardi, C., Palla, F., Evans, N.J., *Ap. J.*, in press

Rydgren, A.E., Schmelz, J.T., Zak, D.S., Vrba, F.J., 1984, *Publ. US Naval Obs.* XXV, Part 1, Washington

Sakashita, S., Hanami, H., 1986, *Publ. Astron. Soc. Japan*, **38**, 879.

Sargent, A.I., Beckwith, S., 1987, *Ap. J.*, **323**, 294.

Shu, F.H., Lizano, S., Ruden, S.P., Najita, J., 1987, *Ap. J.*, in press

Snell, R.L., Bally, J., 1986, *Ap. J.*, **303**, 683.

Snell, R.L., Schloerb, F.P., 1985, *Ap. J.*, **285**, 490.

Snell, R.L., Loren, R.B., Plambeck, R.L., 1980, *Ap. J.*, **239**, L17.

Stocke, J., Hartigan, P., Strom, S.E., Strom, K.M., *et al.* , 1987, *Ap. J.*, in press

Strom, S.E., Strom, K.M., Grasdalen, G.L., Capps, R.W., Thompson, D., 1985, *AJ*, **90**, 2575.

Strom. S.E., Strom, K.M., Edwards, S., 1987a, in *NATO ASI: Galactic and Extragalactic Star Formation*, Eds. R. Pudritz and M. Fich (Dordrecht: Reidel)

Strom, K.M., Strom, S.E., Kenyon, S., Hartmann, L., 1987b, *AJ*, in press

Thompson, R., 1985, in *Protostars and Planets II*, Eds. M.S. Matthews and D.C. Black (Tucson: University of Arizona Press)

Torrelles, J.M., Rodriguez, L.F., Cantó, J., Carral, P., *et al.* , 1983, *Ap. J.*, **274**, 214.

Torrelles, J.M., Cantó, J., Rodriguez, L.F., Ho, P.T.P., Moran, J.M., 1985, *Ap. J.*, **294**, L117.

Wagenblast, R., Bertout, C., Bastian, U., 1983, *Astron. Ap*, **120**, 6.

Walker, M.F., 1972, *Ap. J.*, **175**, 89.

Walker, C.K., Lada, C.J., Young, E.T., Maloney, P.R., Wilking, B.A., 1986, *Ap. J.*, **309**, L47.

Weaver, R., McCray, R., Castor, J., Shapiro, P., Moore, R., 1977, *Ap. J.*, **218**, 377.

Welch, W.J., Dreher, J.W., Jackson, J.M., Terebey, S., Vogel, S.N., 1987, *Science*, **238**, 1550.

Wynn-Williams, C.G., 1982, *Ann. Rev. Astron. Ap.*, **20**, 597.

RADIO AND (SUB)MILLIMETER OBSERVATIONS OF THE INITIAL CONDITIONS FOR STAR FORMATION

Åke Hjalmarson and Per Friberg
Onsala Space Observatory
S-439 00 ONSALA
Sweden

ABSTRACT. Our stellar formation scenario has changed considerably during the last two decades. The very detection of interstellar ammonia in 1968 revealed the existence of a denser cloud population immersed in the dilute HI medium known from extensive observations of the 21 cm hydrogen line. Since then some 80 interstellar molecules have been indentified. These trace molecules are not only curiosities in space. They are our major probes of the physical conditions and processes inside the dense star formation regions. Moreover, the molecular lines provide the cooling necessary for cloud evolution towards stellar formation.

Our current knowledge of the small and large scale structure of molecular clouds as well as their large scale distribution in galaxies is reviewed with emphasis on the initial conditions for stellar formation. We also discuss cloud formation processes, support against gravitaional collapse as well as bipolar outflows indicating extensive mass loss during early stellar evolution.

1. INTERSTELLAR MOLECULES - PHYSICS COUPLED WITH CHEMISTRY

[References: Kutner 1984, Hjalmarson 1985a, Irvine *et al* 1985, 1987]

The interstellar molecules - about 80 have now been identified (Table 1) mainly via observation of >2000 spectral lines in the frequency range $0.7 - 3800$ GHz ($\lambda = 43$ cm $- 77$ μm) - are more than chemical curiosities in "empty" space as summarized by Tables 2 and 3. Since the dominant constituents of the molecular clouds, H_2 and He, cannot radiate at the low cloud temperatures we rely on their collisions with trace molecules (CO, H_2O, $HCN...$), which then transfer energy via their radiative (normally rotational) transitions. While CO is "only" $\sim 10^4$ times less abundant than H_2 most other trace molecules have $10^2 - 10^6$ times lower concentrations than CO.

The cloud masses deduced from molecular line observations are very large. Although the volume filling factor of these clouds in our Galaxy is only ~ 1 %, the total H_2 mass ($\sim 3 \times 10^9$ M_0) is as large as the HI mass (from 21-cm observations). While 90 % of the total H_2 mass is found

A. K. Dupree and M. T. V. T. Lago (eds.), Formation and Evolution of Low Mass Stars, 65–92.
© *1988 by Kluwer Academic Publishers.*

Table 1.   INTERSTELLAR MOLECULES

Simple molecules:

H_2	CO	NH_3	CS
$HC\ell$	SiO	SiH_4 *	SiS
C_2	HNO ?	CH_4 *	OCS
PN ?	SO_2	H_2O	H_2S
$NaC\ell$ *	$A\ell C\ell$ *	$KC\ell$ *	AlF * ?

Nitriles, acetylene derivatives, and related molecules:

HCN	$HC\equiv C-CN$	$H_3C-C\equiv C-CN$	H_3C-CH_2-CN
H_3CCN	$H(C\equiv C)_2-CN$	$H_3C-C\equiv C-H$	$H_2C=CH-CN$
CCCO	$H(C\equiv C)_3-CN$	$H_3C-(C\equiv C)_2-H$	HNC
CCCS	$H(C\equiv C)_4-CN$	$H_3C-(C\equiv C)_2-CN$	HNCO
$HC\equiv CH$ *	$H(C\equiv C)_5-CN$		HNCS
$H_2C=CH_2$ *			

Aldehydes, alcoholes, ethers, ketones, amides, and related molecules:

$H_2C=O$	H_3COH	$HO-CH=O$	H_2CNH
$H_2C=S$	H_3CCH_2OH	$H_3C-O-CH=O$	H_3CNH_2
$H_3C-CH=O$	H_3CSH	$H_3C-O-CH_3$	H_2NCN
$NH_2-CH=O$	$H_2C=C=O$		
HC_2CHO ?	$(CH_3)_2C=O$?		

Ions:

CH^+	HCS^+	H_2D^+ ?	$HCNH^+$
$HN_2{}^+$	SO^+ ?	$HOCO^+$	HOC^+ ?
H_3O^+ ?	HCO^+		

Cyclic molecules:

C_3H_2	SiC_2 *	$c-C_3H$

Radicals:

CH	C_2H	CN	HCO
OH	$\ell-C_3H$	C_3N	NO
	C_4H	NS	SO
	C_5H	C_2S	
	C_6H		

* Detected only in envelopes around evolved stars

inside the solar circle only 30 % of the HI mass occurs here. Most of the H_2 mass appears in giant molecular clouds (GMC) of masses $>10^5$ M_0. Figure 1 illustrates the coupling between the physical and chemical evolutions of a molecular cloud (cf. Prasad et al 1987).

After these introductory notes we will try to review our current knowledge of the small and large scale initial conditions for stellar formation - inside giant molecular clouds and dark clouds (§3.1) and in galaxies (§3.2). But first we need some simple excitation and radiative transfer theory in order to understand problems and uncertainties.

Table 2. WHY ARE INTERSTELLAR MOLECULES INTERESTING ?

□ Pure chemistry (which[1], where, how much[2], why)

□ Spectroscopy (Identification & molecular physics)

□ Cloud probing (of physical conditions & processes)[3]

□ Cloud energetics (cooling by molecular lines)[4]

1) See Table 1.

2) Also physics:
   i) Abundances of some isotopic variants (>50 are known) provide
       information on stellar evolution - *nucleosynthesis*.
   i) Abundances of molecular ions tell us about *ionization level* in
       clouds. Necessary to estimate influences of *magnetic field* on
       cloud evolution and star formation.
   iii)   Abundances govern *cloud energetics*, see 4.

3) See Table 3.

4) We need accurate knowledge of chemical abundances to compute
   cooling/heating rates which govern the *cloud evolution* towards
   protostar or small scale clump formation.

Table 3. MOLECULES AS PROBES

□ Have demonstrated the *existence* of dense ("molecular") clouds - dense
   and massive enough to support gravitational collapse towards star
   formation.

□ Allow observations of the *physical conditions* deep inside optically
   dark nebulae and in circumstellar envelopes (mass, density,
   temperature, velocity... distribution) — provides information on:

   a) *Initial conditions for star formation*

   b) *Early stellar evolution* (Physics and chemistry of *mass loss
       envelopes* around C rich as well as O-rich stars). Return of
       processed matter to ISM.

□ Allow studies of the *large scale distribution of dense clouds* in
   galaxies - provides information on:

   a) *Initial conditions for large scale star formation* - we need to
       know how the dense clouds are distributed and how the dense
       massive clouds are formed (and disrupted).

   b) *Galactic morphology*

   c) *Galactic dynamics*

   d) *Galactic evolution* (includes discussions of spiral arm structure,
       velocity field, cloud and star formation rates, density wave
       theory, stochastic selfpropagating star formation)

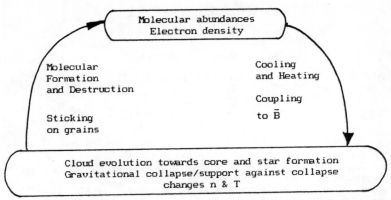

Figure 1. Cloud physics coupled with chemistry.

2. MOLECULAR EXCITATION AND RADIATIVE TRANSFER CONSIDERATIONS.

[References: Kutner 1984, Irvine *et al* 1987]

2.1. Molecular Excitation - a Simplistic Approach

We will here only consider the balance between radiative and collisional processes in a 2-level system, see Figure 2.

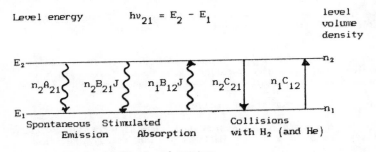

Figure 2. Two-level system

Equilibrium between upward and downward processes means that

$$n_2(A_{21} + B_{21}J + C_{21}) = n_1(B_{12}J + C_{12}) \qquad (1)$$

When collisions dominate (C_{12} & C_{21}, very large) this balance corresponds to a Boltzmann distribution at the <u>cloud kinetic temperature</u> T_k, i.e.,

$$\frac{C_{12}}{C_{21}} = \frac{n_2}{n_1} = \frac{g_2}{g_1} e^{-T_o/T_k} \qquad (2)$$

where $T_0 = h\nu_{21}/k = 4.8 \times 10^{-2} \nu[\text{GHz}] = 14.4/\lambda[\text{mm}]$ and $g_{1,2}$ are statistical weights. Similarly we introduce the <u>excitation temperature</u> T_{ex} to describe the real population distribution,

$$\frac{n_2}{n_1} = \frac{g_2}{g_1} e^{-T_o/T_{ex}}$$ (3)

Since $B_{12}/B_{21} = g_2/g_1$ the equilibrium condition (1) may be reformulated as

$$e^{-T_o/T_{ex}} = \frac{C_{21}e^{-T_o/T_k} + B_{21} J}{C_{21} + A_{21} + B_{21}J}$$ (4)

We note that

i) $T_{ex} \to T_k$, if $C_{21} \to \infty$ ("thermalization")

ii) $T_{ex} \to T_{bg} = 2.7$ K, if $C_{21} \to \emptyset$, we then have balance with the cosmic background radiation $J = (e^{T_0/T_{bg}} - 1)^{-1} \cdot A_{21}/B_{21}$.

We find for optically thin lines and neglecting the background $(J = \emptyset)$

$$T_{ex} = \frac{T_k}{1 + \frac{T_k}{T_o} \ln \left(1 + \frac{A_{21}}{C_{21}}\right)}$$ (5)

We may use (5) or (4) to estimate which C_{21}/A_{21} is required to produce a certain Tex in a cloud of temperature Tk.

The collision rate C_{21} can be expressed as

$$C_{21} \approx \sigma n \langle v \rangle$$ (6)

where n is the cloud number density and

$$\langle v \rangle \approx 10^4 \sqrt{T_k} \ [\text{cm s}^{-1}]$$ (7)

is the mean velocity between H_2 and the heavier molecule. The collision cross section σ may be estimated (very approximately) from the molecular dimensions, i.e. $\sigma \approx 10^{-16} - 10^{-15}$ cm^2.

The minimum cloud density required for observable emission is

$$n \geq n_{min} = \frac{10^{11}A_{21}}{\sqrt{T_k}} \quad (\text{for } C_{21} \geq A_{21})$$ (8)

where we have used $\sigma = 10^{-15}$ cm^2.

Table 4. EXCITATION REQUIREMENTS

Molecule	Frequency (GHz)	$A(s^{-1})$	$n_{min}(cm^{-3})$
H	1.4	3×10^{-15}	2×10^{-5}
OH	1.7	8×10^{-11}	1
NH$_3$(1,1)	24	2×10^{-7}	2×10^3
CO (1-$\emptyset$)	115	7×10^{-8}	7×10^2
CS (2-1)	98	2×10^{-5}	2×10^5
CS (7-6)	343	9×10^{-4}	1×10^7

According to relation (5) we would need a density as high as $n(\min) \cdot Tk/T_0$ to achieve $Tex = Tk/2$. It thus appears that different molecules (and transitions) require very different cloud densities for excitation as illustrated by Table 4 (calculated for 100 K using $\sigma = 10^{-15} \ cm^2$).

In case of high opacity τ (see § 2.2) the spontaneously emitted photons are reabsorbed (trapped) in the cloud and we should use $A_{21},eff \approx A_{21}/\tau$ ($\tau > 1$) in relations (5) & (8). This lowers the required cloud density, especially for CO where normally $\tau \geq 10$ (since it is a very abundant molecule). We should also remember that

$$A_{21} \sim \nu^3_{21}, \tag{9}$$

Hence, for a specific molecule $n(\min)$ is 10^3 times larger at $\lambda = 0.3 \ mm$ than at 3 mm.

We now understand that the very detections of NH_3 and CO emission imply the existence of dense ($\geq 10^3 \ cm^{-3}$) and hence massive clouds. The cloud cores probed e.g. by CS (J=2-1) must have densities in excess of $10^5 \ cm^{-3}$.

More accurate excitation considerations would require multi-level statistical equilibrium analysis based upon measured or quantum mechanically estimated collision rates, cf. Flower (1987). Further discussions will appear in § 3.1.3.

2.2. Radiative Transfer Problems

If we observe a transition at frequency $\nu = (Eu - E\ell)/h$ from a homogeneous cloud the antenna temperature of the line above the background — observed by a loss-less antenna above the atmosphere and perfectly coupled to the source — is given by

$$Ta = [J(Tex) - J(Tbg) - Tcon](1 - e^{-\tau}), \tag{10}$$

where

$$J(T) = \frac{T_0}{e^{T_0/T} - 1} \tag{11}$$

$J(T) \approx T$ for $T \gg T_0 = h\nu/k = 4.8 \times 10^{-2} \ \nu(GHz) = 14.4/\lambda(mm)$, $Tbg = 2.7 \ K$ (cosmic background) and Tcon is the additional continuum background observed by the antenna. At longer wavelengths Tcon may be due to free-free emission and synchrotron radiation while at short millimeter wavelengths dust emission may be significant.

A number of special cases should be considered:

i) For $\tau \gg 1$ (optically thick emission/absorption) we can directly determine Tex, which for sufficient excitation is a measure of Tk.

ii) For $|\tau| \ll 1$ (optically thin emission) we find, ignoring the background,

$$T_a = \frac{T_o \tau}{e^{T_o/T_{ex}} - 1} \tag{12}$$

The optical depth is ($N_{u/\ell} = \int n_{u/\ell} \ ds$ are column densities along the line of sight)

$$\tau = \frac{h\nu}{c}\,\phi_\nu\,\left[B_{\ell u}N_\ell - B_{u\ell}N_u\right] = \frac{h\nu}{c}\,\phi_\nu\,B_{u\ell}N_u\left[\frac{B_{\ell u}}{B_{u\ell}}\cdot\frac{N_\ell}{N_u} - 1\right] \qquad (13)$$

where $\dfrac{B_{u\ell}}{B_{\ell u}}\cdot\dfrac{N_\ell}{N_u} = e^{T_o/T_{ex}}$, ϕ_ν the line profile function, $\int\phi_\nu d\nu = 1$)

Using (13) and $A_{u\ell} = B_{u\ell}\cdot 8\pi h\nu^3/c^3$ relation (12) my be rewritten

$$Ta = (hc^2\phi_\nu/8\pi k\nu)\,A_{u\ell}\,N_u \text{ (independent of Tex)} \qquad (14)$$

Integrating over the line we find

$$\int Tad\nu = (hc^3/8\pi k\nu^2)A_{u\ell}\,N_u \qquad (15)$$

or in terms of convenient units

$$N_u[cm^{-2}] = 2\times10^3\nu^2[GHz]\int Tad\nu\ [K\ km\ s^{-1}]/A_{u\ell}[s^{-1}] \qquad (16)$$

In the optically thin limit we are just counting photons produced by spontaneous emission.

iii) $J(Tex) < J(Tbg)+Tcon$ & $\emptyset < \tau \ll 1$ - absorption lines

iv) In case Tex $\approx$ Tbg = 2.7 K (insufficient collisional excitation) we may still observe absorption lines if only Tcon $\neq \emptyset$ (and if the molecule in question exists at low cloud density).

v) For Tex $< \emptyset$ and $\tau \ll -1$ we find exponential amplification [cf Elitzur 1987]. This case of large population inversion is found in OH, H_2O & SiO masers.

vi) $\tau \approx 1$ - WE HAVE TO LEARN HOW TO USE IT ! We here observe variations of Tex and τ. Observations of rarer isotopic variants (for which $|\tau| \ll 1$) might be needed in order to separate the effects, but do not contain all information and are not strong enough to map.

Our assumption of a homogeneous cloud often is far from the observed situation. Self absorption due to lower excitation (foreground) gas may dramatically distort optically thick line profiles as illustrated in Figure 3. Here the importance of observing rarer isotopic variants is clearly demonstrated. Other examples of severe self-absorption in dark clouds are given by Guélin et al (1982) and Irvine and Schloerb (1984). Also in warm giant molecular clouds such effects are indeed troublesome (cf. HCO^+: Wotten et al 1984; CO: Loren et al 1981, Jaffe et al 1987)

2.3. Estimates of Population Distributions and Optimum Observing Frequencies

If we assume that all energy levels of a molecule are populated according to a Boltzmann distribution at a single temperature T we may relate the total population N_{tot} to N_u via

$$N_{tot} = \Sigma\,N_i = \frac{N_u}{g_u}\,e^{E_u/kT}\,Q(T) \qquad (17)$$

where we have introduced the partition function $Q(T) = \Sigma g_i e^{-E_i/kT}$.

From (17) and (15) we now find the useful relation (for $|\tau| \ll 1$)

$$\ln\left[\frac{8\pi k}{hc^3} \cdot \frac{v^2}{g_u A_{u\ell}} \int T_A dv\right] = \ln\frac{N_{tot}}{Q(T)} - \frac{E_u}{k_T} \qquad (18)$$

We may use relation (18) to estimate Trot (the rotational distribution

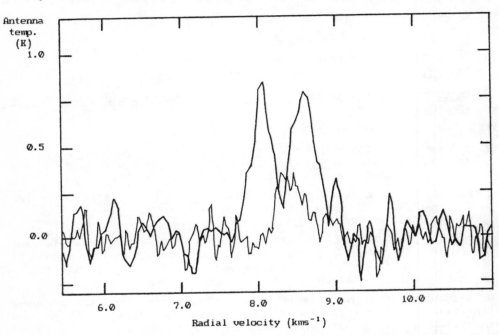

Figure 3. Self absorption spectrum of HCO$^+$ towards B335. The H^{13}CO$^+$ spectra (light) proves that it is self absorption and not two cloud components.

temperature) by plotting the left hand side quantity against the upper state energy Eu for a number of observed transitions of a specific molecule. Trot is determined by the slope of the line while Ntot/Q(Trot) is found for Eu = 0. Ntot may then be accurately determined if only Q(Trot) can be estimated. Trot will be a good measure of the cloud temperature if the cloud density is sufficient for thermalization.

For a <u>linear</u> molecule the energy of the Jth state is

$$E_J = hBJ(J + 1) \qquad (19)$$

and the transition frequencies are

$$v_{J \to J-1} = E_J - E_{J-1} = 2BJ, \qquad (20)$$

where B is the molecular rotation constant.

In this case we may obtain a simple expression for Q(T), since

$$Q(T) = \Sigma(2J + 1) e^{-hBJ(J + 1)/kT} \qquad (21)$$

can, for kT $\gg$ hB, be evaluated as

$$Q(T) \approx \int_0^\infty (2J + 1)e^{-hBJ(J + 1)/kT}dJ = \frac{kT}{hB} = 20.8 \frac{T[K]}{B[GHz]} \qquad (22)$$

Since

$$N_J = (2J + 1)e^{-hBJ(J + 1)/kT}/Q \qquad (23)$$

we may now determine which state has <u>maximum fractional population</u>, viz.

$$J_{max.pop} \approx \sqrt{Q/2} \qquad \left(\frac{dN_J}{dJ} = 0\right) \qquad (24)$$

We should also try to find <u>the state emitting the strongest signal</u> using (15) and (23). After introducing

$$A_{J\to J-1} \sim \upsilon^3\frac{J}{2J + 1} \qquad (25)$$

and (21) we arrive at

$$J_{max.intensity} \approx \sqrt{Q} \qquad \left(\frac{dI}{dJ} = 0\right) \qquad (26)$$

which indicates the optimum transition to observe at a certain cloud temperature (for optically thin emission). In the following Table 5 we have estimated Q, J(max intensity) and optimum observing frequencies for CO, HC_3N and HC_5N. (B = 57.6, 4.55 and 1.33 GHz, respectively).

Table 5. Q, $J_{max\ intensity}$ AND υ_{max} VS. TEMPERATURE

T(K)	=	10.	30.	100.	1000.
CO	Q	3.6	10.8	36.	360.
	Jmax	2	3	6	19
	υ(GHz)	230.	345.	690.	2200. (140 μm)
HC_3N	Q	46.	137.	457.	
	Jmax	7	12	21	
	υ(GHz)	64.	109.	191.	
HC_5N	Q	156.	469.	1560.	
	Jmax	12	22	39	
	υ(GHz)	32.	59.	104.	

It is interesting to note that CO has been observed up to 3800 GHz (77 μm), with maximum intensity for J=19 (140μm) in a shock heated region caused by the Orion KL bipolar outflow (Watson et al 1985, cf Watson 1985). Also note that the low frequency CO lines are very optically thick. Here the information should be more relevant for ^{13}CO and $C^{18}O$. The large Q:s for complex, heavier species seriously limit their detectability (especially in high temperature regions) since the populations are spread over the many available energy levels, resulting in low signal intensities, cf relation (18)

2.4. Determination of H_2 Column Density and Molecular Abundances
[References: Irvine *et al* 1985, 1987, van Dishoeck and Black 1987]

In § 2.2 - 2.3 we indicated how the antenna beam averaged total column density $N(X)$ of a molecule X can be estimated. Molecular abundance ratios can then be estimated in terms of N-ratios. To obtain fractional abundances with respect to H_2 we may proceed via

$$\frac{N(X)}{N(H_2)} = \frac{N(X)}{N(CO)} \cdot \frac{N(CO)}{N(H_2)} , \tag{27}$$

where the fractional abundance of CO rests upon

$$\frac{N(CO)}{N(H_2)} = \frac{N(CO)}{A_v} \cdot \frac{A_v}{N(H_2)} \tag{28}$$

The ratio $N(CO)/A_v$ has been estimated from dark cloud observations of ^{13}CO and $C^{18}O$ together with star counts to determine the visual extinction (A_v), cf Frerking *et al* (1982) and Cernicharo and Guélin (1987). The ratio $N(H_2)/A_v = 0.94 \times 10^{21}$ molecules $cm^{-2} mag^{-1}$ has been determined from observations by the Copernicus satellite (Bohlin *et al* 1978, cf Spitzer 1985) but only for $A_v < 2$ mag.

Assuming a constant gas-to-dust ratio also for higher extinctions we arrive at $N(C^{18}O)/N(H_2)=2 \times 10^{-7}$ and $N(CO)/N(H_2) \approx 10^{-4}$. For further discussions see van Dishoeck and Black (1987).

2.5 Determination of Cloud Masses
[References: Scoville and Sanders 1987, van Dishoeck and Black 1987]

Cloud masses may be estimated from mapping of $N(H_2)$ across the clouds. However this method requires at least CO and ^{13}CO data and hence would be very time consuming for observations of the large scale distribution of molecular clouds (via CO) in our own and nearby galaxies. Instead empirical relations have been deduced for H_2 column density,

$$N(H_2) = (3 \pm 2) \times 10^{20} I_{12} cm^{-2}, \tag{29}$$

and H_2 surface density,

$$\sigma(H_2) = (5 \pm 3) M_\odot pc^{-2} I_{12} \tag{30}$$

where $I_{12} = \int T_a(CO)dv$ [K km s^{-1}] is the velocity integrated CO $(J = 1-0)$ intensity.

The underlying statistics as well as inherent uncertainties are discussed in considerable detail by Scoville and Sanders (1987), van Dishoeck and Black (1987), and Solomon and Rivolo (1987). For virial equilibrium it appears that the proportionality constants in (29) and (30) scale with $\sqrt{n(H_2)}/Tk$, where $n(H_2)$ and Tk are the average density and temperature of the cloud. This reflects some of the uncertainty in the scaling laws. However, still larger uncertainties may be due to underlying assumptions (cf. γ-ray results discussed by Bloemen 1987 vs Osborne *et al* 1987).

3. PROBING THE INITIAL CONDITIONS FOR STELLAR FORMATION

We will first (§ 3.1) discuss the physical conditions in giant molecular clouds (GMCs) and dark clouds to provide some insighty into the initial conditions for star formation. The huge step from cloud to star –

Table 6. SPATIAL RESOLUTION VS BEAM SIZE AND DISTANCE

Beam size =	1"	5"	25"	2'	10'	source size (pc)
Source						
Dark cloud d ∼ 100 pc	0.0005	0.002	0.01	0.06	0.3	⟨0.1-20
GMC d ≥ 500 pc	0.002	0.01	0.06	0.3	1.5	⟨0.1-100
Galatic center d ≈ 8.5 kpc	0.04	0.2	1	5	25	100
SMC & LMC d ≈ 50 kpc	0.2	1	6	30	150	5000
Spiral galax d ≥ 700 kpc	3	17	85	400	2000	10000
M51 d ∼ 10 Mpc	50	240	1200	6000	30000	10000

Table 7. PHASES OF THE INTERSTELLAR MEDIUM

Name	$n(cm^{-1})$	$T(K)$	filling factor
Hot intercloud gas	3×10^{-3}	10^6	0.4-0.7
Warm clouds	0.1	10^3-10^4	0.4
Diffuse clouds	⟩10	50-100	
Dark clouds	⟩10^3	10	0.01-0.05?
GMC:s	⟩10^3	15-40	
HII regions	⟩10	10^4	
SNR	⟩1	10^4-10^7	

compression by $\sim 10^{20}$ or linear contraction by 10^7 - has been much studied but still is very much unknown. Another, perhaps even less known step is the formation of molecular clouds from dilute HI gas - compression by only 10^3 - which we will dwell on in connection with observations of molecular cloud distributions in galaxies (§ 3.2).

Table 6 translates relevant angular resolutions into spatial resolutions for nearby dark clouds, molecular clouds and nearby galaxies. While the smallest beam sizes may be achieved only with (millimeter wave) interferometers, a large mm wave telescope has a resolution of ~ 25". A $10'$ antenna beam characterizes a 1 m antenna operating at 100 GHz (3 mm), or a 100 m antenna at 1 GHz (21 cm). Hence the interiors of nearby clouds can be studied in considerable detail, but only in the most nearby galaxies individual clouds are discernible. In Table 7 we summarize our current knowledge of the phases of the interstellar medium (cf Jura 1987 and Shull 1987).

3.1. Giant Molecular Clouds and Dark Clouds
[References: Goldsmith 1987, Myers 1987, Blitz 1987, Falgarone and Pérault 1987, Fuller and Myers 1987, Walmsley 1987]

3.1.1. Classificitation of Cloud Regions

Table 8, adopted from Goldsmith (1987), summarizes our current knowledge of the initial conditions for stellar formation and also gives Jeans conditions and free fall time formulas.
The "dark clouds" are nearby, cold, quiescent, lower mass molecular clouds visible as dark areas in the sky due to their obscuration of the star light. We note that the observed core sizes and masses are comparable to the Jeans conditions for gravitational collapse and that the line widths found in the dark cloud cores are very close to the thermal widths at 10K.

3.1.2. Temperature
[References: Goldsmith 1987, Walmsley 1987]

Since the CO(J=1-0 and 2-1) transitions i) are thermalized at low cloud densities (Table 4) and ii) are optically thick these lines have become major probes of cloud temperature. cf § 2.2 - the observed CO/^{13}CO intensity ratio is only $\sim$ 2-10, while the isotopic abundance ratio is 40-90. However, this temperature estimate may only apply to the near surface of the cloud, where already the CO emission becomes optically thick if the velocity field of the cloud is microturbulent. Self-absorption may also cause problems. In case of considerable small scale clumping or large scale motions the CO temperature measure may apply to an interior region (cf Kwan and Sanders 1986).

Because of its large number of transitions (around $\lambda \sim 1.3$ cm), sensitive to a wide range of excitation conditions, the symmetric top molecule NH$_3$ has become another important probe of cloud temperature and density (cf. Ho and Townes 1983). Corrections to the derived population distribution temperatures are required at higher temperatures (cf Kuiper

Table 8. MOLECULAR REGIONS IN THE INTERSTELLAR MEDIUM

		Giant Molecular Cloud	Dark Cloud	
C				
O	Size(pc)	20-80	6-20	
M	Density (cm^{-3})	100-300	100-1000	
P	Mass (M_0)	$8 \times 10^4 - 2 \times 10^6$	$10^3 - 10^4$	
L	Linewidth (km/s)	6-15	1-3	
E	Temperature (K)	7-15	~10	
X				
C	Size (pc)	3-20	0.2-4	
L	Density (cm^{-3})	$10^3 - 10^4$	$10^2 - 10^4$	
O	Mass (M_0)	$10^3 - 10^5$	5-500	
U	Linewidth (km/s)	3-12	0.5-1.5	
D	Temperature (K)	15-40	8-15	
C	Size	0.5-3	0.1-0.4	$\approx \lambda$ Jeans
O	Density (cm^{-3})	$10^4 - 10^6$	$10^4 - 10^5$	
R	Mass (M_0)	$10 - 10^3 \geq M$ Jeans	0.3-10	$\approx M$ Jeans
E	Linewidth (km/s)	1-3	0.2-0.4	~ thermal
	Temperature (K)	30-100	~10	width
C	Size (pc)	<0.5	0.05	
L	Density (cm^{-3})	>10^6	10^5	
U	Mass (M_0)	$10 - 10^3$	~ 1	
M	Linewidth (km/s)	4-15	0.2	
P	Temperature (K)	30-200		

$$\lambda_{Jeans} = 0.27 \sqrt{\frac{T/10}{h/10^4}} \ pc; \quad M_{Jeans} = 4M_\odot \sqrt{\frac{(T/10)^3}{n/10^4}}$$

$$t_{ff} = 4 \times 10^5 / \sqrt{n/10^4} \ years$$

1987). Takano (1986) has compared Tk(NH₃) with Tk(CO) in molecular clouds containing bipolar outflow sources. He finds Tk(NH₃) $\approx$ Tk(CO) for lower temperatures, while Tk(NH₃) > Tk(CO) for high luminosity outflows. This may demonstrate the influence of internal heating sources on the interior cloud temperatures. The opposite case of temperature decrease from > 25 K at the cloud edges to 12-15 K in the clump cores has been derived from NH₃ (J, K = 1,1 and 2,2) observations of the very nearby (300 pc) molecular cloud IC348 (Bachiller et al 1987). In this case Tk (CO) was estimated to be about 20 K (at the near surface?). Photoelectric heating due to the radiation field from the Per OB2 cluster is proposed to be responsible for the temperature enhancement at the cloud edges.

In the Orion A molecular cloud there appear to be interior cores (15-50 K) considerably cooler than the near surface cloud temperature (Tk(CO) ≈ 60-100 K) consistent with heating from the near side HII region. There are, however, also regions close to the Kleinmann-Low nebula where the temperature is high. In the Orion A case relevant temperature information has been obtained from NH_3, CH_3C_2H, CH_3CN (all symmetric tops) analysis as well as from multi-transition population distribution observations of CH_3OH (asymmetric top), cf Andersson (1985) and Hjalmarson (1985b).

Peaking of cloud temperatures in compact regions ("hot cores") near infrared sources has been demonstrated by means of NH_3 observations for a number of clouds (e.g. Mauersberger et al 1986, 1987, Henkel et al 1987, Keto et al 1987). Keto et al advocate that they have found the collapse of a hot rotating core in the G10.6-0.4 area.

Similarly Jaffe et al (1987) find evidence for warm 80-150 K, dense cloud cores via CO (J=7-6) observations. The energy of the J=7 level is 155 K above the ground state. We note that our Table 5 indicates maximum intensity for J=6 at T=100 K.

3.1.3. Density
[References: Goldsmith 1987, Walmsley 1987]

Cloud densities can be estimated in a variety of ways, depending upon the purpose of the investigation:

i)   ⟨n⟩ cloud = Mass/Volume
ii)   ⟨n⟩ column = $N(H_2)$/size along line of sight
iii) Two-level excitations analysis, cf § 2.1 and Table 4
iv)   Statistical equilibrium model fit to multi-line observations.

Since different molecular transitions probe different density regions (cf Table 4) we may get rather different estimates from iii). We also understand that the observed cloud sizes will vary with the density requirement, see e.g. the CO(n)10^3 cm^{-3}) and HCN (n)10^5 cm^{-3}) Orion maps of Schloerb and Loren (1982).

Statistical equilibrium modeling of CS(J=2-1,3-2,5-4, and 6-5) mapping data for the cloud cores of M17, S140 and NGC 2024 has been performed by Snell et al (1984), who find that the column density across the sources varies by an order of magnitude while the estimated cloud density stays almost constant ∿ 4-9x10^5 m^{-3}. An LVG (large velocity gradient) model was employed to correct for photon trapping. These results have been confirmed by analysis of optically thin lines from $C^{34}S$ (Mundy et al 1986a) and by a multi-transition mapping of H_2CO (Mundy et al 1987a). The best interpretation of the data appears to be that the emission arises from numerous dense clumps with a beam filling factor changing from 100% at the center to 10% at the core edges. The modeling relies heavily upon good quantum mechanical estimates of the collision rates. However, it should be noted that constant cloud temperatures (from CO observations) have been entered into the models.

VLA mapping of 6 cm H_2CO in absorption in the S140 molecular cloud seems to indicate a clumpy structure in this region with 0.2 pc clumps of line width ∿ 1 km s^{-1} and with a velocity dispersion among the clumps of 1.3 km s^{-1} (Evans *et al* 1987). This would be consistent with the clumpy structure indicated by the modeling of the single dish multi-transition mapping of CS and H$_2$CO.

3.1.4. H$_2$ Column Density and Cloud Mass

According to § 2.4 we may just use N(H$_2$) $\approx$ 5x10$^6 \cdot$N(C^{18}O), where N(C^{18}O) can be estimated by relations (16) and (17). But the C^{18}O lines sometimes may be optically thick and the population distribution temperature is not always well determined by Tex(CO). Cloud masses may be estimated by integrating the H$_2$ column densities - derived from C^{18}O, or, less accurately, from ^{13}CO - over the cloud extent. The optically thick emission from ^{12}CO together with the emperical relations in § 2.5 may be used to determine the mass distribution on a large scale in our own and other galaxies. Virial mass estimates are also used in this context. In all cases distance uncertainties may contribute to large mass error bars.

3.1.5. Velocity Field
[References: Goldsmith 1987, Myers 1987, Fuller and Myers 1987]

Many details about the velocity field still remain uncertain and problematic. In the most quiescent cold dark cloud cores line widths are close to the thermal widths at 10 K, but in general the observed line-widths are highly supersonic. Modeling in terms of turbulence has been quite successful and there is some evidence for magnetically supported clouds in virial equilibrium, where the linewidths are caused by Alfvén waves. Large scale motions and rotation are sometimes found, but there are no unambiguous observations of large rotating disks. Contracting clouds, or cloud cores, have not (yet) been discovered, while high velocity outflow sources are frequently observed.

3.1.6. Bipolar Outflows - Mass Loss During Early Stellar Evolution
[References: Lada 1985, Snell 1987]

This topic is covered by Bertout, Lada and Mundth at this meeting. A few additional comments may be useful.

i)   No large rotating accretion disks have unambiguously been observed although such interpretations of observations do exist.

ii) From an observational point of view it is still unclear whether bipolarity is intrinsic to the way stars form or is caused by channeling by inhomogeneities in the ambient cloud.

iii) Increased search sensitivity leads to discoveries of many more outflow sources as demonstrated clearly by the Onsala observations of the NGC1333 region (Liseau *et al* 1987, presented at this meeting). An apparent question is whether all stars undergo an early mass loss-phase.

3.1.7. Cloud Morphology and Clumping
[References: Goldsmith 1987, Blitz 1987, Falgarone and Pérault 1987, Fuller and Myers 1987, Myers 1987, Wilson 1985]

We will here provide some examples of large and small scale structures:

i) Orion A (distance ~ 500 pc)
Starting at the very large scale we note that Maddalena et al (1986) have surveyed in the CO(1-0) line an area of 850 deg^2 centered on the Orion and Monoceros molecular cloud complexes using the Columbia 1.2 m telescope (9' beam). Emission from elongated, "clumpy" and filamentary structures was detected from ~ 1/8 of the region yielding a total mass of ~ $4x10^5 M_0$. Bally et al (1987) have mapped in the $^{13}CO(1-0)$ line an 8 deg^2 area around Orion A using the Bell Laboratories 7 m antenna (1.5' beam). Also at this scale the overall structure of the cloud is filamentary and clumpy. The total amount of gas is ~ $5x10^4 M_0$, while the Orion A "clump" contributes with ~ $2x10^3 M_0$. Zooming in towards the cloud core a number of studies exists concerning an area of a few to 10^2 $arcmin^2$ around Orion KL. There is a large overall velocity gradient along the cloud. Small scale clumping is indeed apparent with clump sizes of ~ 10-50" (0.02-0.1 pc) and estimated clump masses and densities in the range 10-100 M_0 and 10^5-10^7 cm^{-3}. These numbers are based upon Owens Valley aperture syntesis - including Onsala zero spacing maps - of CS(2-1) emission (beam size 7.5": Mundy et al 1987b, c) and Effelsberg NH_3 multi-transition mapping (beam size ~ 43"; Batrla et al 1983). For further discussions of the Orion KL source structure, including the massive outflow, we refer to Masson et al 1987, Johansson et al 1984, Hjalmarson 1985a, Irvine et al 1985, 1987, and the many references therein.
  It may be important to point out that recent $C^{18}O$ (1-0 and 2-1) maps, observed with 20-25" beams, show a much smoother structure (IRAM 30 m antenna: Wilson et al 1986; FCRAO 14 m telescope: Schloerb et al 1987). The dense clumps presumably are "hidden" in a lower density halo which contributes to a large portion of the $C^{18}O$ emission. The temperature/density structure of the core/halo configuration in Orion A and other molecular clouds is at present modeled by means of extensive CH_3CN, CH_3C_2H and CH_3OH multi-transition mapping at Onsala (first results were presented by Andersson 1985).One of the dense clumps is warm (140 K) and our CH_3OH lines show increasing line width with increasing level energy, which may be indicative of contraction (cf Hjalmarson 1985b).

ii) Sgr B2 (distance 8.5 kpc)
A brief source description is given by Irvine et al (1987). Recent high angular resolution studies of this massive (~ $3x10^6 M_0$) molecular cloud complex, only ~ 200 pc from the Galactic Center, have identified conspicuous chemical differentiation, small-scale structure, hot condensations and outflows. These conclusions are based upon Nobeyama observations (beam size 15") of HC_3N, HNCO, SO and OCS (Goldsmith et al 1987, cf discussion in Irvine et al 1987) and VLA multi-transition NH_3 aperture synthesis (beam size 5"x3") by Vogel et al 1987. Sizes of 0.2 pc and masses in the range 50-5000 M_0 are estimated for the hot cores.

iii) TMC-1 (distance ∿ 100 pc)

Because of its large abundance of linear carbon chain molecules TMC-1 is a comparatively well studied dark cloud fragment in the Taurus region (cf Irvine *et al* 1985, 1987). Ungerechts and Thaddeus (1987) have surveyed, with an angular resolution of 0.5^0, a 750 deg^2 region including the dark nebulae in Perseus, Taurus and Auriga in the CO(1-0)line, see Figure 4. Emission was detected from nearly 50% of the area The total mass is estimated to be ∿ 2×10^5 M_0 of which the nearby Taurus - Auriga clouds contain ∿ 3×10^4 M_0. Murphy and Myers (1985) have mapped with 8' resolution (corresponding to 0.3 pc) 11.5 deg^2 towards the Taurus complex in the CO(1-0) line. Again filaments and clumps are delineated. The mass of Heiles cloud 2 (of which TMC-1 is a part) is estimated to be ∿ 800 M_0. More detailed studies of Heiles cloud 2 have been performed by Schloerb and Snell (1984) using ^{13}CO (and $C^{18}O$). Their map of 3 deg^2 (grid spacing of 1-2') suggests a rotating ring structure in which several dense condensations have been identified - among them TMC-1 with an estimated (virial) mass of 35 M_0. The total mass in their map is ∿ 700 M_0 of which ∿ 400 M_0 resides in the ring. Cernicharo and Guelin (1987) have mapped this region, with 5' resolution, in the J=1-0 lines of CO, ^{13}CO, $C^{18}O$, HCO^+ and $H^{13}CO^+$. Their interpretation of the data is in terms of two bending filaments (instead of a ring) of mass ∿ 300 M_0, embedded in an envelope of similar mass.

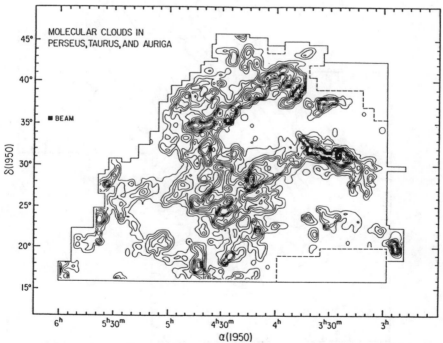

Figure 4. Integrated CO intensity in Perseus, Taurus, and Auriga. From Ungerechts and Thaddeus (1987)

High spatial resolution (1', i.e. 0.03 pc) maps of HC_3N (J=5-4, 9-8 and 12-11) lines towards TMC-1 trace mainly the high density gas (Schloerb *et al* 1983). The data suggest a total mass of $\sim$ 30 M_0 in an elongated structure, subdivided into several smaller fragments containing a few M_0. This is also indicated by the recent $C^{18}O$ data of Friberg, see Figure 5. We note, however, that this clumpy $C^{18}O$ integrated intensity structure is not easily interpreted in its details. The $C^{18}O$ line might be optically thick and hence the real column density variations might be bigger than what is observed. The situation is further complicated by pronounced velocity structure in the cloud as observed in the $C^{18}O$ lines and even more so in Onsala SO data (similar to the SO data for L134N shown in Figure 6). Similar comments apply to the clumping in the C_3H_2 map of Guélin (1987). Simultaneous modeling of

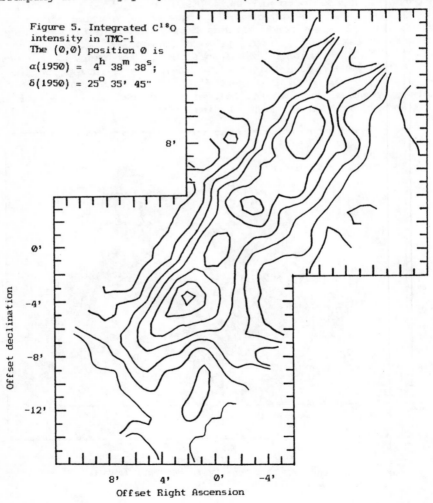

Figure 5. Integrated $C^{18}O$ intensity in TMC-1
The (0,0) position 0 is
$\alpha(1950) = 4^h\ 38^m\ 38^s$;
$\delta(1950) = 25^o\ 35'\ 45"$

Offset declination

Offset Right Ascension

multi-transition maps of several isotopic variants of well selected molecules ultimately will lead to unique interpretations in terms of physical as well as chemical structure variations. These cautionary notes are indeed applicable also to data for another rather well studied (but not too well understood) dark cloud, i.e.,

iv) L134N (L183, distance $\sim$ 100 pc)

From the $NH_3(1,1)$ observations of Rydbeck et al (1977) it appeared that this was a very quiescent cloud. The observed linewidth was only 0.27 km s^{-1} while the thermal width would be 0.16 km s^{-1} in a 10 K cloud. It was also concluded that only $\sim$ 15 % of the Onsala antenna beam (2.5' at 24 GHz) was covered with NH_3 clouds (The observed signal strength was only 15% of that expected for an excitation temperature equal to a cloud temperature of 10 K at the high optical depth $\tau \sim 2.4$). The inferred small scale clumping was subsequently mapped in NH_3 with the Effelsberg 100 m telescope ($\sim$ 40" beam) by Ungerechts et al (1980), who also derived a uniform cloud temperature of 9 K from comparison of the NH_3 (2,2) and (1,1) lines. The intrinsic line widths were as low as 0.2 kms^{-1} with small velocity changes across the cloud. The two observed clumps have sizes $\sim$ 0.1 pc, density of 10^4-10^5 cm^{-3}, and their masses are only $\sim$ 1 M_0 - slightly lower than the Jeans mass (4 M_0 at 10 K and 10^4 cm^{-3}), suggesting gravitational equilibrium rather than collapse. Since then NH_3 surveys of dark clouds by e.g. Myers and Benson (1983) have indicated cold, dense clumps of similar characteristics in many

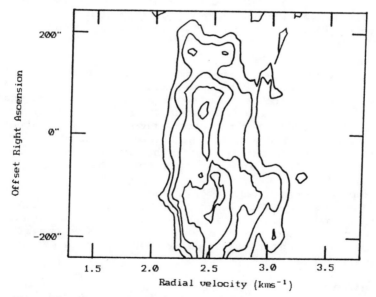

Figure 6 Right Ascension velocity map of SO 3_2-2_1 in L134N. The (0,0) position is $\alpha(1950)$ = 15^h 51^m 32.7^s; $\delta(1950)$ = -2^0 42' 51". The spatial resolution is 40" and the velocity resolution is 0.05 kms^{-1}

dark clouds, cf Fuller and Myers (1987). The IRAS survey has been used by Beichman et al (1986) to search 95 nearby dense cloud cores for evidence of newly forming stars. About 50 % of these clumps have associated IRAS sources, many of which have been further studied in the wavelength range 0.4 to 20 μm (optical and near IR) by Myers et al (1987), confirming that these IRAS sources are indeed stars (of T Tauri or pre - T Tauri type). The NH_3 lines appear significantly broader in cores with stars as discussed by Fuller and Myers (1987).

To complicate a seemingly rather clearcut situation we now have to discuss other structure probes. Extensive mapping of L134N to ultimately disentangle variations of physical characteristics from chemical gradients has been performed by Swade and collaborators at the University of Massachusetts. Some of the results have been reported by Swade et al (1985) and were discussed from a chemical point of view by Irvine et al (1987). A $C^{18}O(1-0)$ map of the core region leads to an estimated mass of 15 M_0 and masses of 0.1-0.2 pc clumps of a few M_0. However, the integrated brightness distributions of NH_3, C_3H_2, $H^{13}CO^+$, CS and SO show different structures and clumps compared with $C^{18}O$ and with each other. The situation is rather unclear since optical depth and excitation effects may be mixed with abundance variations, cf Irvine et al (1987). The very small scale spatial vs velocity structure variations apparent in Onsala $SO(3_2-2_1)$ data of Friberg (Figure 6) further complicates the situation. It seems appropriate to stress that the brightness clumps delineated by e.g. NH_3 may not necessarily outline mass concentrations very well. Elongated structures claimed to indicate large accretion disks hence should be regarded with respectful caution. The fact that the chemical equilibrium time scale is as large as 10^7 years should be a warning here (Prasad et al 1987; Millar et al 1987)

v) Some further references on clumping and hierarchical fragmentation
Observational aspects of small scale clumping in molecular clouds have been discussed by Wilson (1985). Fragmentation on all scales ranging from cloud complexes to protostellar cores is reviewed by Falgarone and Pérault (1987), who also discuss recent models of clumped molecular clouds. Scalo (1985, 1987) presents observational as well as theoretical work on fragmentation, hierarchical structure and turbulence, cf also Dickman (1985) and the "classical" paper by Larson (1981). Norman and Silk (1980) present a theory for low mass clump formation due to T Tauri winds.

Star formation from hierarchical cloud fragmentation has been discussed by Zinnecker (1984). The recent study of Jeans collapse in a turbulent medium by Bonazzola et al (1987) indicates how low mass cores may collapse in otherwise stable clouds. Turbulence is one of several mechanisms working to support classically Jeans unstable clouds against collapse, cf § 3.1.9.

3.1.8. Ionization Level and Magnetic Fields

The free electron abundance determines the degree to which magnetic fields will influence cloud contraction towards clump and star formation. Here interstellar chemistry shows its importance and (present) limitations (Hjalmarson 1985a, Langer 1985). The fractional electron abundance must be higher than that of the most abundant ion HCO^+, i.e. $>10^{-8}$. But its upper limit has become uncertain due to laboratory measurements demonstrating that the cruxial $e+H_3^+$ reaction rate was 100 times lower than "expected". This has shifted the observational upper limit from 10^{-7} to 10^{-5} (!), although arguments involving (uncertain) estimates of the cosmic ray ionization rate may lead to $\leq 3x10^{-7}$ in dense regions.

Observations of interstellar magnetic fields have been critically reviewed by Heiles (1987). Recent estimates by Crutcher et al (1987) lead to a working hypothesis that the total magnetic field strengths are of order 30 and 120 μG in cool dark clouds (10^3-10^4 cm^{-3}) and warm molelcular clouds (10^4-10^6 cm^{-3}), respectively. Zeeman splitting observed in OH maser fragments (10^6-10^8 cm^{-3}) leads to field strengths in the range 2000 - 10 000 μG. In low density (0.1 - 100 cm^{-3}) regions, on the other hand, fields of 1 - 10 μG have been estimated.

The importance of magnetic fields in connection with cloud contraction and star formation has been discussed e.g. by Langer (1978), Nakano (1984), Mestel (1985), and Mouschovias (1987), and with more emphasis towards star formation by Lizano and Shu (1987), Shu et al (1987) and Shu at this meeting. A model for ionization - regulated star formation in magnetized molecular clouds has been presented by Pudritz and Silk (1987).

3.1.9. Cloud Energetics and Support Against Gravitational Collapse

Some useful references on cloud cooling by molecular and atomic lines and thermal balance are Goldsmith and Langer (1978), Takahashi et al (1985), Falgarone and Puget (1985, 1986) and the review papers by Watson (1985), Phillips (1987) and Black (1987). Hjalmarson (1985a) points out the (large!) abundance uncertainties for important coolants like CO, CI, O_2, H_2O and HCℓ.

Only 20 years ago a main problem in stellar formation modeling was the accumulation of enough mass in a limited volume to invoke collapse - in an HI cloud (10 cm^{-3} at 75 K) the Jeans' mass is 10^4 M_0. With the discovery of the dense molelcular clouds the situation became rather the opposite. While the observed star formation rate (SFR) in our Galaxy is 3 - 10 M_0/year the rate estimated from the total H_2 mass/free fall time $= 2x10^9 M_0/4x10^6$ years (for n $\sim$ 100 cm^{-2}) = 500 M_0/year. Hence some kind of support against collapse is needed. A number of alternative processes have been proposed (cf Turner 1984, Downes 1987, Lizano and Shu 1987, Myers 1987):

i) Rotation - is not observed in most clouds
ii) Winds — may suffice (Lada at this meeting; Norman and Silk 1980)
iii) Supersonic turbulence (caused by ii, iv or v?)
iv) Magnetic fields — may be very important since observed density vs size and turbulent velocity vs size relations agree well with the idea of magnetically supported clouds in virial equilibrium, where the observed supersonic linewidths would be explained by "Alfvén wave turbulence" (Myers 1987, Lizano and Shu 1987).

v) "Star cloud turbulence" or "Gravito-turbulence", originating (and replenished) by some cascading mechanism from differential galactic rotation (see Turner 1984, cf also Henriksen 1986).

3.1.10. Cloud Ages
[References: Casoli and Combes 1982, Kwan and Valdes 1983, Turner 1984, Lo et al 1987]

For giant molecular clouds the following age limits (among others) may be argued for (?):

$t < 10^9$ years	(Gas mass/SFR; $5 \times 10^9 M_0 / 5 M_0 yr^{-1}$)
$t < 3 \times 10^7$ years	(disruption by OB star formation)
$t < 10^8$ years	(arm crossing time)
$t > 10^8$ years	(interarm crossing time)
$t > t_{ff} = 4 \times 10^6$ years	(for $n = 100$ cm^{-3})

In case of dark clouds the situation is even less certain, as will appear from our brief discussion of the formation of dense clouds, cf. § 3.2.3.

3.2. Molecular Clouds in Spiral Galaxies

Our main question here is: *What can we learn about the formation of dense clouds and stars from studies of cloud distributions rather than cloud interiors?* Because of limited space (and time!) available we will here give very fragmentary notes on this issue. Comprehensive reviews on molecular cloud distributions (and "statistical" cloud properties) in our own (Solomon and Sanders 1985, Solomon and Rivolo 1987, Scoville and Sanders 1987) and nearby galaxies (Rydbeck 1985, Young 1987a,b) are available.

3.2.1. Radial Distribution

Although the total HI and H_2 masses may be similar the radial distributions are conspicuously different. While the molecular cloud surface density tends to be centrally peaked and/or to delineate massive "rings" of molecular clouds in the inner galaxy regions, the main part of the dilute HI gas seems to reside in outer envelopes. In the Milky Way about 90 % of the H_2 mass occurs inside the solar circle, but only 30% of the HI mass is found here (for uncertainties see § 2.5 and Osborne et al 1987). The good correlation of blue light luminosity with that of CO has led to the conclusion that the efficiency of star formation (starformation rate/available gas mass in a unit volume) is rather constant.

Wyse (1986) has provided a "recipe" for producing molecular cloud distributions very similar to the observed ones (from CO) from observed HI distributions and rotation curves. This purely phenomenological model uses HI cloud – cloud collisions dynamically driven through galactic rotation. We note that for M51 the agreement between the model produced radial CO distribution and the observed one becomes even more striking when the CO distribution is deconvolved with the antenna beam response (cf. Rydbeck et al 1985).

3.2.2. Azimuthal Distribution and Velocity Field

Since spiral arm patterns of bright stars and HII regions often are conspicuous two outstanding question are: i) Do spiral arms of dense molecular clouds exist?, and ii) Do spiral density waves play a role?

Only very recently we have found some clues in this direction. In the Milky Way a population of warmer molecular clouds delineates spiral arms in a background disk population of cold clouds (cf. Solomon and Sanders 1985, Solomon and Rivolo 1987, Scoville and Sanders 1987). In M51 pronounced molecular cloud spiral arms – and velocity shifts across arms in the sense predicted by density wave theory (cf. Sundelius *et al* 1987) – have been discovered via Onsala CO observations deconvolved with maximum-entropy and by Owens Valley aperture synthesis (Rydbeck *et al* 1985, 1987a,b,c; Lo *et al* 1987). Although our statistical sample is small (cf. Rydbeck *et al* 1987a) it seems safe to state that

i) density waves do trigger the formation of giant molecular clouds
ii) it is not (yet) clear to what extent density waves trigger massive
   star formation (e.g. via cloud – cloud collisions, cf. Scoville *et al* 1986, Rydbeck *et al* 1987b,c).

3.2.3. Cloud Formation and Destruction

From the observational indications that:
i) giant molecular clouds are found mainly (or only?) in spiral arms,
   with velocity shifts as predicted by density wave theory, and
ii) dark clouds are found everywhere,
we may choose between the following cloud formation alternatives:

a) GMC's form from/in dilute HI (super) clouds in density wave
   spiral arms ("various instabilities")
b) GMC's form by spiral density wave organization of already
   existing dense clouds ("orbit crowding")
c) Dark clouds form when GMC's are disrupted by O,B stars and
   SNR's
d) Dark clouds form due to density waves in arms and/or everywhere
   (spontaneously ?).

In this context the results from the coordinated CO and HI observations (Nobeyama and VLA) by Lada *et al*.1987 of a small part of M31 are very suggestive. The GMC's are situated inside the HI arm and the arm-inter-arm contrast is considerably smaller for HI than for CO. Lada *et al*. hence argue that GMC's are formed and destroyed over relatively short times within the HI spiral arm feature, our alternative a).
Some useful, more recent references on cloud formation and destruction processes (on various scales) are e.g. Casoli and Combes (1982), Kwan and Valdes (1983), Combes and Gerin (1985), Wyse (1986), Roberts and Stewart (1987), Elmegreen (1985a,b,c, 1987a,b,c,d) and Elmegreen and Elmegreen (1986), cf also Tenorio-Tagle *et al*.(1987), and Franco *et al* at this meeting.

4. CONCLUDING REMARKS

This review contains a, hopefully, balanced mixture of observational facts, estimates (based upon assumptions and models), interpretations, and perhaps wishes. Here the wise words of Harlow Shapley are useful to contemplate:
"A hypothesis or theory is clear, decisive and positive but is believed by no one but the man who created it. Experimental findings, on the other hand, are messy, inexact things which are believed by everyone except the man who did the work".
Raising these cautionary sentences in Viana do Castelo Frank Shu reminded us about the equally wise words of Eddington: "Never believe in observations until they have been confirmed by theory". Experiments paired with theory – and our dreams – do indeed constitute the ultimate driving force in science.

5. ACKNOWLEDGEMENTS

To the organizers of this meeting for the invitation to produce this lecture, to colleagues in many countries for stimulating discussions and collaboration and to the Swedish Natural Science Research Council (NFR) for financial support.

6. REFERENCES

6.1. Recent Review Papers and Books:

[1] Winnewisser et al: Astrophysics of Interstellar Molecules, in Modern Aspects of Microwave Spectroscopy, ed. G.W. Chanty, Academic Press 1979, p. 313

[2] Kutner: Probing Molecular Clouds, Fund. Cosm. Phys. (1984) 9, 233.

[3] Lucas et al (eds.): Birth and Infancy of Stars, North Holland 1985 (Les Houches course).

[4] Black and Matthews (eds.): Protostars and Planets II, U. Arizona Press 1985.

[5] Shaver and Kjär (eds.): ESO-IRAM-Onsala Workshop on (Sub)Millimeter Astronomy, ESO Conf. & Workshop Proc. No. 22, 1985.

[6] Diercksen et al (eds.): Molecular Astrophysics-State of the Art and Future Directions, Reidel 1985 (NATO ASI)

[7] Gahm (ed.): Astrophysical Aspects of the Interstellar Medium and Star Formation, Physica Scripta (1985) T11 (Proc. Crafoord Symposium in Honour of Lyman Spitzer Jr., Stockholm).

[8] van Woerden et al (eds.): The Milky Way Galaxy, IAU Symp. No. 106, Reidel 1985

[9] Chiosi and Renzini (eds.): Spectral Evolution of Galaxies, Reidel 1986.

[10] Peimbert and Jugaku (eds.): Star Forming Regions, IAU Symp. No. 115, Reidel 1987.

[11] Lonsdale Persson (ed.): <u>Star Formation in Galaxies</u>, NATO Conf. Publ. 2466, 1987 (Pasadena).
[12] Gilmore and Carswell (eds.): <u>The Galaxy</u>, Reidel 1987.
[13] Hollenbach and Thronson, Jr. (eds.): <u>Interstellar Processes</u>, Reidel 1987 (Wyoming).
[14] Morfill and Scholer (eds.): <u>Physical Processes in Interstellar Clouds</u>, Reidel 1987 (NATO ASI).
[15] Vardya and Tarafdar (eds.): <u>Astrochemistry</u>, IAU Symp. No. 120, Reidel 1987.
[16] Nakano: <u>Contraction of Magnetic Interstellar Clouds</u>, Fund. Cosm. Phys. (1984) <u>9</u>, 139.
[17] Turner: <u>How Stars Form – A Synthesis of Modern Ideas</u>, Vistas in Astronomy (1984) <u>27</u>, 303.
[18] Appenzeller: <u>Star Formation and Pre-main-sequence Stellar Evolution</u>, Fund. Cosm. Phys. (1982) <u>7</u>, 313.

6.2. Text References

Andersson, M. 1985, in Ref. 5, p. 353.
Bachiller, R., Guilloteau, S. and Kahane, C. 1987, Astron. Astrophys. <u>173</u>, 324.
Bally, J., Langer, W.D., Stark, A.A. and Wilson, R.W. 1987, Astrophys.J. <u>312</u>, L45
Batrla, W., Wilson, T.L., Bastien, P. and Ruf. K. 1983, Astron. Astrophys. <u>128</u>, 279.
Beichman, C.A., Myers, P.C., Emerson, J.P., Harris, S., Mathieu, R., Benson, P.J. and Jennings, R.E. 1986, Astrophys.J. <u>307</u>, 337.
Black, J.H. 1987, in Ref. 13, p. 731.
Blitz, L. 1987, in Ref. 14, p. 35.
Bloemen, H. 1987, in Ref. 13, p. 143.
Bohlin, R.C., Savage, B.D., and Drake, J.F. 1978, Astrophys.J. <u>224</u>, 132 (cf. Spitzer Jr. 1985, in Ref. 7, p. 5)
Bonazzola, S., Falgarone, E., Heyvaerts, J., Pérault, M. and Puget, J.L. 1987, Astron. Astrophys. <u>172</u>, 293.
Casoli, F. and Combes, F. 1982, Astron. Astrophys. <u>110</u>, 287.
Cernicharo, J. and Guélin, M. 1987, Astron. Astrophys. <u>176</u> 299.
Combes, F. and Gerin, M. 1985, Astron. Astrophys. <u>150</u>, 327.
Crutcher, R.M., Kazés, I. and Troland, T.H. 1987, Astron. Astrophys. <u>181</u>, 119.
Dickman, R.L. 1985, in Ref. 4, p. 150.
Downes, D. 1987, in Ref. 10, p. 93.
Elitzur, M. 1987, in Ref. 13, p. 763.
Elmegren, B.G. 1985a, in Ref.7, p.48
                1985b, in Ref.3, p.215
                1985c, in Ref.4, p.33.
Elmegren, B.G. 1987a, in Ref. 10, p. 457
                1987b, in Ref. 13, p. 259
                1987c, in Ref. 14, p. 1
                1987d, Astrophys.J. <u>312</u>, 626
Elmegren, B.G. and Elmegren, D.M. 1986, Astrophys.J. <u>311</u>, 554.
Evans, N.J., Kutner, M.L. and Mundy, L.G. 1987, preprint.

90

Falgarone, E. and Puget, J.L. 1985, Astron. Astrophys. <u>142</u>, 157
1986, Astron. Astrophys. <u>162</u>, 235.
Falgarone, E. and Pérault, M. 1987, in Ref. 14, p. 59.
Flower, D.R. 1987, in Ref. 13, p. 745.
Frerking, M.A., Langer, W.D. and Wilson, R.W. 1982, Astrophys.J. <u>262</u>, 590.
Fuller, G.A. and Myers, P.C. 1987, in Ref. 14, p. 137.
Goldsmith, P.F. 1987, in. Ref. 13, p. 51
Goldsmith, P.F. and Langer, W.D. 1978, Astrophys.J. <u>222</u>, 881.
Goldsmith, P.F., Irvine, W.M., Hjalmarson, Å. and Elldér, J. 1986, Astrophys.J. <u>310</u>, 383.
Goldsmith, P.F., Snell, R.L., Hasegawa, T., and Ukita, N. 1987, Astrophys.J. <u>314</u>, 525.
Guélin M. 1987, in Ref. 15, p. 171.
Guélin, M., Langer, W.D. and Wilson, R.W. 1982, Astron. Astrophys. <u>107</u> 107.
Heiles, C. 1987, in Ref. 13, p. 171 and in Ref. 14, 429.
Henkel, C., Wilson, T.L. and Mauersberger, R. 1987 Astron. Astrophys. <u>182</u>, 137.
Henriksen, R.N. 1986, Astrophys.J. <u>310</u>, 189.
Hjalmarson, Å. 1985a, in Ref. 5, p. 285.
Hjalmarson, Å. 1985b, in Ref. 7, p. 59.
Ho, P.T.P. and Townes, C.H. 1983, Ann. Rev. Astron. Astrophys. <u>21</u>, 239.
Irvine, W.M. and Schloerb, F.P. 1984, Astrophys.J. <u>282</u>, 516.
Irvine, W.M., Schloerb, F.P., Hjalmarson, Å. and Herbst, E. 1985, in Ref. 4, p. 579.
Irvine, W.M., Goldsmith, P.F. and Hjalmarson, Å. 1987, in Ref. 13, p. 561.
Jaffe, D.T., Harris, A.I. and Genzel, R. 1987, Astrophys.J. <u>316</u>, 231.
Johansson, L.E.B. et al 1984 (the Onsala spectral scan), Astron. Astrophys. <u>130</u>, 227.
Jura, M. 1987, in Ref. 13, p. 3.
Keto, E.R., Ho, P.T.P. and Haschick A.D. 1987, Astrophys.J. <u>318</u>, 712.
Kuiper, T.B.H. 1987, Astron. Astrophys. <u>173</u>, 209.
Kutner, M.L. 1984, see Ref. 2.
Kwan, J. and Valdes, F. 1983, Astrophys.J. <u>271</u>, 604.
Kwan, J. and Sanders, D.B. 1986, Astrophys.J. <u>309</u>, 783.
Lada, C.J. 1985, Ann. Rev. Astron. Astrophys. <u>23</u>, 269.
Lada, J.C., Margulis, M., Sofue, Y., Nakai, N. and Handa, T. 1987, preprint.
Langer, W.D. 1978, Astrophys.J. <u>225</u>, 95
Langer, W.D. 1985, in Ref. 4, p. 650.
Larsson, R.B. 1981, Mon. Not. R. Astr. Soc. <u>194</u>, 809.
Liseau, R., Sandell, G. and Knee, L. 1987, Astron. Astrophys., in press.
Lizano, S. and Shu, F.H. 1987, in Ref. 14, p. 173.
Lo, K.Y., Ball, R., Masson, C.R., Phillips, T.G., Scott, S. and Woody, D.P. 1987, Astrophys.J. <u>317</u>, L63.
Loren, R.B., Plambeck, R.L., Davis, J.H. and Snell, R.L. 1981, Astrophys.J. <u>245</u>, 495.

Maddalena, R.J., Morris, M., Moscowitz, J. and Thaddeus, P. 1986, Astrophys.J. 303, 375.

Masson, C.R., Lo, K.Y., Phillips, T.G., Sargent, A.I., Scoville, N.Z. and Woody, D.P. 1987, Astrophys.J. 319, 446.

Mauersberger, R., Henkel, C., Wilson, T.L., Walmsley, C.M. 1986, Astron. Astrophys. 162 199.

Mauersberger, R., Henkel, C. and Wilson. T.L. 1987, Astron. Astrophys. 173, 352.

Mestel, L. 1985, in Ref. 4, p. 320 (cf. also Ref. 7, p. 53).

Millar, T.J., Leung, C.M., Herbst, E. 1987, Astron. Astrophys. 183, 109-117.

Mouschovias, T.Ch. 1987, in Ref. 14, p. 453.

Mundy, L.G., Scoville, N.Z., Bååth, L.B., Masson, C.R. and Woody, D.P. 1986a, Astrophys.J. 304, L51.

Mundy, L.G., Snell, R.L., Evans, N.J., Goldsmith, P.F., and Bally, J. 1986b, Astrophys.J. 306 670.

Mundy, L.G., Evans, N.J., Snell, R.L. and Goldsmith P.F. 1987a, Astrophys.J. 318, 392.

Mundy, L.G. Cornwell, T.J., Masson, C.R., Scoville, N.Z., Bååth, L.B. and Johansson, L.E.B. 1987b, OVRO preprint 1987:12.

Murphy, D.C. and Myers, P.C. 1985, Astrophys.J. 298, 818.

Myers, P.C. 1987, in Ref. 13, p. 71.

Myers, P. and Benson, P. 1983, Astrophys.J. 266, 309.

Myers, P.C., Fuller, G.A., Mathieu, R.D., Beichman, C.A., Benson, P.J., Schild, R.E. and Emerson, J.P. 1987, Astrophys.J. 319, 340.

Nakano, T. 1984, see Ref. 16.

Norman, C. and Silk, J. 1980, Astrophys.J. 238, 158.

Osborne, J.L., Parkinson, M., Richardson, K.M. and Wolfendale, A.W. 1987, in Ref. 14, p. 81.

Phillips, T.G. 1987, in Ref. 13, p. 707.

Prasad, S.S., Tarafdar, S.P., Villere, K.R. and Huntress Jr, W.T. 1987, in Ref. 13, p. 631.

Pudritz, R.E. and Silk, J. 1987, Astrophys.J. 316, 213.

Roberts Jr, W.W. and Stewart, G.R. 1987, Astrophys.J. 314, 10.

Rydbeck, G. 1985, in Ref. 5, p. 169.

Rydbeck, O.E.H., Sume, A., Hjalmarson, Å., Elldér, J., Rönnäng, B.O. and Kollberg, E. 1977, Astrophys.J. 215, L35.

Rydbeck, G., Hjalmarson, Å. and Rydbeck, O.E.H. 1985, Astron. Astrophys. 144, 282.

Rydbeck, G., Hjalmarson, Å., Johansson, L.E.B. and Rydbeck, O.E.H. 1987a, in Ref. 10, p. 535.

Rydbeck, G., Hjalmarson, Å., Johansson, L.E.B., Rydbeck, O.E.H. and Wiklind, T. 1987b, in Ref. 11, p. 315.

Rydbeck et al 1987c, in preparation, and presented at the University of Massachusetts Conference "Molecular Clouds in the Milky Way and Nearby Galaxies".

Scalo, J.M. 1985, in Ref. 4, p. 201.

Scalo, J.M. 1987, in Ref. 13, p. 349.

Schloerb, F.P. and Loren, R.B. 1982, in Symposium on the Orion Nebula to Honor Henry Draper, eds. A.E. Glassgold, P.J. Huggins and E.L. Schucking, Ann.N.Y.Acad.Sci. 395, 32.

Schloerb, F.P., Snell, R.L. and Young, J.S. 1983, Astrophys.J. 267, 163.

Schloerb, F.P. and Snell, R.L. 1984, Astrophys.J. 283, 129.

Schloerb, F.P., Snell, R.L. and Schwartz, P.R. 1987, Astrophys.J. 319, 426.

Scoville, N.Z. and Sanders, D.B. 1987, in Ref. 13, p. 21.

Scoville, N.Z., Sanders, D.B. and Clemens, D.P. 1986, Astrophys.J. 310, L77.

Shu, F.M. Lizano, S. and Adams, F.C. 1987, in Ref. 10, p. 417, see also Ann. Rev. Astron. Astrophys. (1987) 25.

Shull, J.M. 1987, in Ref. 13, p. 225.

Snell, R.L. 1987, in Ref. 10., p. 213

Snell, R.L., Mundy, L.G., Goldsmith, P.F., Evans, N.J. and Erikson, N.R. 1984, Astrophys.J. 276, 625.

Solomon, P.M. and Sanders. D.B. 1985, in Ref. 4, p.59.

Solomon, P.M. and Rivolo, A.R. 1987, in Ref. 12, p. 105.

Sundelius, B., Thomasson, M., Valtonen, M.J. and Byrd, G.G. 1987, Astron. Astrophys. 174, 67.

Spitzer, Jr.L. 1985, in Ref. 7, p. 5

Swade, D.A., Schloerb, F.P., Irvine, W.M. and Snell, R.L. 1985, in Masers, Molecules and Mass Outflows in Star Forming Region, ed. A Haschick (NEROC Haystack Observatory), p. 73.

Takahashi, T., Hollenbach, D.J. and Silk, J. 1985, Astrophys.J. 292, 192.

Takano, T. 1986, Astrophys.J. 303, 349.

Tenorio-Tagle, G., Franco, J., Bodenheimer, P. and Rozycka, M. 1987, Astron. Astrophys. 179, 219.

Turner, B.E. 1984, see Ref. 17.

Ungerechts, H., Walmsley, C.M., and Winnewisser, G. 1980, Astron. Astrophys. 88, 259.

Ungerechts, H. and Thaddeus, P. 1987, Astrophys.J. Suppl. Ser. 63, 645.

van Dishoeck, E.F. and Black, J.H. 1987, in Ref. 14, p. 241.

Vogel, S.N., Genzel, R. and Palmer, P. 1987, Astrophys.J. 316 243.

Walmsley, C.M. 1987, in Ref. 14, p. 161.

Watson, D.M. 1985, in Ref. 7, p. 33.

Watson, D.M., Genzel, R., Townes, C.H. and Storey, J.W.V. 1985, Astrophys.J. 298, 316 (cf. Watson 1985).

Wilson, T.L. 1985, Comments Astrophys. 11, 83, see also Ref. 5, p. 401.

Wilson, T.L., Serabyn, E., Henkel, C. and Walmsley, C.M. 1986, Astron. Astrophys. 158, L1.

Wootten, A., Loren, R.B., Sandqvist, Aa., Friberg, P. and Hjalmarson, Å. 1984, Astrophys.J. 279, 633.

Young, J.S. 1987a, in Ref. 10, p. 557.

Young, J.S. 1987b, in Ref. 11, p. 197.

Wyse, R.F.G. 1986, Astrophys.J. 311, L41.

Zinnecker, H. 1984, Mon. Not. R. Astr. Soc. 210, 43.

Infrared Energy Distributions and the Nature of Young Stellar Objects

Charles J. Lada
Steward Observatory, University of Arizona, Tucson, Arizona, USA
85721

The formation and early evolution of stars in our galaxy occurs in dense regions of molecular clouds, where newly formed stars are physically associated with varying amounts of gas and dust that alter their observed appearance. Indeed, the actual star formation process and the youngest stellar objects themselves are rendered completely invisible by the obscuration of opaque circumstellar dust. As a consequence, a significant fraction of the luminous energy of a young stellar object is radiated in the infrared portion of the electromagnetic spectrum. Infrared observations therefore offer a powerful tool for studying the earliest phases of stellar evolution and perhaps even star formation itself. Recent studies of the infrared energy distributions of young objects have resulted in a new and clearer understanding of the processes involved in the formation and early evolution of low mass stars. In this paper some of the more interesting results of these investigations are reviewed.

1. INTRODUCTION

The formation and early evolution of stars in the galaxy takes place within the dust enshrouded dense cores of molecular clouds. In these circumstances we expect that young stars will be physically associated with varying amounts of gas and dust, which will affect their observed appearance. Indeed, the youngest stellar objects should be rendered completely invisible by the obscuration of opaque circumstellar dust. Thus the process of star formation is veiled from direct observation at optical wavelengths. Moreover, there is mounting evidence to suggest that many visible young stellar objects are surrounded by circumstellar disks which themselves are luminous either because they absorb and reprocess substantial amounts of radiation from their central stars or because they radiate away large amounts of gravitational potential energy as a result of accretion processes. As a consequence of these effects a significant fraction of the luminous energy of a young stellar object is radiated in the infrared portion of the electromagnetic spectrum. Infrared observations therefore offer a powerful tool for studying the earliest phases of stellar evolution and perhaps even star formation, itself.

A. K. Dupree and M. T. V. T. Lago (eds.), Formation and Evolution of Low Mass Stars, 93–109.

Indeed, infrared astronomy has provided the first glimpses of objects in the earliest stages of stellar evolution and, has contributed significantly to our understanding of star formation over the last two decades. Most of what we have learned from infrared observations about the nature of embedded objects in molecular clouds has been derived from broad–band photometric measurements. Until a few years ago, the majority of such observations were obtained with ground–based telescopes. Because of the opacity of the terrestrial atmosphere, these observations provided a relatively limited coverage in infrared wavelength, typically 2–10 μm, corresponding to emitting temperatures of about 300 to 1200 Kelvins. The launch and successful mission of the Infrared Astronomical Satellite (IRAS) nearly five years ago, however, enabled the extension of infrared studies out to wavelengths of 100μm, and thus provided systematic measurements of the cold (30–300K) component of dust emission from around young stellar objects. The combination of ground–based observations with IRAS data enables one to make a nearly complete accounting of the luminous energy radiated by young stellar objects.

The shape of the broad–band infrared spectrum of a young stellar object (YSO) depends both on the *nature* and *distribution* of the surrounding material. Clearly then, we expect that the shape of the spectrum will be a function of the state of evolution of a YSO. The earliest (protostellar) stages, during which an embryonic star is surrounded by large amounts of infalling circumstellar matter, should have a very different infrared signature than the more advanced (pre–main sequence) stages, where most of the original star forming material has already been incorporated into the young star itself. Recent observations of YSOs (e.g., Lada and Wilking 1984; Rucinski 1985; Myers *et al.* 1987; Wilking, Lada and Young 1988) have largely confirmed this expectation and shown that YSOs can be meaningfully classified by the shapes of their optical–infrared spectral energy distributions (e.g., Lada 1987). Moreover, such classifications can be arranged in a more or less continuous evolutionary sequence which is well modeled by at least one physically self-consistent theory of star formation and early stellar evolution (Adams, Lada and Shu 1987). In this contribution I will briefly discuss some of the interesting results obtained from recent infrared studies of embedded YSOs. Because of their utility for investigating star formation and early stellar evolution, it is useful to begin this lecture with a brief review of the salient properties of stellar and YSO energy distributions.

2. INFRARED ENERGY DISTRIBUTIONS OF YOUNG STELLAR OBJECTS

2.1. Some Useful Properties of Source Energy Distributions

Broad–band photometric observations are important for the study of the overall energetics of astronomical sources. In particular, they are essential for the investigation of objects which emit radiation over a wavelength range significantly broader than that emitted by a blackbody, or equivalently, of objects whose spectra cannot be characterized by a single emitting temperature. Most YSOs are in this latter category and to be fully understood require knowledge of their emergent spectra over a broad range of wavelength. The construction of the energy distribution of a source is particularly useful in this regard. An energy distribution is the plot of observable quantities λF_λ vs λ (or equivalently, νF_ν vs ν) in log–log space. The quantity νF_ν is useful because it is proportional to the actual energy flux (as opposed to the spectral energy flux density, i.e., F_ν) radiated in each logarithmic frequency or wavelength interval. Thus a flat distribution of νF_ν when plotted against $\log \nu$ represents equal energy radiated in each logrithmic frequency interval of the spectrum. This is obvious by noting that:

$$\int F_\nu d\nu = \int \nu F_\nu \, d\ln\nu$$

Thus, the peak of a source's energy distribution occurs at the frequency where the greatest emergent energy is located. Moreover, the blackbody function acquires certain useful properties when plotted as an energy distribution. For example, the shape of a blackbody curve is invariant to translations in energy flux and frequency on an energy distribution plot. In addition, one can demonstrate that the following relations hold for blackbody curves:

(1) $\qquad\qquad [\nu B_\nu]_{max} = 1.3283 \times 10^{-5} T^4 \quad$ ergs s^{-1} cm^{-2} ster^{-1}

(2) $\qquad\qquad \int\limits_0^\infty B_\nu(T) \, d\nu \; = \; \frac{\sigma}{\pi} T^4 \; = \; 1.3586[\nu B_\nu]_{max}$

(3) $\qquad\qquad \theta \; = \; 6.384 \times 10^7 \; \dfrac{[\nu B_\nu]_{max}^{\frac{1}{2}}}{T^2} \qquad$ arcsec

(4) $\qquad\qquad\qquad \lambda_{max} \; = \; \dfrac{.3670}{T} \quad$ cm

where B_ν is the Planck function, i.e.,

$$B_\nu(T) \; = \; \frac{2h\nu^3}{c^2} \frac{1}{\exp(\frac{h\nu}{kt}) - 1}.$$

2.2. Observations: An Empirical Classification for YSO Energy Distributions

If one defines a spectral index $\alpha = d\log\lambda F_\lambda / d\log\lambda$, then the spectral energy distributions (SEDs) observed between 1–100 μm of most known YSOs fall into three distinct morphological classes (e.g., Lada and Wilking 1984; Adams, Lada and Shu 1987; and Lada 1987). These are illustrated in Figure 1. Class I sources have SEDs which are broader than a single blackbody function and for which α is positive. These sources are often assumed to be protostellar in nature and are sometimes called "infrared protostars" (Wynn-Williams 1982). Class II sources have SEDs which are also broader than a single blackbody function but have values of α which are negative. Class III sources have SEDs which are characterized by negative values of α but have widths that are comparable to those of single blackbody functions, consistent with the energy distributions expected from purely

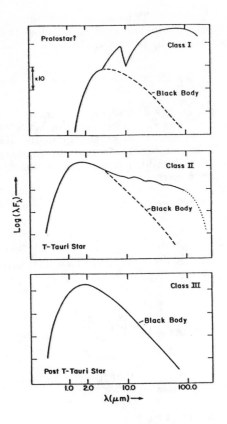

Figure 1. Classification scheme for YSO energy distributions (Lada 1987).

reddened photospheres of young stars. Class I sources derive their steep positive spectral slopes from the presence of large amounts of circumstellar dust. These sources are usually deeply embedded in molecular clouds and rarely exhibit detectable emission in the optical band of the spectrum (e.g., Lada and Wilking 1984; Myers *et al*, 1987). However, nearly all known Class II sources can be observed optically as well as in the infrared. When classified optically Class II sources are usually found to be T Tauri stars or FU Ori stars (see Rucinski (1985) for some good examples of T-Tauri star SEDs). Their negative spectral indices indicate that Class II YSOs are surrounded by considerably less circumstellar dust than Class I sources. Class III sources are usually optically visible with no or very little detectable excess emission at near- and mid-infrared wavelengths, and therefore little or no circumstellar dust. Class III objects include both young main sequence stars and pre-main sequence stars, such as the so-called "post"-T Tauri stars (e.g., Lada and Wilking 1984) and the recently identified "naked"- T Tauri stars (e.g., Walter

1987). It is apparent from existing studies of YSOs that there is a more or less continuous variation in the shapes of SEDs from Class I to Class III (e.g., Myers *et al.*,1987). This is illustrated in Figure 2 which displays the frequency distribution of the spectral indices observed for the population of embedded YSOs in the core of the Ophiuchi dark cloud.

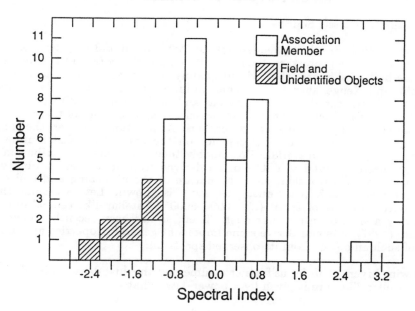

Figure 2. *Frequency Distribution of Spectral Indices for the SEDs of 57 sources in the Ophiuchi dark cloud. (from Wilking, Lada and Young 1988).*

Although most YSO energy distributions can be classified into one of the three classes in Figure 1, there are a few sources which exhibit considerable, often well defined (e.g., double humped), structure in their SEDs and they cannot be characterized by a single spectral index in the infrared. Typically, a source with a double humped energy distribution will exhibit a high frequency hump at around $1\mu m$ wavelength and a low frequency hump at around $60\mu m$ wavelength. The energy distributions of such double humped sources can usually be classified as sub-classes of either Class II or III (e.g., Wilking, Lada and Young 1988). A Class II–D source has a double peaked energy distribution whose high frequency peak is characterized by a spectral index similar to that of a T–Tauri star. Such a source is likely intermediate between pure Class I and pure Class II. A Class III–D source is characterized by a high frequency peak with a spectral index similar to that of a purely reddened photosphere. Such SEDs are usually associated with massive, early type (B) stars which have reached the zero age main sequence, and have sufficient luminosity to heat relatively distant and otherwise cold dust to temperatures which produce observable emission at 60–$100\mu m$. In some situations it is possible that a complex energy distribution could result from confusion and the superposition of more than one source in an observer's beam. Such confusion is most likely to occur at long wavelengths and for distant sources where the effective spatial resolution may be relatively poor.

3.0 INTERPRETATION: THE NATURE OF EMBEDDED YSOS

3.1. The Nature of Class I Sources

Because of their deeply embedded nature, it has often and long been argued that Class I sources are protostars, that is, embryonic stellar cores in the process of acquiring the bulk of the mass they will ultimately contain as main sequence stars (e.g., Becklin and Neugebauer 1967). Their luminosity is assumed to be derived almost entirely from accretion of infalling gas and dust. Although direct evidence for such a hypothesis is lacking, analysis of the energy distributions of these objects tends to lend strong support to such contentions. As mentioned earlier, the shape of the energy distribution of a YSO depends on the nature and distribution of surrounding circumstellar material. Presumably then, the density distribution of circumstellar matter around a YSO can be derived from modeling its observed spectrum or energy distribution, once the nature of the absorbing and emitting material that surrounds it (e.g., dust, gas, etc.) is known. Let us assume that a Class I source consists of a hot (i.e., 3000–5000 K) stellar–like central object, surrounded by a shell of interstellar dust. Then, as has been recently shown by Myers *et al.* (1987), one can derive some important general properties about the structure of such an object from its observed spectrum.

Following Myers *et al.* let us further assume that the dust density distribution around the stellar–like core is given by a power–law. That is:

$$(5) \qquad\qquad n(r) = br^{-p}$$

The observed extinction to the core object is determined by:

$$(6) \qquad\qquad A_v = 10^{-21} \int_{r_1}^{r_2} n(r)\, dr$$

where the limits of integration are the inner and outer radii of the dust shell. If the outer radius of the dust shell extends to a size typical of a star forming core in the Taurus dark cloud, then the constant b can be estimated from the observed size and average density of such a core (i.e., 0.05 pc and 3×10^4 cm^{-3}; Myers and Benson 1983) for a given value of the power law index p (see Myers *et al.* 1987). Then, with the specification of an inner radius, equation 6 can be integrated to give the extinction to the underlying stellar–like core. For example if one sets p = 1.5 and sets r_1 to equal the dust destruction radius (about 0.1 AU for a low mass

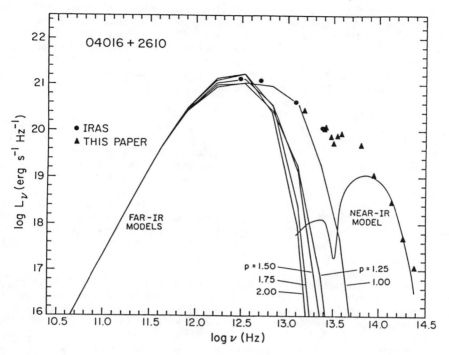

Figure 3. Two component model fits of Myers et al. (1987) to the spectrum of a Class I source in Taurus.

star) then A_v is found to be about 1400 magnitudes! The near infrared spectrum of such an object would be impossible to detect, even in the nearest sources. Yet Class I sources are readily detected between 2–5 μm.

On the other hand, if the central object in a Class I source is assumed to be similar in size and temperature to a T Tauri star, then the total extinction to that object cannot exceed 30–90 magnitudes at visual wavelengths in order to be consistent with typically observed near–infrared fluxes of Class I sources (Myers et al. 1987). For a power–law density distribution, then, we can solve equations 5 and 6 for r_1, the inner radius of the dust shell. For values of p $\approx$ 1.5, Myers et al. found that r_1 ranged between 10 and 100 AU and concluded that the stellar source at the core of a typical Class I object is surrounded by a cavity 10 to 100 AU in extent. The far–infrared spectrum from a Class I source presumably arises from the reprocessing of the light emitted by the central source and absorbed by the dust shell. With the parameters of the dust shell known, one can now model and predict the far–infrared emission spectrum of the source and compare it to actual observations. In addition, one can also model the expected spectrum of the extincted central source and together with the far–infrared model predict the complete spectrum of a Class I source. Figure 3 shows such a two component model made by Myers et al. in an attempt to fit the spectrum of a Class I source in the Taurus dark cloud. Although the model can match the far–infrared peak and the near–infrared points quite well it produces a large deficit of radiation at mid–infrared (i.e., 8–20 μm) wavelengths. In retrospect, the existence of this deficit is not suprising, since emission in this wavelength range corresponds to emitting temperatures of around 150–300 K. The equilibrium temperature of dust around a low luminosity YSO would reach such values only at distances between 10–100 AU from the star. This is just in the region of the inferred cavity.

Evidently, the central regions of this Class I object contain, in addition to a hot central star, a larger and cooler source of luminosity which produces a substantial amount of mid–infrared radiation. This luminosity source must be comprised of material with a density distribution that departs significantly from spherical symmetry in order to avoid completely extinguishing the emission from the central star. The presence of a highly flattened circumstellar disk could provide the missing luminosity and, if not viewed edge–on, simultaneously produce little, if any, extinction to the central star. Together these simple considerations suggest a model for Class I sources which consists of a central stellar–like object, surrounded by: 1) a luminous, spatially flat, disk 10–100 AU in extent, 2) a more or less spherically symmetric cavity of the same dimensions, and 3) a more or less spherically symmetric dust shell with a power law density gradient extending from the outer edge of the cavity to distances on the order of 0.1 parsec from the central star.

The physical characteristics of Class I sources outlined above, interestingly, turn out to be predictions of a recent physically self–consistent theory of protostars developed by Shu (1977) and Adams and Shu (1986). In this picture the evolution of a magnetically supported molecular cloud leads to the formation of dense cores through the ambipolar diffusion of local fields (see Shu, Adams and Lizano 1987). These cores are assumed to have density distributions similar to that of an isothermal sphere, that is:

$$(7) \qquad \rho(r) \; = \; \frac{a^2}{2\pi G} r^{-2}$$

where a is the speed of sound, G the gravitational constant and ρ the mass number density. Shu (1977) showed that the collapse of an isothermal sphere can proceed in a self–similar manner and be characterized by a single parameter, the mass infall rate, given simply by:

$$(8) \qquad \dot{M} \; = \; \frac{.975 a^3}{G}$$

This collapse proceeds in a nonhomologous, inside–out manner with a flattening of the initial isothermal density gradient within the infalling region (i.e., $\rho(r) \propto r^{-1.5}$). In the center, the collapse is arrested at a dense protostellar core which grows in mass linearly with time, that is, M(t) = $\dot{M}$t The protostellar core becomes increasingly luminous as it radiates away the gravitational potential energy lost by the infalling gas as it is arrested and thermalized in an accretion front at the protostar's surface, i.e.,

$$(9) \qquad L_*(t) \; = \; \frac{G M(t) \dot{M}}{R_*}$$

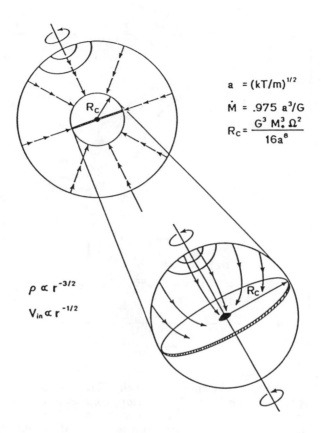

$$a = (kT/m)^{1/2}$$

$$\dot{M} = .975 \, a^3/G$$

$$R_c = \frac{G^3 \, M_*^3 \, \Omega^2}{16a^8}$$

$$\rho \propto r^{-3/2}$$

$$V_{in} \propto r^{-1/2}$$

Figure 4. Schematic diagram of a rotating and collapsing protostar based on the models of Terebey, Shu and Cassen (1984) and Adams and Shu (1986).

Since this collapse solution specifies the density distribution of the collapsing core, one can calculate the emergent spectrum of the protostar with knowledge of the properties of the dust in the infalling envelope. Adams and Shu (1986) used such self-similar collapse models together with the best available calculation of dust opacities (i.e., Draine and Lee 1984) to predict the emergent spectra of protostars. However, because these particular collapse models did not naturally contain large cavities or disks in their central regions, they were unable to produce enough near and mid–infrared radiation to match the observations of Class I sources.

By introducing a slight amount of rotation, Terebey, Shu and Cassen (1984) showed that collapsing isothermal spheres could, as a consequence of conservation

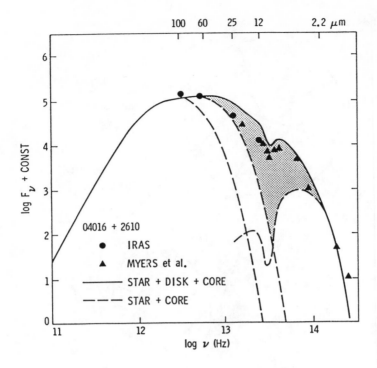

Figure 5. *Model fit to the spectrum of the Class I source shown in Figure 3 using the the model of the rotating, collapsing isothermal sphere (Adams, Lada and Shu 1987). The presence of a luminous disk and a centrifugally evacuated cavity conspire to fit the data remarkably well and remove the mid–infrared deficit of the two component model as shown (adapted from Beichman 1988).*

of angular momentum, produce infalling density distributions with central cavities as well as flattened central disk–like structures. A schematic diagram of such a collapsing isothermal sphere is shown in Figure 4. In the outer regions the infall velocities are nearly radial and virtually similar to the non–rotating case. However, as material falls closer to the center, conservation of angular momentum produces curved infall trajectories and, in the equatorial plane, a centrifugal barrier whose radius is given by:

$$(10) \qquad R_c = \frac{G^3 M^3 \Omega^2}{16a^8}$$

here Ω is the rotation rate and M the mass of the central star. When R_c is larger than the radius of the central stellar core, a disk is formed as infalling material falls onto the equatorial plane. Moreover, the curved (parabolic) trajectories within R_c produce a region of significantly reduced density compared to the non–rotating solutions. Using the density distributions derived from such models and taking in to account the luminosity which would be radiated by the central disks as well as the protostellar core, Adams and Shu (1986) and Adams, Lada and Shu (1987) were able to successfully model the observed spectra of numerous Class I sources. For example, Figure 5 shows the model fit to the source initially depicted in Figure 3. Clearly the deficit of mid–infrared emission is now accounted for. In addition, the model provides a satisfying and self–consistent physical explanation of the simultaneous presence of a disk and a cavity of reduced dust density around the central source.

This theory has two additional virtues. First, the emergent spectrum or energy distribution of a model protostar is essentially completely specified by two parameters (a and Ω) which in principle can be *independently* determined from observation of molecular gas. Second, the model can successfully account for a wide variety of observed Class I energy distributions. For example, the depth of the silicate absorption feature at 10 μm in Class I SEDs is predicted to be inversely proportional to Ω and the breadth of the source energy distribution in frequency space. This is because the larger Ω and therefore R_c, the larger the inner cavity and the lower the observed extinction to the central protostar. Fits to the energy distributions of a variety of Class I sources in Taurus and Ophiuchus appear to confirm this prediction (Adams, Lada and Shu 1987). The robust nature of this theory in its ability to fit the energy distributions of a variety of Class I sources, appears to lend strong credence to the contention that Class I objects are true protostars.

3.2. The Nature of Class II Sources

As mentioned earlier Class II sources are usually associated with T Tauri stars. Indeed, observational studies of the energy distributions of T Tauri stars suggest that their energy distributions can be characterized by power–law slopes longward of 2μm wavelength (see Rucinski 1985; Rydgren and Zak 1987). For these stars the average value of the spectral index, α, is about -0.6, although individual sources exhibit a significant range in α (between values of 0 and -1.3). This departure from a black-body spectrum at infrared wavelengths is often referred to as the infrared excess of T Tauri stars. More than a decade ago Lynden-Bell and Pringle (1974) predicted that T Tauri stars would display such energy distributions if they were surrounded by luminous accretion disks. If we assume that a pre–main sequence star is surrounded by a disk, one can show that if the disk is optically thick and spatially thin it will radiate everywhere like a black-body and be characterized by a spectral index

$$(12) \qquad \alpha \; = \; \frac{2}{n} \; - \; 4$$

where

$$(13) \qquad T_{disk}(r) \; \propto \; r^{-n}$$

If $n = \frac{1}{2}$ then $\alpha = 0$; for $n = 0.75$, $\alpha = -\frac{4}{3}$.

A viscous accretion disk is predicted to produce a temperature gradient characterized by n=0.75, corresponding to an α of -1.33 (Lynden–Bell and Pringle 1974). Recently Adams and Shu (1986) and Friedjung (1985) showed that a flat *passive*

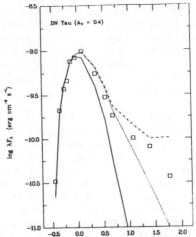

Figure 6. Energy distribution of the T Tauri star DN Tau, with three different models fit to the data: the solid line is a reddened photosphere, the dotted line a reddened star plus a flat, passive disk, and the dashed line a reddened star with a flaring passive disk. The horizontal axis plots log(λ). From Kenyon and Hartmann (1987).

disk which derives all its luminosity from the reprocessing and re-radiation of light it has absorbed from the central star, also has an equilibrium temperature gradient characterized by n = 0.75. However, most Class II sources have spectra characterized by slopes which are less steep than those predicted for either a passive disk or an accretion disk. Kenyon and Hartmann (1987) have argued that a passive, reprocessing disk can produce a more shallow temperature distribution and a more shallow spectral slope if the disk is flared (see Figure 6). Most observed Class II energy distributions can probably be explained with such a model. However, it may also be possible that the shallow temperature gradients required for the disks arise from some nonviscous accretion process (Adams, Lada and Shu 1988). In particular, flat spectrum sources are not easily fit with flaring disks and the nature of their luminosity generation is unclear. However, it is likely that these sources, which include the star T Tauri itself, are characterized by disks which are *active* and have an intrinsic luminosity source in addition to the reprocessed radiation from the central star. In any event, as many authors have argued (e.g., Rucinski 1985; Beall 1987; Adams, Lada and Shu 1987 and Kenyon and Hartmann 1987), the most likely interpretation of the nature of the Class II sources is that they represent pre–main sequence stars which are surrounded by disks. Perhaps the most compelling evidence to support the disk hypothesis comes from spectroscopic measurements of optical and infrared absoption lines in the spectrum of FU Ori by Hartmann and Kenyon (1987). Comparision of the velocity widths of the optical and infrared lines indicate that the infrared rotational velocity is smaller than the optical rotational velocity as would be expected for a differentially rotating structure such as a disk in Keplerian motion around a central star. Moreover, the energy distribution of FU Ori is Class II and is well modeled by a disk with a temperature power-law index of n = 0.75 (Adams, Lada and Shu 1987).

3.3. The Sequence Of Formation And Early Evolution For Low Mass Stars

The variation in the shapes of YSO energy distributions from Class I to III represents a variation in the amount and distribution of luminous circumstellar dust around an embedded YSO. It is natural therefore to hypothesize that the empirical sequence of spectral shapes is a sequence of the gradual dissipation of gas and dust envelopes around newly formed stars and is therefore an evolutionary sequence (see Lada 1987). Recently, Adams, Lada and Shu (1987) have been able to theoretically model this empirical sequence as a more or less continuous sequence of early stellar evolution from protostar to young main sequence star using the self-consistent physical theory of rotating collapsing isothermal spheres as the initial condition. In this theoretical picture Class I sources are true protostars, objects undergoing accretion and assembling the bulk of the mass they will ultimately contain when they arrive on the main sequence. They consist of central embryonic cores steadily gaining mass from accretion and infall of surrounding matter. As a result of rotation they are surrounded by luminous disks, which themselves are contained in cavities 10–100AU in extent. The cavities are surrounded by large shells of infalling matter which absorb and reprocess most of the radiation from the central star and disk. When this outer infalling matter is either all incorporated into the central star or removed by some agent the veil of obscuration is also removed and the YSO appears as Class II source.

To evolve from a Class I to a Class II object requires the removal or dissipation of the infalling envelope. This is most likely accomplished by an intense sellar wind or outflow (e.g. Lada 1988) coupled with the gradual depletion of the surrounding infalling shell as more and more of the initial cloud mass is incorporated into the central protostar. It is expected that as the infalling envelope is depleted and removed the Class I energy distribution will become significantly modified. As the inner regions become excavated by the wind and the outer regions depleted by the infall, residual infalling gas and dust produce less and less extinction and contribute emission to the source energy distribution only at the longest infrared wavelengths. A double–humped energy distribution should result, with a peak at short wavelengths (high frequencies) due to the less extincted central star and disk and a second peak at long wavelengths due to emission from the remnant infalling envelope. To evolve from a Class II to III source requires the further depletion and removal of the circumstellar disk.

Figure 7 displays the results of theoretical models calculated by Adams, Lada and Shu (1987; 1988) to fit observations of YSOs in each of the primary evolutionary states. The top pannel shows the model fit to the well known Class I source L1551 IRS 5. Here the solid curve represents a model which consists of a 1 $M_\odot$ accreting stellar core surrounded by a rotating luminous disk and infalling envelope. The mass infall rate needed to produce this fit was $\approx 10^{-5}$ $M_\odot$ per year. This source is therefore about one hundred thousand years old. The middle pannel shows the transition object, VSSG 23, which has been self-consistently fit by a reddened stellar photosphere surrounded by a luminous, spatially thin but optically thick circumstellar disk and an outer infalling shell which emits optically thin infrared radiation. The infalling shell has an inner radius of 80 AU and accounts for all the observed extinction to the star. However, its density has suffered a depletion of a factor of 100 compared to that of the initial cloud core. The luminous disk is passive in the sense that it derives its luminosity from purely reprocessing radiation absorbed from the central star. The bottom pannel shows a theoretical fit to a

106

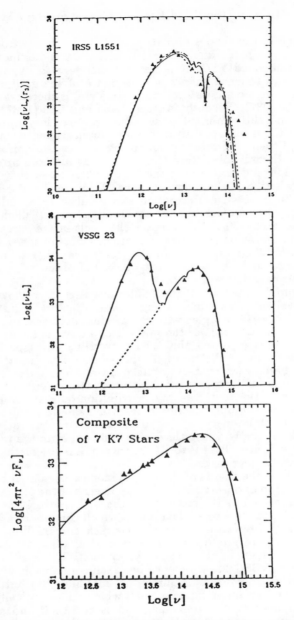

Figure 7. Theoretical model fits to the SEDs of three well known YSOs in different stages of evolution. Note that the energy distributions are plotted with frequency rather than wavelength on the horizontal axis. (from Adams, Lada and Shu 1987 and 1988).

composite energy distribution of 7 well studied T Tauri stars. The theoretical fit consists of a model spectrum for a K7 star surrounded by a circumstellar disk which derives a significant portion of its luminosity from accretion (Adams, Lada and Shu 1988). There is no evidence for residual infalling material in this SED. As is evident from these fits, the simple theory outlined above appears to be able to account for the observed variation in the spectral shapes of YSO SEDs extremely well.

If indeed the sequence of SED shapes corresponds to a sequence in evolution, then study of the frequency distributions of SED types, their relation to YSO luminosities and their relation to outflows should reveal much about the star formation process. For example, consider the frequency distribution of the spectral indices of the 57 sources in the core of the Ophiuchi dark cloud shown in Figure 2. Although, the energy distributions in this cloud display a nearly continuous variation in shape, there appear to be two peaks in the distribution corresponding to Class I and Class II sources. There are very few Class III sources in this sample of 57 stars. An additional 44 sources have been identified in the region which may be associated with the cloud but for which sufficient data did not exist to permit classification. These sources are most likely field stars or very low luminosity (i.e., $\leq 0.1 L_\odot$) members of the embedded population, some of which could be Class III sources. The fact that there are nearly equal numbers of Class I and Class II sources in this cloud suggests that a typical YSO spends roughly equal amounts of time in each stage during its early life. Since the oldest T Tauri stars in the Ophiuchi cloud have estimated ages of a few million years (Cohen and Kuhi 1979), the embedded sources probably spend no more than a million years or so as Class I objects. Thus if Class I sources are growing protostars, they must accumulate their entire main sequence mass during this period. Since the embedded stars in the Ophiuchi cloud are thought to have masses comparable to that of the sun (Lada and Wilking 1984), the infall rate for the Class I sources in the cloud must be greater than or roughly equal to a few times $10^{-6} M_\odot$ per year. The model fits of Adams, Lada and Shu (1987) suggest a mass accretion rate of about $10^{-5} M_\odot$ per year for a selected group of Class I sources in Ophiuchus, suggesting a lifetime of a few hundred thousand years for the Class I stage, roughly consistent with the above estimate. Only 5 of the 57 sources studied had double-humped SEDs, suggesting that this transition phase has a duration of only about 10% of that of either Class I or II.

Figure 8 shows the derived luminosity function for the 57 sources considered in Figure 2. Within the uncertainties, the distribution is consistent with what one would expect from a population of stars whose masses were distributed similar to the IMF. What is interesting in this luminosity function is the apparent segregation of Class I and Class II objects in luminosity. Of the 16 sources with luminosities greater than $5 L_\odot$, 11 have Class I SEDs, 4 are double–humped and 2 are Class III. If we eliminate the three main sequence B stars in the sample, then there are only 3 sources in the group of 13 pre-main sequence objects which are not Class I objects. For luminosities less than $5 L_\odot$, Class II objects clearly dominate the luminosity function. This observation also hints at an evolutionary difference between the two classes of YSOs. Apparently, a star is more luminous during the infall or protostellar stage of evolution than in the T Tauri stage of evolution. Could this additional luminosity be observational evidence for accretion around Class I sources? Or is it the result of a star forming process peculiar to the Ophiuchi cloud (such as the sequential formation of progressively more massive stars with time)? The answer to this question will be a fundamental clue to understanding the details of the star formation process in the Ophiuchi dark cloud. The above

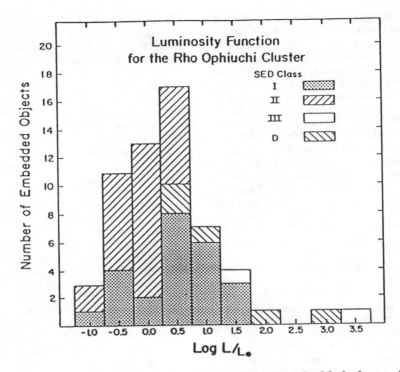

Figure 8. The luminosity function for the embedded cluster in the Ophiuchi dark cloud (from Wilking, Lada and Young 1988).

observations of the embedded population of the Ophiuchi cloud clearly illustrate the potential power of employing infrared energy distributions to decipher the mysteries of star formation and early evolution. Similar studies of the embedded populations of other molecular clouds and the refinement of the classification system through high resolution infrared spectroscopy of YSOs promise to significantly increase our knowledge of the earliest stages of stellar evolution.

Acknowledgements. I thank Bruce Wilking for useful discussions in preparing this review. I am grateful to the Conference Organizing Committee for inviting me to this stimulating meeting and providing the basic financial support which made my participation possible. I also thank the University of Arizona Committee For Foreign Travel for assistance with travel support.

REFERENCES

Adams, F.C. and Shu F.H. 1986,*Ap.J.*,**308**,836.
Adams, F.C., Lada, C.J., and Shu, F.H. 1987,*Ap.J.*,**213**,788.
Adams, F.C., Lada, C.J., and Shu, F.H. 1988,*Ap.J.*,(in press).
Beall, J.H. 1987, *Ap.J.*,**316, 227**
Becklin, E.E., and Neugebauer, G. 1967, *Ap.J.*,**147**, 799.
Beichman, C.A. 1988, IPAC Preprint No. 0030.
Cohen, M. and Kuhi, L.V. 1979,*Ap.J.Suppl.*,**41**, 743.
Draine, B., and Lee, H.M. 1984, *Ap.J.*,**285**, 89.
Friedjung, M. 1985, *Astr.Ap.*,**146**, 336.
Hartmann, L. and Kenyon, S.J. 1987,*Ap.J.*,**312**, 243.
Kenyon, S.J. and Hartmann, L. 1987,*Ap.J.*,**323**, 714.
Lada, C.J. 1987,*in IAU Symposium No. 115: Star Forming Regions*, eds. M. Piembert and J. Jugaku (Dordrecht: Reidel), p.1.
Lada, C.J. 1988, in *Galactic and Extragalactic Star Formation*, eds. R. Pudritz and M. Fich, (Dordrecht: Reidel).
Lada, C.J. and Wilking, B.A. 1984,*Ap.J.*,**287**, 610.
Lynden-Bell, D., and Pringle, J.E. 1974, *M.N.R.A.S.*, **168**, 603.
Myers, P.C. *et al.* 1987,*Ap.J.*,**319**, 340.
Myers, P.C. and Benson P.J. 1983, *Ap.J.*,**266**, 309.
Rucinski, S.M. 1985,*A.J.*,**90**, 2321.
Rydgren, A.E. and Zak, D.S. 1987, *Publ.Astr.Soc.Pacific*,(in press).
Shu, F.H. 1977,*Ap.J.*,**214**, 488.
Shu, F.H., Adams, F.C., and Lizano, S. 1987, *Ann.Rev.Astr.Ap.* ,**25**, 23.
Terebey, S., Shu, F.H., and Cassen P. 1984, *Ap.J.*,**286**, 529.
Walter, F.M. 1987,*Publ.Astr.Soc.Pacific*,**99**, 31.
Wilking, B.A., Lada, C.J. and Young E.T. 1988, preprint.
Wynn–Williams, C.G. 1982,*Ann.Rev.Astr.Ap.* ,**20**, 587.

Near-Infrared Array Images of Low-Luminosity IRAS Sources in Dark Clouds

H. Zinnecker

Max-Planck-Institut für Physik und Astrophysik
Institut für extraterrestrische Physik
8046 Garching b. München, Fed. Rep. Germany

ABSTRACT

Dense cores in dark clouds with obscured young stellar objects in them (showing up as IRAS sources) are excellent targets for near-infrared imaging with the new 2-dimensional infrared arrays such as IRCAM at UKIRT (58 x 62 pixels, 0.6 arcsec/pixel). Here we report some of the first results with this instrument obtained for embedded sources in the Taurus and Ophiuchus clouds (L1489, L1495; Elias 21, Elias 22). We also discuss the promise of future observations.

1. INTRODUCTION

The development of two-dimensional infrared detector arrays similar in kind to optical CCD devices has opened up exciting new opportunities to observe young low-mass stars still embedded in their parent molecular cloud. Low-mass stars are much more numerous than high mass stars, therefore we can find many more examples of nearby low-mass objects than high-mass objects. Consequently, at a given spatial resolution (typically of order 1") for IRCAM, young low-mass stars and their environments can be studied in greater detail than the more distant young high-mass objects. Furthermore, the early evolution of low-mass objects is much slower than that of high-mass young stellar objects which allows to investigate better the circumstellar environment as a function of evolutionary age (characterized by the shape of the spectral energy distribution; see Lada 1986 and in this volume). For example, studies of the surface brightness distribution of the scattered near infrared light will be useful to understand the evolution of the dust density distribution and grain growth around young low-mass stars where planetary systems are likely to form. After all,

111

A. K. Dupree and M. T. V. T. Lago (eds.), Formation and Evolution of Low Mass Stars, 111–121.
© *1988 by Kluwer Academic Publishers.*

the sun is a low-mass star and the formation of the solar system is far from understood.

Apart from studies of the circumstellar environment of single stars, the discovery of binary and multiple embedded systems (associated with a single IRAS point source) is greatly facilitated with the use of infrared arrays. In fact searching the error box of the IRAS beam with single aperture raster scans will not only be tedious but will fail to detect the multiplicity of sources with component separations smaller than 5 arcsec or so. We estimate the chances of an embedded object to be double in the critical range of separation between 2-5 arcsec to be of the order of 20% for objects in nearby dark clouds. Separations smaller than 2 arcsecs must be probed by single detector infrared slit scans or infrared speckle interferometric methods (e.g. Chelli et al. 1988, Zinnecker and Perrier 1988). New detections of embedded binary objects on infrared array images will complement optical CCD studies of pre-main sequence binaries (see Reipurth, this volume) and will allow a comparison with the binary data of solar-type main-sequence stars (Abt and Levy 1976).

Other (extended) IRAS sources for IR-array imaging might include embedded star clusters similar in kind to the embedded cluster of young low-mass stars in the core of the Rho Ophiuchi dark cloud (Lada and Wilking 1984). Here the high resolution of the IR array helps to overcome crowding and confusion problems. Moreover very faint sources can be seen, allowing us to probe for the existence of very low mass objects, possibly even brown dwarfs (note that the limiting flux density for a detection at 2 μm is some ten thousand times fainter than the 2 μm flux of T Tau). The faint end of the 2 μm luminosity function in a low-mass cluster will have important implications on many aspects of star formation theory (cf. the Orion cluster 2.2 μm luminosity function; McLean, McCaughrean, Aspin, and Rayner 1988).

Here, we will report only about extended single sources and binary objects but not about embedded clusters. The images that will be presented have not been fully exploited; in particular the images have not been deconvolved with the point spread function. Therefore the images will give mostly morphological information and only very little photometric information. No detailed physical modelling has been attempted. Nevertheless the images should give a reasonable first impression of what we may expect to see in the years to come. We draw attention to the Proceedings of the Workshop "Infrared Astronomy with Arrays" held in March 1987 in Hilo, Hawaii (eds. Wynn-Williams and Becklin), in which many more infrared images of a variety of sources are shown.

2. DESCRIPTION OF THE IR CAMERA AT UKIRT

The imaging observations presented below were taken on the 3.8 m United Kingdom Infrared Telescope on Mauna Kea, Hawaii using the newly commissioned two-dimensional infrared camera called IRCAM (McLean et al. 1986). The camera operates in the 1-5 microns infrared windows. The usual broad band filters J (1.2 μm), H (1.6 μm), K (2.2 μm), L (3.6 μm), and M (4.8 μm) are available as well as several narrow band filters (e.g. 2.12 μm, 2.17 μm, and 3.07 μm for H_2, B_γ, and ice band observations, respectively). Briefly, the camera consists of a LHe cryostat that cools the detector array and associated optical components to an operating temperature between 35 and 50 K. The detector array is a 62 x 58 pixel Indium-Antimonide (InSb) direct voltage read-out device provided by Santa Barbara Research Corporation. The read out noise is 400 e, and there is a dark current of a few hundred electrons per sec. Three plate scales can be selected: 0.6 arcsec/ pixel, 1.2 arcsec/pixel, and 2.4 arcsec/pixel. The corresponding on sky fields of view (FOV) are roughly 0.6 arcmin x 0.6 arcmin (smallest FOV), 2.4 arcmin x 2.4 arcmin (largest FOV), and 1.2 arcmin x 1.2 arcmin (intermediate FOV). The highest spatial resolution will give the best photometric accuracy. The instrument is common-user at UKIRT now and is controlled via two dedicated LSI 11/23+ micro processors from the main Vax 11/730 mini-computer. Data acquisition can be performed using individual on-chip exposure times from 65ms to about 300s with on-line coaddition of multiple exposures. Final images are transferred to the Vax computer where they can be immediately displayed and manipulated using an extensive in-house data reduction/display package (written by C. Aspin). Normally, the dark current is first subtracted from the images. The dark current (which is a strong function of temperature) is measured for several exposure times using cold blank filters instead of the narrow or broad band filters. A stable temperature is essential. The flat field images are taken on blank patches of sky and are also dark current subtracted prior to the flat fielding procedure. Flat fielding similar to that utilized with optical CCD images is used, dividing the dark current subtracted object image by the dark current subtracted flat field image and rescaling to the mean of a "clean" area of the flat field image. Often the images are Gaussian-smoothed over a few pixels. However, no information can "spill over" from one pixel to another as is sometimes the case for optical CCD arrays. The basic difference between infrared arrays and optical arrays, apart from the number of pixels, lies in the read out process.

The sensitivity of IRCAM in the 0.6 arcsec/pixel mode is of the order 20 mag/arcsec^2 in the J-band, 19.5 mag/arcsec^2 in the H-band, and 19 mag/ arcsec^2 in the K-band assuming a 5 min integration (source + sky frames) and a signal-to-noise ratio = 10 (McCaughrean 1987; calculated figures which have more or less been confirmed in actual observations). Somewhat different figures hold for the 1.2 and 2.4 arcsec/pixel modes.

3. CURRENT OBSERVATIONS

(a) in Taurus

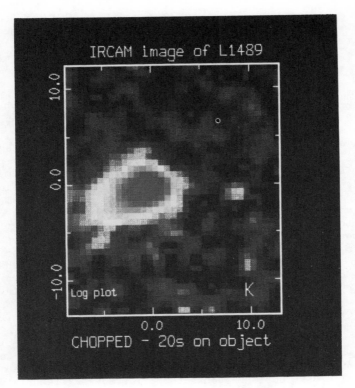

Fig. 1: IRCAM image of L1489 (see text); scale is in arcsec.

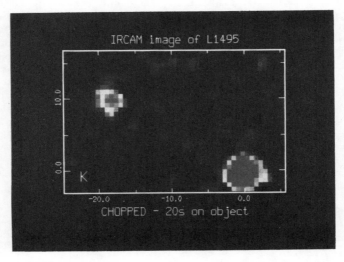

Fig. 2: IRCAM image of L1495 (see text); scale is in arcsec.

Fig. 1 and 2 show infrared images of the IRAS sources L1489 and L1495 listed in Beichman et al. (1986). [Integrated IRAS luminosities of these sources are 2.9 and 0.6 $L_\odot$, respectively]. L1489 was found to be a single object with extended infrared emission, both in the H and K band. The extended emission is interpreted as scattered emission rather than thermal emission. The extension is very unsymmetrical: to the NW the HWHM is 1.5", whereas to the SE the HWHM is 2.5". To the NW the one sigma sky is reached at 3" from the peak, whereas to the SE the one sigma sky is not reached until about 9". The H image (not shown) has a morphology and total extend similar to the K image. L1489 is known to be associated with a CO outflow (Myers et al. 1988). It is also interesting to note that the NIR source is offset from the centre of their NH_3 contour map (see their Fig. 1b).

L1495 was found to be a double source, albeit with a rather large separation of 21 arcsec (already reported in Zinnecker, McLean, and Coulson 1987). It could have been discovered without the use of an infrared array since the separation is so large but it was not. Whether the double source is an optical or a physical pair cannot really be said, however statistical arguments would seem to argue that it is a physical pair (if so, the linear separation is of order 3000 AU - a wide but still not a very wide binary). Such binaries may be formed by fragmentation during the isothermal phase of the collapse of a rotating cloud (see Bodenheimer, this volume).

(b) in Ophiuchus

Fig. 3 shows Elias 21 (= GSS 30), a most spectacular infrared reflection nebula (L = 35 $L_\odot$; Wilking and Lada 1983). It was first discovered by Elias (1978) who carried out NIR aperture photometry. Later, Castelaz et al. (1985a) presented a raster scan map using a 6 arcsec beam and gave a thorough discussion and interpretation. They gave surface brightness cuts in several radial directions and also presented infrared polarimetric maps at K. They provided evidence in favour of a protostellar disk with bipolar outflow lobes perpendicular to the disk. Furthermore, 10 μm data were also supplied (Castelaz et al. 1985b). Although the 10 μm structure (nebulosity extended over 10") might be due to thermal emission from a warm disk, scattered emission is a more likely interpretation. Such scattering would point to the existence of grains much larger than normally assumed, i.e. to grain growth in a dense circumstellar environment.

The Elias 21 nebula is quite extended in the NE/SW direction. There is a second source in the NE (#2) about 15 arcsec from the peak of the main source whose association with the infrared nebula is not clear. The elongation of the nebula is probably due to scattering of 2 μm photons off the dust in the bipolar lobes. CO-, CS-, and NH_3 observations done with the Nobeyama 45 m dish have revealed that indeed a bipolar flow exists together with a dense gaseous disk (Tamura et al. 1988 and priv. commun.). It is claimed that the bipolar structure is elongated along the local magnetic field direction (Sato

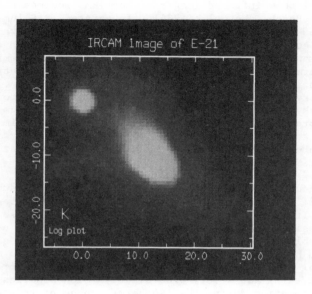

Fig. 3: K-image of the infrared reflection nebula Elias 21.

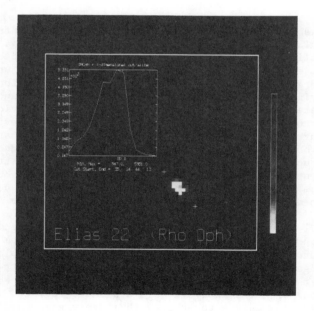

Fig. 4: K-image of the young binary star Elias 22. The insert displays
a cut of the intensity along a line joining the two crosses.

et al. 1988). If this is true, then infrared imaging could offer the possibility of tracing the magnetic field direction (in case the image exhibits an elongated structure).

Our image is a 40 sec exposure, taken in rather windy conditions, and the source #2 (the presumed point source to the NE) has a FWHM of 2.5 arcsec. We have used source #2 which is known to have K= 10.2 in a 6 arcsec aperture (measured by Castelaz et al. 1985a) as a reference to infer K = 8.85 ± 0.10 in a 6 arcsec (software) aperture in the IR-image.

Fig. 4 shows a K-image of Elias 22 (=GSS31, also named Do-Ar 24E) which is somewhat degraded by the effects of seeing, windshake, and perhaps an RA oscillation of the telescope. The object is definitely a double point source. From the present image its projected separation is estimated to be 1.5 ± 0.2 arcsec and its brightness ratio is 1.37 ± 0.06 (the northern component is 0.35 mag brighter). Elias 22 would have been bright enough to be resolved in one 0.4 sec exposure so we could have avoided the degradation of the image quality caused by too long an exposure. We mention that this object had originally been discovered to be double in a 2 μm slit scan experiment at the ESO 3.6 m telescope (Zinnecker et al. 1987, Chelli et al. 1988) and that it was chosen in order to test the resolving capabilities of the UKIRT infrared camera system. The V-magnitude of the system is 14.8, while K=6.7 so that V-K=8.1. The total luminosity of Elias 22 is about 9 $L_\odot$, and $L_1/L_2 \approx 5$. The primary is visible in the optical (through 6 mag of visual extinction) but the secondary shows up only in the infrared ("infrared companion").

Other Rho Oph IRAS sources (e.g. Elias 29, WL 16, and WL 22; see Young, Lada, and Wilking 1986) were also imaged with IRCAM but the data have not been fully evaluated (all the objects quoted seem to be single).

4. FUTURE OBSERVATIONS

One promising direction for future IRCAM observations of young low-mass stars will be the detection and investigation of infrared reflection nebulae similar to Elias 21 or L1551-IRS5 (the core-halo structure of the former can be seen in the infrared speckle data of Zinnecker, Chelli, and Perrier 1986 obtained with the ESO speckle system, while the extended nature of the latter is demonstrated in the infrared images of Moneti et al. 1988 obtained with the Rochester IR-array).

It is the density distribution and the grain properties of these infrared reflection nebulae that lend themselves to array studies in the near infrared. The density distribution will result from the deconvolved radial surface brightness distribution and radial infrared colour distributions (and comparison with a grid of theoretical pre-

118

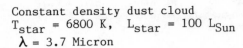

Constant density dust cloud
T_{star} = 6800 K, L_{star} = 100 L_{Sun}
λ = 3.7 Micron

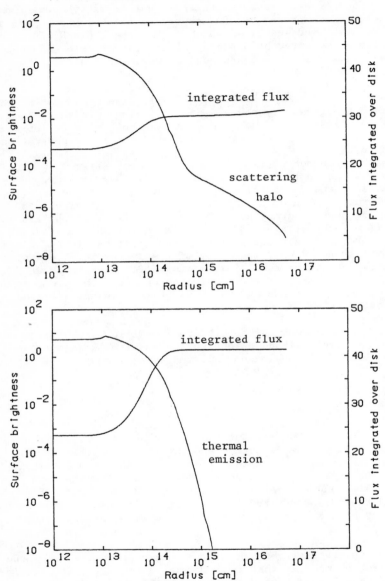

Fig. 5: Monochromatic radial surface brightness profile at 3.7 microns of a star embedded in a constant density cloud (A_v = 30 mag). Top: scattering included; bottom: scattering omitted.

dictions). The grain properties (e.g. grain sizes may change with distance from the central object and/or evolutionary age of the young star) will be derived from multi-wavelength infrared polarimetric imaging.

It must be stressed that under normal circumstances and assumptions (Wolfire and Churchwell 1987, Dent 1988) the _extended_ infrared surface brightness at JHKL is almost all due to scattered light. There is practically no contribution from thermal emission. The effect of taking scattering into account or omitting it is illustrated in Fig. 5, kindly provided by H.W. Yorke (Fig. 5 was calculated similarly to Fig. 4 in Yorke 1980). The figure shows that thermal NIR emission of embedded Taurus or Ophiuchus sources cannot be resolved by IR-arrays with scales of 1"/pixel while the more extended scattered emission can (10^{16} cm in Figure 5 translate to ≈ 4" at the distance of 150 pc).

The polarimetric data can be used to estimate grain sizes by determining the wavelength where the polarisation reaches a maximum (Savage and Mathis 1979). Furthermore, these data can be used to infer the scattering optical depth, that is whether single or multiple scattering prevails. Finally, and perhaps most importantly, polarimetric vector patterns offer the best opportunity to locate the position and constrain the geometry of an embedded source.

Compared to single detector raster scanning employed in the past the use of 2D imaging arrays will not only allow to probe scattered radiation much closer to the star but will also help to better determine the profile of differential extinction across the face of the source and therefore allow a better derivation of the intrinsic properties of the nebula (see again Castelaz et al. 1985a describing the procedure to analyse the data). Regarding polarimetry, the high spatial resolution of IRCAM will enable one to see a less diluted degree of polarisation without the degradation due to averaging over large areas of the source.

5. DISCUSSION

A big advantage of studying young low-mass stars over young high-mass stars lies in the fact that low-mass stars are far less disruptive to their parent molecular cloud than high mass stars, thus by combining infrared array images with VLA interferometric maps of molecular emission (e.g. NH_3, CS, HCN, etc.) of similar resolution it should be possible to establish to a high degree of precision the coincidence or the offset of the position of the embedded star from the density peak of the molecular gas (cf. Clark 1987 who investigated the positions of IRAS point sources with respect to the positions of ammonia cores). Then we can hope to infer more precisely some of the initial conditions for the formation of a particular star and for low-mass star formation in general.

Such studies may also enable us to see directly the effects of collimated outflows from a young stellar object on the adjacent dense gas. Then we can hope to answer the question whether or not it is the onset of a stellar wind that ultimately brings an end to protostellar accretion (as suggested by Shu et al. 1987), or whether perhaps differential acceleration between the protostar and its gaseous environment (resulting in supersonic relative motion) plays a role, too.

REFERENCES

Abt, H.A. and Levy, S.G. (1976). Ap. J. Suppl. 30, 273.

Beichman, C., Myers, P.C., Emerson, J.P., Harris, S., Mathieu, R., Benson P.J., and Jennings, R.E. (1986). Ap. J. 307, 337.

Bodenheimer, P. (1988). this volume.

Castelaz, M.W., Hackwell, J.A., Grasdalen, G.L., Gehrz, R.D., and Gullixson, C. (1985a). Ap. J. 290, 261.

Castelaz, M.W., Gehrz, R.D., Grasdalen, G.L., and Hackwell, J.A. (1985b). P.A.S.P. 97, 924.

Chelli, A., Zinnecker, H., Carrasco, L., Cruz-Gonzales, I., and Perrier, C. (1988). Astron. Astrophys. (submitted).

Clark, F.O. (1987). Astron. Astrophys. 180, L1.

Dent, W.R.F. (1988). Ap. J. 325, 252.

Elias, J.H. (1978). Ap. J. 224, 453.

Lada, C.J. (1986). in IAU-Symposium 115 "Star Forming Regions" (eds. M. Peimbert and J. Jugaku, Reidel), p. 1.

McCaughrean, M. (1987). Ph.D. Thesis, Univ. of Edinburgh.

McLean, I.S., Chuter, T.C., McCaughrean, M. J., and Rayner, J.T. (1986). SPIE Vol. 627 Instrumentation in Astronomy VI, p. 430.

McLean, I.S., McCaughrean, M.J., Aspin, C., and Rayner, J.T. (1988). preprint.

Moneti, A., Forrest, W.J., Pipher, J.L., and Woodward, C.E. (1988). Ap. J. (in press).

Myers, P.C., Heyer, M., Snell, R., and Goldsmith, P. (1988). Ap. J. 324, 907

Reipurth, B. (1988). this volume (ESO-preprint 548).

Sato, S., Tamura, M., Nagata, T., Kaifu, N., Hough, J., McLean, I.S., Garden, R. P., and Gatley, I. (1988). MNRAS (in press).

Savage, B.D. and Mathis, J.S. (1979). Ann. Rev. Astron. Astrophys. 17, 73.

Shu, F.H., Adams, F.C., and Lizano, S. (1987). Ann. Rev. Astron. Astrophys. 25, 23.

Tamura, M., Sato, S., Suzuki, H., Kaifu, N., Ukita, N., Hasegawa, T., and Hough, J. (1988). Ap. J. Lett. (submitted).

Wilking, B.A. and Lada, C.J. (1983). Ap. J. 274, 698.

Wolfire, M.G. and Churchwell, E. (1987). Ap. J. 315, 315.

Yorke, H.W. (1980). Astron. Astrophys. 85, 215.

Young, E., Lada, C.J., and Wilking, B.A. (1986). Ap. J. 304, L45.

Zinnecker, H., Chelli, A., and Perrier, C. (1986). in IAU-Symposium 115 "Star Forming Regions" (eds. M. Peimbert and J. Jugaku, Reidel), p. 71.

Zinnecker, H. and Perrier, C. (1988). ESO-Messenger No. 51 (in press).
Zinnecker, H., McLean, I.S., and Coulson, I.M. (1987). in Proc. Hilo
    Workshop "Infrared Astronomy with Arrays" (eds. G. Wynn-Williams
    and E.E. Becklin, Univ. of Hawaii, Honolulu), p. 291.

ACKNOWLEDGEMENT

I thank my friends at UKIRT for valuable discussions and help, in
particular C. Aspin, M. McCaughrean, J. Rayner, and I.S. McLean.
I also thank H.W. Yorke for permission to include Fig. 5.

BEGINNING AND END OF A LOW-MASS PROTOSTAR

Frank H. Shu, Susana Lizano, Fred C. Adams, and Steven P. Ruden
Astronomy Department, University of California, Berkeley, CA 94720

ABSTRACT. We present some new results concerning the first and last stages in the formation of a low-mass star before it becomes visible as an optical object. The first stage involves the slow condensation of a molecular cloud core to the brink of gravitational collapse. We give detailed theoretical models of this process under the assumption that the rate of condensation is governed by the slip of neutrals relative to the ions and magnetic fields that help to support a molecular cloud against its self-gravity. The theoretical calculations compare well with the existing observational information. The last stage involves the onset of a stellar wind that helps to clear away the surrounding placenta of gas and dust, thereby making the young stellar object optically visible. We discuss new observational evidence that the emerging wind is largely *neutral* and *atomic* in low-mass protostars. We then suggest a simple theoretical mechanism for the generation of such powerful neutral winds.

1. THE SITES OF LOW-MASS STAR FORMATION

1.1. Properties of Small Molecular Cloud Cores

Small dense cores of molecular clouds, observed in NH_3 emission, are the sites of low-mass star formation (Myers and Benson 1983). To be able to excite NH_3 collisionally at the measured temperature of ~ 10 K, the ambient density must exceed $\sim 3 \times 10^4$ H_2 molecules per cm^3. Within such an isodensity contour, ammonia cores typically have radii ~ 0.05 pc, masses ~ 1 $M_\odot$, and nearly thermal linewidths. In cores without stars, "turbulent" motions (perhaps due to the presence of nonlinear Alfven waves, see the review of Shu, Adams, and Lizano 1987) constitute only about 45% of the thermal speed of the cosmic mix (Fuller and Myers 1987). Beichman et al. (1986) found that $\sim 50\%$ of the NH_3 cores in dark clouds like Taurus have deeply embedded infrared sources inside them. The infrared spectral energy distributions of these sources correspond well with the theoretical expectations for rotating accreting protostars (Adams and Shu 1986; Adams, Lada, and Shu 1987), with the ages of the sources equal to several hundred thousand years. Since there are approximately twice as many small cloud cores as there are embedded low-mass protostars, the inferred ages of the cores dense enough to excite ammonia emission, must be of the order of 1×10^6 yr. The challenge for any theory of the first stage of low-mass star formation is to understand these basic facts.

1.2 Origin by Ambipolar Diffusion

Many people have suggested that magnetic fields underlie the mechanical support of interstellar clouds (see, e.g., Mestel 1965, Strittmatter 1966, Mouschovias 1976). In a lightly ionized medium such as a molecular cloud, magnetic support is synonymous with ambipolar diffusion, the process by which neutral gas slips past the ions and magnetic field that

A. K. Dupree and M. T. V. T. Lago (eds.), Formation and Evolution of Low Mass Stars, 123–137.
© *1988 by Kluwer Academic Publishers.*

124

provide a relatively long-lasting impediment to gravitational collapse (Mestel and Spitzer 1956, Mouschovias 1978, Nakano 1979, 1984). The process produces a natural mechanism for slowly separating out small dense cores supported primarily by thermal pressure from a more diffuse molecular envelope supported principally by magnetic fields and "Alfvenic turbulence" (Shu 1983, Lizano 1988).

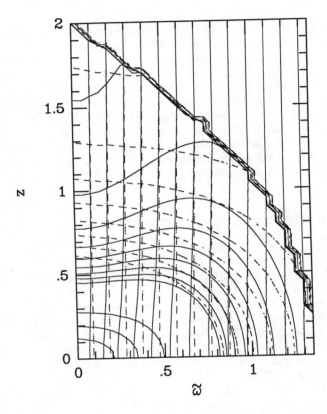

Figure 1. The evolution of a molecular cloud core by ambipolar diffusion. The growing central concentration drags in field lines slightly as indicated by the displacement of the solid vertical curves from the dashed ones. The isodensity contours for the dashed initial state correspond, from top to bottom, to 3, 4, 5, 6, 7, 8, 9, and 10 in units of ρ_0, and the central density is 23. The isodensity contours for the solid curves correspond, from top to bottom to 2, 3, 4, 5, 6, 7, 8, 9, 10, 30, 60, and 120 in units of ρ_0, and the central density is 220. The piling up of the contours at the Roche lobe is an artifact of the plotting routine.

Figure 1 shows the results of a detailed calculation which assumes axial symmetry with respect to the z axis (Lizano and Shu 1988). We consider an infinite periodic chain of identical regions, which are mutually gravitating and spaced a uniform distance apart along z. The calculation for ambipolar diffusion and force balance is carried out for the points interior to the "Roche lobe" of each region, with the magnetic field assumed to have a uniform value B_0 on the boundary. Because no attempt is made to follow the evolution of the material or field outside the Roche lobe (the common envelope), matter is not allowed to cross the Roche surface. The ionization fraction is approximated by an analytic fit to the equilibrium calculations of Elmegreen (1979). We compute the gas kinetic pressure

$$P = a^2 \rho, \tag{3a}$$

where ρ is the mass density and $a \equiv (kT/m)^{1/2}$ is the isothermal sound speed, by assuming a constant temperature T. We simulate "Alfvenic turbulence" by the heuristic inclusion of a "logatropic" pressure, $P_{\text{turb}} \propto \ln \rho$, which has an associated square of the dispersion velocity that satisfies:

$$v_{\text{turb}}^2 \equiv dP_{\text{turb}}/d\rho = K/\rho, \tag{3b}$$

with K taken to be a constant in accordance with the empirical relationship noted by a number of radio observers (e.g., Solomon and Sanders 1985, Dame et al. 1986, Myers and Goodman 1988).

The dimensionless parameters in the specific calculation illustrated in Figure 1 have been chosen so that if $B_0 = 30 \, \mu G$, $T = 10$ K ($a = 0.19$ km s^{-1}), and the unit of density ρ_0 corresponds to 10^3 hydrogen molecules per cm^3, then the spacing between adjacent Roche lobes equals 0.46 pc (a larger spacing would have been more realistic, but it would have restricted the dynamical range over which we could have followed the density in the core). With these choices, the mass interior to each Roche lobe is 7 $M_\odot$. The constant K has been chosen so that v_{turb}^2 is 6 times larger than a^2 at a density ρ_0. For the fiducial choices listed above, this yields a "turbulent" speed that is 45% of the thermal speed at the critical density 3×10^4 cm^{-3} needed to excite ammonia. The dashed curves display a somewhat arbitrarily chosen "initial" state, one in which even the central density is marginally insufficient to excite NH$_3$ emission. The solid curves illustrate the situation a time 1.4×10^6 yr later, when the central density has increased by a factor of 16 over its original value. The magnetic field has diffused outward relative to the neutrals in a Lagrangian sense, but the increasing central concentration of the matter due to its growing self-gravitation has pulled the field lines inward in an Eulerian sense. The central portions of the configuration given by the solid curves have sizes, densities, and velocity widths that resemble the NH$_3$ cores of Myers and Benson (1983). The evolution of the inner regions at this stage proceeds very quickly, with the central density trying to develop a singular cusp on a time scale $\sim 10^5$ yr. The ensuing dynamical collapse cannot be followed with a quasistatic code.

We have also computed cases in which the "turbulence" parameter is appreciably greater than the nominal value 6. For $K = 10$, the time required to produce an ammonia core approaches $\sim 10^7$ yr, but once the ammonia core has formed, the additional time that it takes to reach a state of dynamical collapse is not much different from the case when $K = 6$ (or lower). On the other hand, if the medium is extremely "turbulent" (K in excess of 14), ammonia core formation can be prevented altogether. (This conclusion unrealistically presumes that the high level of "turbulence" can be maintained indefinitely, resulting in clouds that become entirely supported by a combination of such motions and thermal pressure, with the [mean] magnetic field becoming asymptotically straight and uniform [i.e., force-free].)

The conclusion is clear. With interstellar parameters of the magnitude conventionally attributed to dark clouds like the Taurus region, the process of ambipolar diffusion naturally and spontaneously leads to the production of thermally supported entities with the properties of observed ammonia cores. They survive in such a condition about 10^6 yr, evolving quickly at the end toward a state of extreme central concentration and "inside-out" collapse. The isodensity contour defining the critical density for ammonia emission ($\sim 3 \times 10^4$ cm^{-3}) and containing about 1 $M_\odot$ of gas has no particular *dynamical* significance over any other isodensity contour. In other words, the existence of 1 $M_\odot$ ammonia cores does not automatically imply the formation of 1 $M_\odot$ stars. If the infall onto a forming protostar were not reversed by some internal means, we would expect the collapse to proceed to use up not only the entire material in a given "Roche lobe" (7 $M_\odot$ in Fig. 1), but even cause the material in the common envelope to eventually rain down. Gas would be transferred over as well from neighboring "Roche lobes." Given the observed inefficiency of star formation in most molecular clouds, we believe that the termination of the infall process – the end of the protostellar phase – must usually rely on some process other than simply the running out of the raw material for star formation.

2. BIPOLAR OUTFLOWS

2.1 Neutral Protostellar Winds

Growing evidence has accumulated in recent years that the process of protostellar build-up ends with the onset of a powerful stellar wind that often takes a bipolar form (see the reviews of Lada 1985, Welch et al. 1985, Bally 1987, Snell 1987). Thus, *forming stars help to determine their own masses.* The theoretical identification (Shu and Terebey 1984) that bipolar flow sources represent a phase of protostellar evolution combining inflow (along the *equatorial* regions) and outflow (along the rotational *poles*) receives support from the fact that they often have infrared spectral energy distributions that are indistinguishable from the models for pure accretion (Adams, Lada, and Shu 1987; see Figs. 2 and 3). As long as the cones of warm dust pushed away from the poles by the stellar outflow encompasses relatively little total solid angle, the thermal emission from the dust in the ambient (infalling) molecular envelope should remain relatively unaffected by their removal. The disruption of the remnant molecular cloud core as the evacuated polar cones widen in time has been captured especially convincingly in the case of L43 (Mathieu et al. 1988).

Until recently, however, the underlying properties of the hypothesized stellar wind that reverses the protostellar infall have remained largely conjectural. Attempts to find the wind from its signature as an *ionized* gas have generally set upper limits that lie one to two orders of magnitude below the momentum requirement of the observed bipolar outflow of swept-up molecular gas (Rodriguez and Canto 1983, Levreault 1985, Strom et al. 1986). Nevertheless, the discovery of extended far-infrared emission associated with the CO lobes of L1551 (Clark and Laureijs 1986; Edwards et al. 1986) leaves little doubt that there must exist a powerful source of missing mechanical luminosity coming from the central source IRS 5. The stellar photons from L1551 are insufficient to heat the very extended distribution of warm dust; the latter must somehow be heated by a mechanical outward transport of energy, presumably in the form of a *neutral* stellar wind.

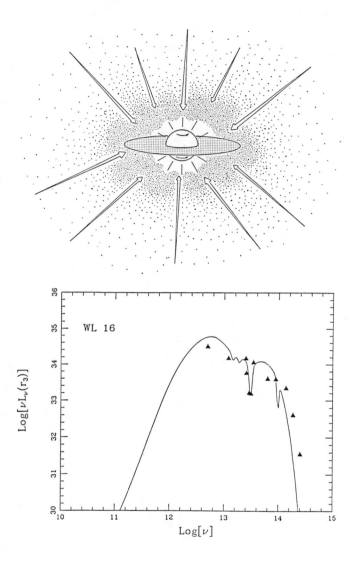

Figure 2. Comparison of theoretical and observed emergent spectral energy distributions of protostellar candidate WL 16 (ν and L_ν in cgs units): data from Wilking and Lada (1983), Lada and Wilking (1984), and Young, Lada, and Wilking (1986); the theoretical model assumes an infall rate $\dot{M} = 1 \times 10^{-5}\ M_\odot\ \mathrm{yr}^{-1}$, a central mass $M = 0.5\ M_\odot$, and a cloud core rotation rate $\Omega = 5 \times 10^{-13}\ \mathrm{s}^{-1}$.

128

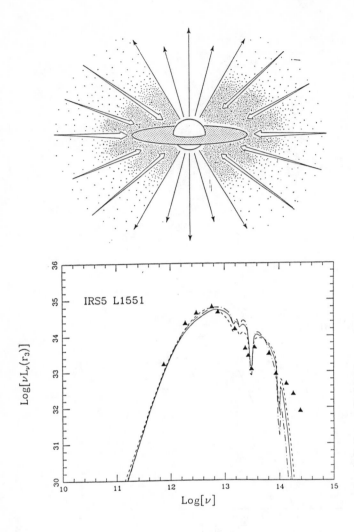

Figure 3. Theoretical and observed emergent spectral energy distributions of bipolar flow source IRS 5 L1551: data from Cohen and Schwartz (1983), Cohen et al. (1984), and Davidson and Jaffe (1984); the theoretical model corresponding to the dashed curve assumes $\dot{M} = 1 \times 10^{-5}\ M_\odot\ \mathrm{yr}^{-1}$, $M = 0.5\ M_\odot$, and $\Omega = 5 \times 10^{-13}\ \mathrm{s}^{-1}$; the dashed-dotted curve, $\dot{M} = 8 \times 10^{-6}\ M_\odot\ \mathrm{yr}^{-1}$, $M = 0.675$ $M_\odot$, and $\Omega = 1 \times 10^{-13}\ \mathrm{s}^{-1}$; the solid curve, $\dot{M} = 1 \times 10^{-5}\ M_\odot\ \mathrm{yr}^{-1}$, $M = 1.0\ M_\odot$, and $\Omega = 1 \times 10^{-13}$ s^{-1}, with stellar and disk dissipation efficiencies of $(\eta_*, \eta_D) = (1.0, 0.5)$ instead of $(0.5, 1.0)$ as in all the other cases. Notice that the most sensitively determined parameter is the infall rate $\dot{M}$ (see Adams, Lada, and Shu [1987] for details).

Observations carried out by Lizano et al. (1988) validate this expectation in the case of the bipolar flow source, HH7-11. Emerging from the central star, SVS 13, is a stellar wind composed of *atomic* hydrogen (see Fig. 4). (A similar situation may also hold for IRS 5 L1551, but the viewing geometry is less favorable – nearly equator-on – so that galactic hydrogen at moderately high velocities offers more confusion.) The H I detected in HH7-11 by the Arecibo telescope reaches speeds as high as ±170 km s^{-1}, with ~ 0.015 $M_\odot$ being contained in the fast moving gas represented by the line wings in Figure 4. Since the profile is roughly triangular, the average line-of-sight velocity is 170 km s$^{-1}/3 \approx 60$ km s^{-1}, yielding a travel time across the (projected) radius represented by the Arecibo beam of ~ 0.3 pc/60 km s$^{-1} \sim 5000$ yr. Thus, the rate of mass loss $\dot{M}_w$ suffered by SVS 13 $\sim$ 0.015 $M_\odot/5000$ yr $\approx 3 \times 10^{-6}$ $M_\odot$ yr^{-1}. The accumulated atomic hydrogen represented by the line core in Figure 4 amounts (after various corrections) to ~ 0.2 $M_\odot$, implying a total outflow time of 0.2 $M_\odot/3 \times 10^{-6}$ $M_\odot$ yr$^{-1} \sim 7 \times 10^4$ yr. Such a stellar wind would more than suffice as the underlying power that drives the extended CO lobes (see, e.g., Edwards and Snell 1984).

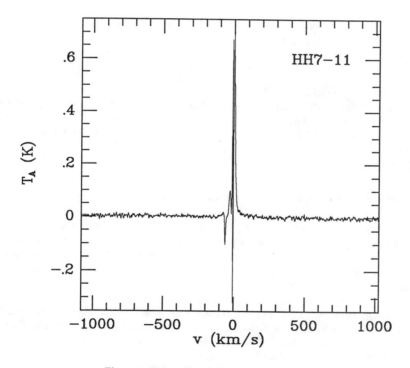

Figure 4. H I profile of HH7-11 taken at Arecibo.

The infrared spectral energy distribution of HH7-11 can be fit by a standard protostellar model. Figure 5 demonstrates that a reasonable fit requires a mass infall rate $\sim 1 \times 10^{-5}$ $M_\odot$ yr^{-1}. This is roughly 3 times greater than the mass outflow rate from the source, so a net accumulation evidently still occurs for the central star.

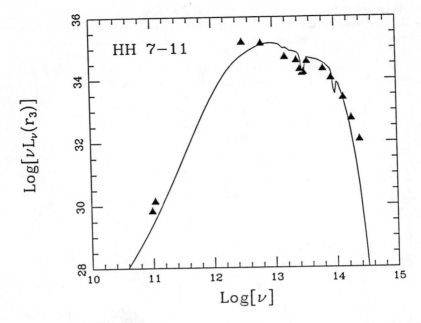

Figure 5. Spectral Energy Distribution of HH7-11. The model fit parameters are a central star of mass $M_* = 1.5\ M_\odot$, a mass infall rate of $\dot{M} = 1\times10^{-5}\ M_\odot\ \mathrm{yr}^{-1}$, and a cloud core rotation rate of $\Omega = 6\times10^{-14}\ \mathrm{s}^{-1}$. In the model, the disk accretion rate $\dot{M}_\mathrm{d}$ is taken in a steady state to equal the infall rate $\dot{M}$.

2.2 Evolutionary Context for Heavy Protostellar Mass Loss

An important property of the stellar wind in HH7-11 is the finding (on the Kitt Peak 12-m dish) that the high-velocity gas (up to $\pm$ 160 km s^{-1}) bears CO in a proportion consistent with all of the elemental carbon being contained in the form of that molecule. The combination CO plus *atomic* hydrogen characterizes matter in cool stellar atmospheres (rather than, say, molecular clouds or molecular disks), and places strong contraints on the possible physical mechanisms that can be responsible for accelerating the stellar wind. We have proposed that the mass loss originates from a magnetized protostar rotating at break-up on its equator because it is being spun up by an adjoining accretion disk (Shu, Lizano, Ruden, and Najita 1988a). The basic mechanism is that of a centrifugally-driven magnetic wind (Hartmann and MacGregor 1982).

The need for an unusual wind mechanism arises because protostars suffer much heavier mass loss than do their normal-star counterparts. Among ordinary cool stars, only supergiants such as the Mira variables have mass loss rates approaching $10^{-6}\ M_\odot\ \mathrm{yr}^{-1}$ (see, e.g.,

Fig. 14 in the review of Dupree 1986). The low-mass protostars lying at the centers of low-luminosity bipolar flow sources probably correspond to cool subgiants. (Stripped of their shroud of infalling material, they would be T Tauri stars on Stahler's [1983] "birthline;" see also Cohen and Kuhi [1979] and Cohen [1984].) However, ordinary cool subgiants have mass loss rates at least two orders of magnitude smaller than cool supergiants. Both have outer convection zones, one of the main ingredients for producing strong magnetic activity at the surface. In comparison with protostars, what the older objects lack is rapid rotation, presumably because they have had a relatively long time to brake magnetically (see, e.g., Schatzman 1962, Mestel 1968, Hartmann 1983). In contrast, a forming star should be rotating nearly at break-up if the infall process produces a circumstellar disk (Terebey, Shu, and Cassen 1984; see also Yuan and Cassen 1985), with a substantial fraction of the mass of the star being accumulated through disk accretion.

To allow the protostar to accumulate material of relatively high specific angular momentum, spin-up by the disk must be exactly balanced by spin-down by an extraordinarily powerful magnetized stellar wind (which works, as we shall see, only when the protostar is rotating almost exactly at "break-up.") The requirement of critical rotation can be shown to yield the result

$$\dot{M}_{\rm w} = f\dot{M}_{\rm d}, \tag{4a}$$

where the fraction f is given by the formula

$$f = \frac{1 - 2b}{\bar{J} - 2b}, \tag{4b}$$

with $\bar{J}$ being the streamline-averaged value of the specific angular momentum carried away asymptotically by the protostellar wind, measured in units of the specific angular momentum of the gas in Keplerian orbit at the equatorial radius R_e of the star, and with b being the total angular momentum of the star $\mathcal{L}_*$ divided by its mass M_* and radius R_e, measured in units of the virial speed $v_0 \equiv (GM_*/R_e)^{1/2}$. To derive equation (4b) we have assumed that the protostar lies on the "deuterium birthline" (see below) and has a nearly completely convective interior. This yields an equatorial break-up speed $v_0 \approx 140$ km s^{-1}, nearly independent of the mass of the star as long as it lies between ~ 0.1 and 2 $M_\odot$. Similarly, the combination of a polytrope of index 1.5 and critical rotation (at a more or less uniform rate throughout the star) yields $b = 0.136$ (James 1964).

For a magnetically torqued wind in which the thermal speed is small in comparison with the virial speed, it can be shown that $\bar{J}$ must be greater than 3/2 to allow matter to be centrifugally flung to infinity. If the Maxwell stresses associated with the magnetic field asymptotically become unimportant, $\bar{J}$ is given in terms of the mean terminal speed $\bar{v}_{\rm term}$ and the equatorial break-up speed v_0 through the formula

$$\bar{J} = \frac{3}{2} + \frac{1}{2}\left(\frac{\bar{v}_{\rm term}}{v_0}\right)^2. \tag{4c}$$

For the neutral wind in HH7-11 the ratio $\bar{v}_{\rm term}/v_0 = 1.2$, yielding a predicted ratio $f \approx 0.38$ from equation (4b) that is in reasonable agreement with the empirically determined values from Figures 4 and 5 of $\dot{M}_{\rm w} = 3 \times 10^{-6}$ $M_\odot$ yr^{-1} and $\dot{M}_{\rm d} = 1 \times 10^{-5}$ $M_\odot$ yr^{-1}. It would be valuable to have additional observational examples in which the predictions of equation (4a)-(4c) can be checked. Even more illuminating would be a measurement of the rotational velocity of a young stellar object of low mass that is optically revealed but still blows a neutral wind. The test would be to see if $v_0 \approx 140$ km s^{-1}.

2.3 The X-celerator Mechanism

In most theories of mass loss, the rate is determined by the conditions that apply at the location where the gas makes a sonic transition. The heavy mass loss $\dot{M}_w$ associated with protostars must originate from near their photospheres (to have sufficient density at the sonic transition). The problem is that the acoustic speed in the photosphere of a cool star forms a very small ratio compared to the escape speed (typically 0.04); therefore, thermal pressure is able to overcome gravity only where the local gravity is relatively weak (cf. an analogous problem in the case of mass-transfer binaries considered by Lubow and Shu 1975). A narrow band where the effective gravity is locally weak will naturally occur on the equator of a protostar rotating at break-up because it is being spun up by accretion from a surrounding nebular disk (that is itself supplied by infall from a rotating molecular cloud core). The equator of a critically rotating star corresponds to an "X-point" of the effective gravitational potential, $\mathcal{V}_{\text{eff}}$, (including the inertial effects of being in a frame that rotates with the star) where $\nabla \mathcal{V}_{\text{eff}} = 0$. Notice, however, that the centrifugal force can bring about a reversal in sign of the effective gravity only if the gas can be kept (nearly) corotating with the star. If the gas pushed outward by the thermal pressure preserves its original angular momentum, it will rotate too slowly to remain in orbit at the larger radius, and inertial effects would deflect it back to the star. Strong magnetic fields (generated, say, by dynamo action after the low-mass protostar has developed an extensive outer convective zone), tied to the deep interior of the star, will try to enforce the requisite corotation of the surface layers. (We assume the coupling between the the neutrals and the small fraction of ions to be good enough to treat the gas as a single conducting medium [see Natta et al. 1988].) The gas pushed out by the thermal pressure in an equatorial band now acquires more than enough angular momentum to maintain itself at the enlarged distance, so it continues to move outward as it is torqued up by the stellar rotation. The process drives the gas eventually to superalfvenic speeds, at which point the motions become essentially ballistic. If the gas speed at the Alfven point exceeds the local escape speed, the flow will develop into a wind that blows to infinity (Fig. 6).

For the mechanism to work, it can be calculated that a field strength of the order of ~ 100 Gauss is needed as an average over the surface of a cool protostar. Such fields are compatible with the estimates of Lago (1984) for RU Lupi, a T Tauri star which corresponds to a later stage of evolution than the protostars being discussed here. In our view, strong magnetic fields in low-mass protostars derive from dynamo action when rapid rotation combines with active convection zones that arise in response to the onset of deuterium burning (see, e.g., Shu and Terebey 1984).

Provided that such magnetic fields do get generated, the mechanism described above (which we call the "X-celerator" because of the crucial role played by the X-point in the effective gravitational potential) can accomodate almost arbitrarily high mass-loss rates. Because of the spin-up provided by the accretion disk, the mass loss involves a *forced* wind, not a free one. If $\dot{M}_w$ is less than required to satisfy equation (4a), the spin-up of the equatorial regions would continue to move the photosphere of the star outward relative to the X-point. The (exponentially) increasing density of the material making a sonic transition would then rise to a value sufficient to give the required steady-state rate of mass loss. Conversely, if $\dot{M}_{\mathrm{w}}$ is greater than the equilibrium rate dictated by equation (4a), the spin-down of the equatorial regions would lower the density at the X-point, decreasing the mass loss rate.

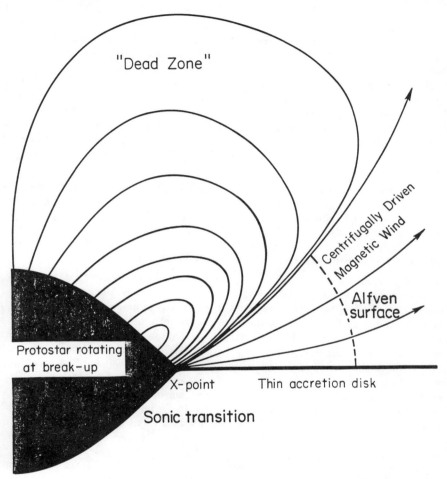

Figure 6. Schematic picture of a magneto-centrifugally driven wind from the equatorial region of a protostar rotating at break-up. The wind makes a sonic transition near the X-point of the effective gravitational potential, but the bulk of the acceleration occurs magneto-centrifugally in the region between the sonic surface and the Alfven surface. Much beyond the Alfven surface, the material acquires nearly ballistic trajectories, whose ability to escape from the system (in the limit of a small ratio for the thermal to virial speeds) depends only on the amount of angular momentum imparted to the gas by the magnetic torque of the star. In realistic circumstances, some of the closed field lines in the "dead zone" may also be opened by the effects of an "ordinary" wind.

A detailed model of the magnetohydrodynamics can be formulated if one assumes that the flow is axisymmetric and steady; in particular, one can solve self-consistently for the amount of spreading that occurs for the streamlines in the meridional plane. A tendency exists for the wind to be deflected toward the rotational poles because of the the magnetic pressure gradient which arises when the toroidal magnetic field is wrapped by the differential rotation of the outflowing gas (Nerney and Suess 1975, Sakurai 1985). Preliminary calculations (Shu, Lizano, Ruden, and Najita 1988b) indicate that the amount of magnetic focusing in the current problem may suffice to generate the geometry observed for bipolar CO flows, but the focusing is insufficiently strong to explain the highly collimated optical and radio jets seen in many young stellar objects of low mass (e.g., Mundt and Fried 1983, Bieging et al. 1984). We believe that such jets may arise from an "ordinary" wind that appears in the zone of the star which is "dead" to the X-celerator mechanism (see Fig. 6). This ordinary wind, which would usually blow in a quasi-spherical manner, is confined to flow toward the rotational poles by the more powerful ("extraordinary") wind driven by the X-celerator mechanism. A case for two component winds in bipolar flow sources has also been made on observational grounds by Stocke et al. (1988) and by Mundt (this volume).

In any case, notice that in our proposal the disk plays no active role in focusing the flow. Indeed, it has never been clear to us how various proposed *thin-disk* mechanisms could *naturally* produce well-collimated outflows. For us, the disk is merely the agent behind the scene that keeps the "flywheel" of the star spinning at break-up speeds. Notice also that once the magnetic field has been generated, the "flywheel" has almost 100% efficiency in converting excess rotational energy into an extraordinary wind.

In the presence of a rotating gravitational infall, we speculate that the breakout of both the ordinary and extraordinary winds occurs most readily along the channels of least resistance, which again lie in the directions of the rotational poles (Shu and Terebey 1984). As time proceeds, more and more of the infall occurs onto the outer parts of the disk rather than onto the star, and the outflow will naturally widen to reverse an ever increasing fraction of the inflow. An accompanying drop in the mass accretion rate through the disk will lead to a decrease in the power of the extraordinary wind, which will tend to decollimate the ordinary wind, giving it a greater lever arm for magnetic braking of the star. This effect should be quite dramatic once the X-celerator mechanism shuts off (when the angular speed of the star drops appreciably below the critical value). The system may then quickly lose the characteristics of a protostar and begin its life as a classical T Tauri star with modest rates of rotation and mass losss (Vogel and Kuhi 1981, Bouvier et al. 1986; Hartmann et al. 1986). It is intriguing to ask what kind of objects might represent the "missing links" of this scenario. We personally would place our bets on the T Tauri stars with flat infrared spectra (Adams, Lada, and Shu 1988; Strom et al. 1988).

This work is funded in part by grants from the National Science Foundation and from the NASA astrophysics program which supports a joint Center for Star Formation Studies at UC Berkeley, UC Santa Cruz, and NASA Ames Research Center. S. L. gratefully acknowledges the support of fellowships from the National University of Mexico and the Amelia Earhart Foundation.

REFERENCES

Adams, F. C., and Shu, F. H. 1986, *Ap. J.*, **308**, 836.

Adams, F. C., Lada, C. J., and Shu, F. H. 1987, *Ap. J.*, **312**, 788.

—————— . 1988, *Ap. J.*, in press.

Bally, J. 1987, *Irish Astr. J.*, **17**, 270.

Bieging, J., Cohen, M., and Schwartz, P. R. 1984, *Ap. J.*, **282**, 699.

Bouvier, J., Bertout, C., Benz, W., and Mayor, M. 1986, *Astr. Ap.*, **165**, 110.

Clark, F. O., and Laureijs, R. J. 1986, *Astr. Ap.*, **154**, L26.

Cohen, M. 1984, *Physics Rep.*, **116(4)**, 173.

Cohen, M., and Kuhi, L. V. 1979, *Ap. J. Suppl.*, **41**, 743.

Cohen, M., and Schwartz, R. D. 1983, *Ap. J.*, **265**, 877.

Cohen, M., Harvey, P. M., Schwartz, R. D., and Wilking, B. A. 1984, *Ap. J.*, **278**, 671.

Davidson, J. A., and Jaffe, D. T. 1984, *Ap. J. (Letters)*, **277**, L13.

Edwards, S., and Snell, R. L. 1984, *Ap. J.*, **281**, 237

Edwards, S., Strom, S. E., Snell, R. L., Jarrett, T. H., Beichman, C. A., and Strom, K. M. 1986, *Ap. J. (Letters)*, **307**, L65.

Dame, T.M., Elmegreen, B. G., Cohen, R. S., and Thaddeus, P. 1985, in *IAU Symposium No. 106, The Milky Way*, ed. H.van Woerden, R. J. Allen and W.B. Burton, (Dordrecht: Reidel), p. 303.

Dupree, A. K. 1986, in *Ann. Rev. Astr. Ap.*, **24**, 377.

Elmegreen, B. G. 1979, *Ap. J.*, **232**, 729.

Fuller, G. A., and Myers, P. C. 1987, in *Physical Processes in Interstellar Clouds*, ed. D. M. Scholer, (Dordrecht: Reidel), p. 137.

Hartmann, L. 1983, in *IAU Symposium No. 102, Solar and Magnetic Fields*, ed. J. O. Stenflo (Dordrecht: Reidel), p. 419.

Hartmann, L., Hewitt, R., Stahler, S., and Mathieu, R. D. 1986, *Ap. J.*, **309**, 275.

Hartmann, L., and MacGregor K. B. 1982, *Ap. J.*, **259**, 180.

James, R. A. 1964, *Ap. J.*, **140**, 552.

Lada, C. J., 1985, *Ann. Rev. Astr. Ap.*, **23**, 267.

Lada, C. J., and Wilking, B. A. 1984, *Ap. J.*, **287**, 610.

Lago, M. T. V. T. 1984, *M. N. R. A. S.*. **210**, 323.

Levreault, R. M. 1985, Ph.D. Thesis, University of Texas.

Lizano, S. 1988, Ph.D. Thesis, University of California, Berkeley.

Lizano, S. and Shu, F. H. 1988, in preparation.

Lizano, S., Heiles, C., Rodriguez, L. F., Koo, B., Shu, F. H., Hasegawa, T., Hayashi, S., and Mirabel, I. F. 1988, *Ap. J.*, in press.

Lubow, S. H., and Shu, F. H. 1975, *Ap. J.*, **198**, 383.

Mathieu, R. D., Benson, P. J., Fuller, G. A., Myers, P. C., and Schild, R. E. 1988, *Ap. J.*, in press.

Mestel, L. 1965, *Quart. J. Roy. Astron. Soc.*, **6**, 265.

_____ . 1968, *M.N.R.A.S.*, **138**, 359.

_____ . 1985, in *Protostars and Planets II*, ed. D. C. Black and M. S. Matthews (Tucson: University of Arizona Press), p. 320.

Mestel, L., and Spitzer, L. 1956, *M. N. R. A. S.*, **116**, 503.

Mouschovias, T. Ch. 1976, *Ap. J.*, **207**, 141.

Mouschovias, T. Ch. 1978, in *Protostars and Planets*, ed. T. Gehrels (Tucson: University of Arizona Press), p. 209.

Mundt, R., and Fried, J. W. 1983, *Ap. J. (Letters)*, **274**, L83.

Myers, P. C., and Benson, P. J. 1983, *Ap. J.*, **266**, 309.

Myers, P. C., and Goodman, A. 1988, *Ap. J.*, in press.

Nakano, T. 1979, *Pub. Astr. Soc. Japan*, **31**, 697.

_____ . 1984, *Fund. Cosmic Phys.*, **9**, 139.

Natta, A., Giovanardi, C., Palla, F., and Evans, N. J. 1988, *Ap. J.*, in press.

Nerney, S. F., and Suess, S. T. 1975, *Ap. J.*, **196**, 837.

Rodriguez, L. F. and Canto, J. 1983, *Rev. Mex. Astr. Astrof.*, **8**, 163.

Sakurai, T. 1985, *Astr. Ap.*, **152**, 121.

Schatzman, E. 1962, *Ann. d'Astrophys.*, **25**, 18.

Shu, F. H. 1983, *Ap. J.*, **273**, 202.

Shu, F. H., Adams, F. C., and Lizano, S.,1987, *Ann. Rev. Astr. Ap.*, **25**, 23.

Shu, F. H., Lizano, S., Ruden, S. P., and Najita, J. 1988a, *Ap. J. (Letters)*, submitted.

_____ . 1988b, in preparation

Shu, F. H., and Terebey, S. 1984, in *Cool Stars, Stellar Systems, and the Sun*, ed. S. Baliunas and L. Hartmann (Berlin: Springer-Verlag), p. 78.

Snell, R. L. 1987, in *IAU Symposium 115, Star Forming Regions*, ed. M. Peimbert and J. Jugaku (Dordrecht: Reidel), p. 213.

Solomon, P. M. and Sanders, D. B. 1985, in *Protostars and Planets II*, ed. D. C. Black and M. S. Matthews (Tucson: University of Arizona Press), p. 59.

Stahler, S. W. 1983, *Ap. J.*, **274**, 822.

Stocke, J., Hartigan, P., Strom, S. E., Strom, K. M., Anderson, E. R., Hartmann, L., and

Kenyon, S. A. 1988, *Ap. J.*, submitted.

Strittmatter, P. A. 1966, *M. N. R. A. S.*, **132**, 359.

Strom, K. M., Strom, S. E., Wolf, S. C., Morgan, J., and Wenz, M. 1986, *Ap. J. Suppl.*, **62**, 39.

Strom, K. M., and Strom, S. E., Kenyon, S. J., and Hartmann, L. 1988, *Ap. J.*, in press.

Terebey, S., Shu, F. H., and Cassen, P. 1984, *Ap. J.*, **286**, 529.

Vogel, S. N., and Kuhi, L. V. 1981, *Ap. J.*, **245**, 960.

Welch, W. J., Vogel, S. N., Plambeck, R. L., Wright, M. C. H., and Bieging, J. H. 1985, *Science*, **228**, 1329.

Wilking, B. A., and Lada, C. J. 1983, *Ap. J.*, **274**, 698.

Young, E. T., Lada, C. J., and Wilking, B. A. 1986, *Ap. J. (Letters)*, **304**, L45.

Yuan, C., and Cassen, P. 1985, *Icarus*, **64**, 435.

COLLAPSE OF A ROTATING PROTOSTELLAR CLOUD

Peter Bodenheimer
Lick Observatory
University of California
Santa Cruz, CA 95064, USA

Michal Różyczka
Warsaw University Observatory
Al. Ujazdowskie 4
PL-00-478 Warszawa, Poland

Harold W. Yorke
Universitäts-Sternwarte Göttingen
Geismarlandstr. 11
D-3400 Göttingen, FRG

Joel E. Tohline
Department of Physics and Astronomy
Louisiana State University
Baton Rouge, Louisiana 70803, USA

ABSTRACT. Two dimensional numerical hydrodynamical calculations with radiative transfer have been performed for the inner regions of collapsing, rotating protostellar clouds of about 1 solar mass. The region of disk formation is investigated in detail. Infrared spectra and isophotes are shown as a function of viewing angle at different stages of collapse. The usually assumed initial conditions for protostar collapse do not lead to the formation of a T Tauri star plus a "solar nebula" of about 0.2 solar masses, if conservation of angular momentum of each mass element is assumed; angular momentum transfer is required to produce such a system.

1. INTRODUCTION

The characteristics of the collapse of a protostellar cloud, starting from a molecular cloud core have been recently reviewed by Shu *et al.* (1987) and Boss (1987). The spectral appearance of the stellar core, its surrounding disk and the infalling cloud has been calculated approximately (assuming an equivalent spherical temperature distribution) by Adams and Shu (1986). The object of the present calculations is first, to determine the conditions required in an interstellar cloud for producing a slowly rotating central star plus a disk of

139

A. K. Dupree and M. T. V. T. Lago (eds.), Formation and Evolution of Low Mass Stars, 139–151.
© *1988 by Kluwer Academic Publishers.*

about 100 AU in radius, and second to determine the spectral appearance of the system during the process of disk formation as a function of viewing angle.

2. CALCULATIONS

The region of calculation is that within 10^{13} and 10^{15} cm from the central star. Although it includes only the inner part of the collapsing protostar, it allows adequate spatial resolution of the disk-forming region with a 50 by 50 grid. The region is dusty and optically thick; the inner boundary corresponds approximately to the dust evaporation point. The density in the region is everywhere high enough so that the magnetic field is not coupled to the matter. Therefore the calculation can be carried out with axial symmetry under the assumption of conservation of angular momentum of each mass element. Other mechanisms for angular momentum transport, such as turbulent viscosity, are not effective during the gravitational collapse phase. The initial cloud is assumed to be at rest, to contain about 1 solar mass, to have a uniform angular velocity distribution, to be isothermal, and to have a power-law density distribution.

The calculations were performed with the two-dimensional explicit hydrodynamic code of Różyczka(1985) modified to include self-gravity, rotation, and radiation transport in the diffusion approximation. The technique for calculating the gravitational potential is modelled after that described by Black and Bodenheimer (1975). Radiative diffusion was calculated fully implicitly using an ADIP approach. The Rosseland mean opacities for the dust were taken from the calculations of Pollack *et al.* (1985). After the hydrodynamic evolution was calculated with this code, frequency-dependent radiative transfer calculations were applied to selected models, following the approach of Bertout and Yorke (1978) and using their grain opacities. This mixture of silicates, graphites and ice-coated grains is consistent with the Rosseland mean opacities and with the opacities of Draine and Lee (1984) at near infrared wavelengths. These calculations, performed on a grid of lines of sight through the object, allow the calculation of the emergent spectrum and isophote maps at selected frequencies, both as a function of viewing angle.

The outer boundary, which has a fixed radius in the Eulerian grid, is assumed to have a constant temperature of 20 K and zero velocity, that is, no material is allowed to enter or leave the grid. At the inner boundary, matter falls into an unresolved central core. The mass in the core is included in the calculation of the gravitational potential. Given the mass and total angular momentum in the core, an approximate model for it is calculated based on the Maclaurin spheroid approximation. From this model, the ratio of rotational to gravitational energy in the core can be calculated, as can its equatorial radius R. The accretion luminosity is then calculated according to $L = G M \dot{M}/R$. The quantity L $\cdot \Delta$ t, where Δ t is the time step, is added to the internal energy of the central zone and is included in the radiative transfer calculation for the model.

3. RESULTS

The particular case to be discussed has density inversely proportional to spherical radius r, $\log \Omega = -10$, where Ω is the initial angular velocity, and a total angular momentum of $\log J = 52.8$ (both Ω and J in cgs units). Density and temperature contour plots with velocity vectors as well as spectra and isophotal contours are shown in the following figures.

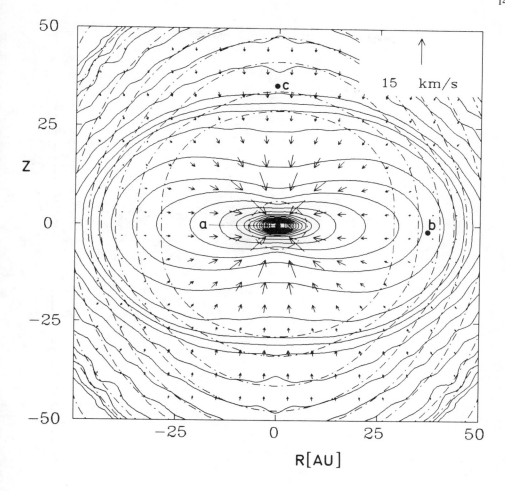

Figure 1. The cloud shortly after the beginning of collapse (110 years), in the (R,Z) plane. The Z axis corresponds to the rotation axis, and the R axis to the equatorial plane. The outer regions are still relatively spherical and the inner regions are beginning to flatten. The solid curves are equidensity contours; densities(in cgs) range from log ρ = -10 at the center to -17 at the outer edge. The dashed lines are isotherms, with temperatures ranging from 2000 K at the center to 60 K in the outer regions. Arrows correspond to velocity vectors. The accretion luminosity at this stage is 1000 solar luminosities, corresponding to a mass accretion rate (in solar masses per year) of log $\dot{M}$ = -2.5. This high rate is a consequence of the high initial density of the protostar.

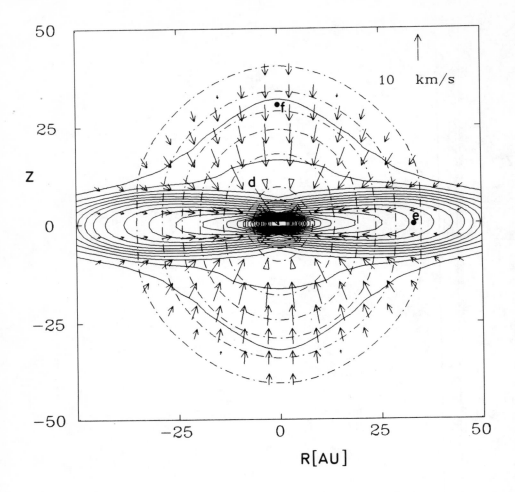

Figure 2. Representation of the protostar in the (R,Z) plane after 189 years. Density and temperature contours are shown, as in Figure 1. The values of density range from log ρ = -12 to -17; the temperature contours from 1000 K to 30 K. Most of the low angular momentum material that is able to fall into the core has already done so; the material with higher angular momentum is now beginning to form the disk. Therefore the accretion rate onto the core has decreased to log $\dot{M}$ = -6.4 and the accretion luminosity to 0.23 solar luminosities. Because of the lower accretion luminosity, which forms the main energy input for the disk, that region is beginning to cool by radiative diffusion. About 0.64 solar masses now has fallen into the core.

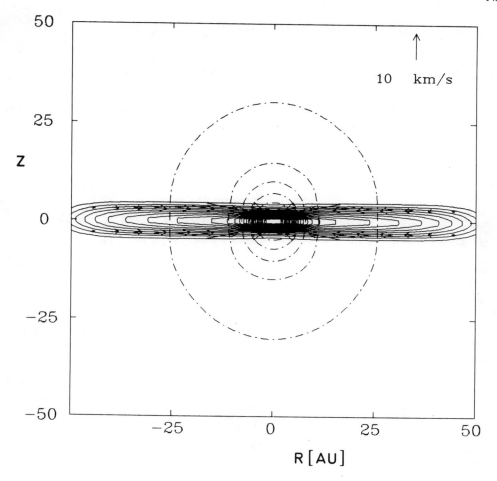

Figure 3. The protostar in the (R,Z) plane after 222 years, showing the well-developed disk. Density and temperature contours as well as velocity vectors are shown, as in Figure 1. Inflow along the equatorial plane is still taking place , so the disk is not yet in equilibrium, but the accretion rate onto the core has been reduced to a negligible value. The maximum temperature on this diagram is about 600 K; the maximum density about log ρ = -12. In Figures 1-3 the contour intervals are Δ log ρ = 0.3 and Δ log T = 0.15. The core itself has a mass of 0.66 solar masses, and it is rapidly rotating, with an estimated ratio of rotational energy to gravitational energy of 0.4. Non-axisymmetric instabilities are likely to occur there.

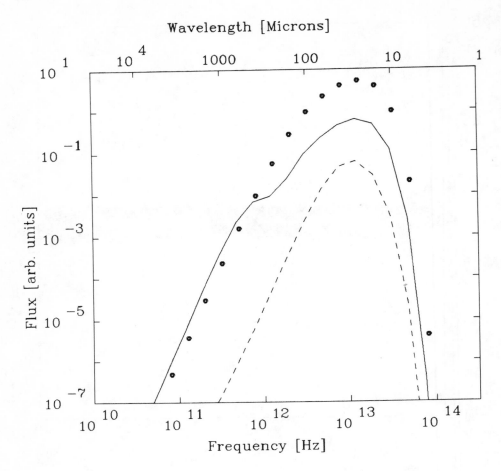

Figure 4. The infrared spectrum of the model shown in Fig. 1, as viewed from the pole. The symbols give the total spectrum, which peaks here at 25 microns, assuming that the beam includes the entire protostar. The solid line gives the spectrum that would be obtained with a very thin beam directed at the pole, while the dashed line gives the spectrum, again with a thin beam, along a line of sight offset by about 33 AU from the pole but parallel to the rotation axis. The vertical scale is in arbitrary units; the solid and dashed curves are plotted on the same vertical scale, which is different from that used for the total spectrum.

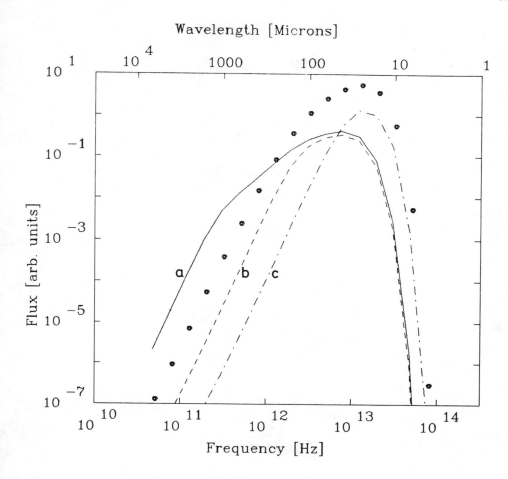

Figure 5. The infrared spectrum of the model shown in Fig. 1, as viewed from the equator. The overall spectrum (symbols) is very similar to that shown in Figure 4, with a peak at 25 microns. The flux is reduced by a factor 2 from that received from the pole. Most of the radiation comes from the outer cool layers, which at this time are fairly spherically symmetric. The spectra along specific lines of sight are given by the curves labelled a, b, and c, where the labels refer to the points marked on Figure 1. The three curves use the same (arbitrary) vertical scale, which is offset from that used for the overall spectrum. Curves a and b are views directly into the heavily obscured equator which show a shift in the maximum intensity toward the red as compared with the overall spectrum.

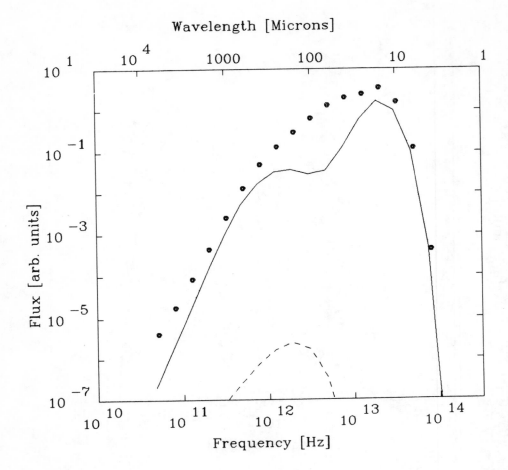

Figure 6. The infrared spectrum of the model shown in Fig. 2, as viewed from the pole. The total spectrum (symbols) peaks at 16 microns, slightly blueward of the peak shown in Fig. 4. The protostar is now relatively flattened, and the view toward the pole now receives radiation from the warm regions just outside the core. The line of sight to the pole (solid line) shows both a warm and a cool component. The line of sight offset by 33 AU from the pole (dashed curve) has a highly reduced flux with a peak longward of 100 microns. As in Figures 4 and 5, the vertical scale is arbitrary, and the two curves are on the same scale which, however, differs from that used for the symbols.

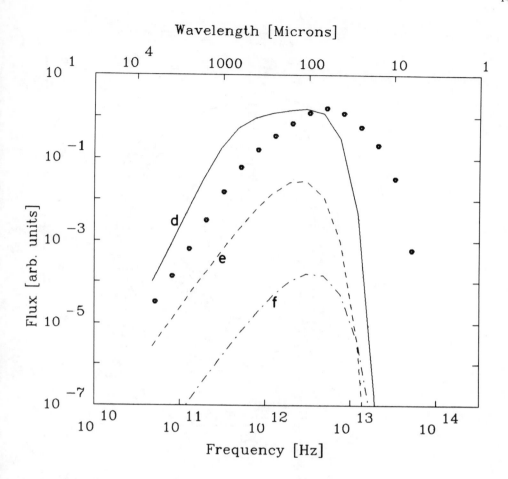

Figure 7. The infrared spectrum of the model shown in Fig. 2, as viewed from the equator. The peak of the total spectrum (symbols) is now at 63 microns, notice-ably shifted from that of the corresponding spectrum at the earlier time (Figure 5) and also shifted from that at the pole (Figure 6). The spectra along specific lines of sight are given by the curves labelled d, e, and f, where the labels re-fer to the points marked on Figure 2. These three curves use the same (arbi-trary) vertical scale, which is offset from that used for the overall spectrum. Note that the maximum intensity of all of these curves is shifted to the red rel-ative to the total spectrum. In all cases cool or heavily obscured regions are being observed, either in the equatorial plane or well above it. The warm radia-tion evident in the total spectrum is emitted from the region near the pole, but just above the plane.

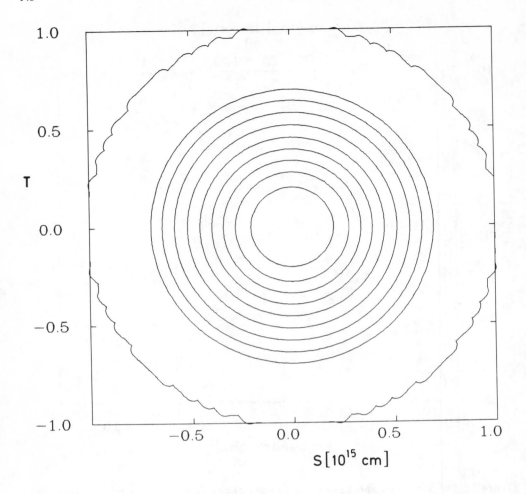

Figure 8. Isophotes for the model shown in Fig. 1 at a wavelength of 40 microns, near the peak of the overall spectrum. The coordinate axes refer to the plane of the sky in a view directly toward the pole. The linearly spaced values of the contours are normalized to the central value and correspond to intensities of 0.9, 0.8, 0.7, etc. times that at the center. The total luminosity over a full sphere that would be deduced from the view at this angle is about 700 solar luminosities. The view from the equator in this model would give about 350 solar luminosities. These values are somewhat less than the accretion luminosity generated at the core, 1000 solar luminosities, because some of the energy goes into dissociation and heating in the envelope.

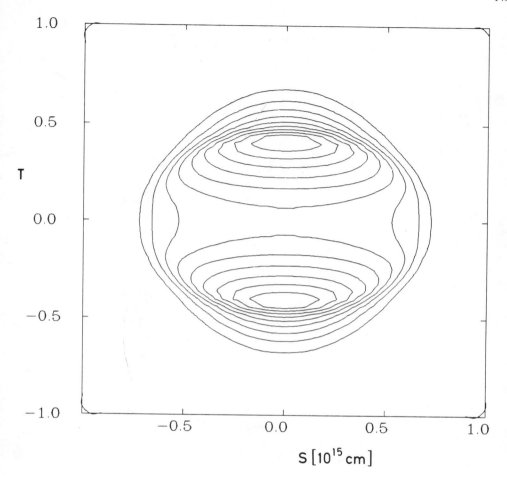

Figure 9. Isophotes for the model shown in Fig. 1 at a wavelength of 40 microns, assuming an equator-on view. The linearly spaced contour levels are normalized to the maximum value which in this case is shifted to points above and below the plane. The relative contour levels are the same as in Fig. 8. At shorter wavelengths (10 microns) the optical depth effects result in spatial shifts of the maximum intensity to even greater distances away from the plane, while at 100 microns, the two maxima, while still present, are very close to the plane. The same qualitative effect (not shown) occurs in the more disk-like model plotted in Fig. 2.

150

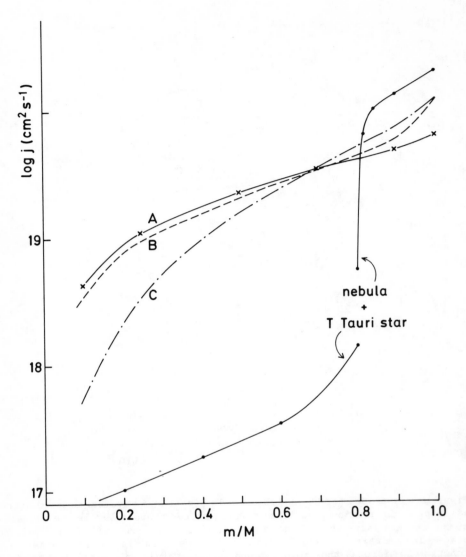

Figure 10. This figure illustrates the significance of the assumed initial distributions of density and angular momentum. The angular momentum per unit mass (j) is plotted as a function of the cylindrical mass fraction for several different configurations, all of which have the same total total mass and angular momentum. The solid line represents a T Tauri star of .8 solar masses, a radius of 4 solar radii and rotational velocity of 50 km /sec, surrounded by a Keplerian disk of 0.2 solar masses and a radius of 20 AU. The other curves give the angular momentum distributions in uniformly rotating spheres (A) of uniform density, (B) with the density distribution of an n=1.5 polytrope, and (C) with a density proportional to the inverse square of the radius.

4. CONCLUSIONS

Calculations of the collapse of rotating protostars have been performed for the phase during which disk formation occurs. The standard case is described in Figures 1-9; other cases with different density distributions and different total angular momenta have also been considered. Over a relatively small parameter range a disk with radius of about 100 AU is formed. The energy input to the disk region is provided mainly by the accretion luminosity onto the protostellar core. When the supply of low angular momentum material is exhausted, this luminosity drops sharply. Infrared spectra of the system as a function of viewing angle have been obtained. During disk formation there is a considerable shift in the wavelength of maximum intensity between the pole-on view and the equator-on view. The central core of the protostar, inside 1 AU, is very rapidly rotating and is probably unstable to non-axisymmetric perturbations. Therefore the standard initial conditions, even with highly peaked density distributions, do not lead to a slowly rotating central star plus a surrounding nebula of relatively low mass, if the collapse proceeds with conservation of angular momentum of each mass element. Substantial angular momentum transport must occur during or immediately after the collapse; gravitational torques induced by non-axisymmetric instabilities in the central regions are a likely possibility.

Acknowledgements: This work was supported through the NSF grant AST-8521636, DFG grants Yo 5/4 and Yo 5/3-2, and under the auspices of a special NASA astrophysics theory program which supports a joint center for star formation studies at NASA/Ames Research Center, UC Berkeley, and UC Santa Cruz. PB acknowledges a Senior U.S. Scientist Award from the Alexander von Humboldt Foundation.

REFERENCES

Adams, F.C., and Shu, F.H. 1986. *Ap.J.* **308**, 836.
Bertout, C., and Yorke, H.W. 1978. In *Protostars and Planets*, ed. T. Gehrels (Tucson: Univ. of Arizona Press), p. 648.
Black, D.C. and Bodenheimer, P. 1975. *Ap.J.* **199**, 619.
Boss, A.P. 1987. In *Interstellar Processes*, eds. D. Hollenbach and H. Thronson (Dordrecht: Reidel), p. 321
Draine, B.T. and Lee, H.M. 1984. *Ap.J.* **285**, 89.
Pollack, J.B., McKay, C., and Christofferson, B. 1985. *Icarus* **64**, 471.
Różyczka, M. 1985. *Astron. Ap.* **143**, 59.
Shu, F.H., Adams, F.C., and Lizano, S. 1987. *Ann. Rev. Astron. Ap.* **25**, 23.

ACCRETION DISCS

J E Pringle
Institute of Astronomy
Madingley Road
Cambridge, CB3 0HA
United Kingdom

ABSTRACT. The role of accretion discs in the formation of protostars is reviewed and discussed. It is argued that most stars form via a protostellar accretion disc and that accretion disc remnants around stars should therefore be common.

1. INTRODUCTION

It is becoming increasingly evident that accretion discs play a major role in the star formation process. From a theoretical point of view this is not surprising, since a major obstacle to forming a star from the interstellar medium is getting rid of angular momentum. Accretion discs provide a known and plausible mechanism for overcoming this problem. It has long been known that accretion discs are of immediate relevance to the formation of the solar system and all the early accretion disc papers were written with this in mind. A detailed review of the role of accretion discs in the dynamical origin of the solar system is given by Lin and Papaloizou (1985). Larson (1983) discusses protostellar discs and the angular momentum problem, and protostellar discs are also reviewed by Cassen, Shu and Tereby (1985). From the observational point of view evidence for remnant discs around recently formed stars is being discovered - for example around HL Tau (Sargent and Beckwith 1987) and in T Tauri stars (Bertout, Basri and Bouvier, 1987; Adams, Lada and Shu, 1987; Kenyon and Hartmann, 1987), and the dust discs around low mass mainsequence stars (see the review by Wolstencroft and Walker 1987).

2. THE STANDARD THIN ACCRETION DISC

If the pressure of the disc gas is dynamically unimportant then the gas in the disc is in circular orbits about the central point such that at each point centrifugal force acting on the orbiting particles balances the internally directed gravitational pull. For a disc orbiting about a central point mass, M this implies that the angular velocity, Ω, of the gas at radius R is given by

A. K. Dupree and M. T. V. T. Lago (eds.), Formation and Evolution of Low Mass Stars, 153–162.
© 1988 by Kluwer Academic Publishers.

$$R\Omega^2 = GM/R^2 \tag{2.1}$$

which implies

$$\Omega = (GM/R^3)^{\frac{1}{2}}. \tag{2.2}$$

Because $d\Omega/dR \neq 0$ this is a shear flow and hence at each point in the disc it is possible to tap this shear energy. A simple viscosity present in the disc would lead to local dissipation of shear energy and generation of heat. On the global scale this heat must come ultimately from the gravitational potential energy of the particles involved and so local dissipation must also lead to inflow of material through the disc.

Suppose the disc is in a steady state such that the rate of inflow of material, $\dot{M}$, through each radius, R, is a constant equal to, perhaps, the rate at which matter is fed into the disc by some external mechanism at large radius. Then consider a particular annulus in the disc at radius R and of width ΔR. Then the rate, $\dot{E}$, at which potential energy is lost by $\dot{M}$ flowing through ΔR is

$$\dot{E} \propto M. \frac{GM}{R}. \frac{\Delta R}{R} \tag{2.3}$$

If, as is true for a thin disc, the energy liberated is radiated away locally, we can use $\dot{E}$ to define an effective temperature, T_e, in terms of the energy loss rate per unit area (remember the disc radiates from two sides)

$$\dot{E} \propto 4\pi R.\Delta R.\sigma\, T_e^4 \tag{2.4}$$

which gives

$$T_e^4 \propto GM\dot{M}/R^3 \tag{2.5}$$

This is a basic property of steady accretion discs and depends on nothing more than energy conservation.

In fact, when the calculation is done in full (e.g. Pringle 1981, Frank, King and Raine 1985) the formula turns out to be

$$T_e^4 = \frac{3GM\dot{M}}{8\pi\sigma R^3} \left[1 - (\frac{R_*}{R})^{\frac{1}{2}} \right], \tag{2.6}$$

where R_* is the inner disc radius. From this one finds that the total disc luminosity is

$$L_{disc} = \int_{R_*}^{\infty} \sigma T_e^4 4\pi R\, dR$$

$$= GM\dot{M}/2R_* \tag{2.7}$$

Thus the energy emitted by the disc is only one half of the gravitational potential energy available. This implies that the accretion disc is at best only half the story and that there can be as much radiation emitted again as the inner disc material comes to rest on the stellar surface in what is called the "boundary layer".

If each element of the disc does in fact radiate like a black body then the disc spectrum can be calculated straightforwardly,

viz.

$$S_\nu \propto \int_{R_*}^{R_{out}} B_\nu[T_e(R)]2\pi R dR.$$

$$(2.8)$$

If the ratio of outer to inner disc ratio is large enough, $R_{out}/R_* \gtrsim 10^3$, then much of the spectrum looks like a power law of the form $\nu S_\nu \propto \nu^{4/3}$. At high frequencies there is an exponential cutoff corresponding to the hottest temperature present in the disc, and at low frequencies the spectrum is $\nu S_\nu \propto \nu^3$ corresponding to the Rayhigh-Jeans tail of the coolest temperature present. At present using this blackbody approximation to discs around protostars appears to be adequate, but it should be remembered that as the accuracy of the observations improves, and, for example, broadband fluxes become spectra, more thought may be required.

As a final comment on the steady state disc, we should note that the kinematic viscosity, ν, in such a disc is required to have the form

$$\nu = \frac{\dot{M}}{3\pi\Sigma}\left[1 - \left(\frac{R}{R}*\right)\right] \quad ,$$

$$(2.9)$$

where Σ is the disc surface density. It should be remembered that the laws of physics may not permit the viscosity to be of this form, that is, may not permit the disc to be steady state.

This problem becomes more acute when we come to consider the time-dependent accretion disc. The time dependence of the disc surface density is given by the equation

$$\frac{\partial\Sigma}{\partial t} = \frac{3}{R}\frac{\partial}{\partial R}\left\{R^{\frac{1}{2}}\frac{\partial}{\partial R}(\nu\Sigma R^{\frac{1}{2}})\right\} \quad .$$

$$(2.10)$$

This is basically a diffusion equation. The physical reason for this is that material cannot move inwards until it has lost angular momentum. Thus loosely the role of the viscosity is to separate out matter which flows inwards and angular momentum which flows outwards. The timescale, t_ν, on which viscosity acts, and so on which the surface density changes can be seen to be

$$t_\nu \sim R^2/\nu$$

$$(2.11)$$

and so is given directly by the viscosity.

Little is known about the viscosity in accretion discs, other than the fact that it is present. The clearest discussion of the possible viscous processes present in accretion discs is given by Shakura and Sunyaev (1973) who proposed a way of parametrizing our ignorance. They argued that the most likely forms of viscosity would be hydrodynamic or hydromagnetic turbulence, and that the same parametrization could be applied to both. From a simple mixing length standpoint, the effective viscosity of turbulence is

$$\nu \sim v_t . l \qquad (2.12)$$

where l is the size and v_t the turnover velocity of the largest eddies. Since one expects $l \lesssim H$, the disc semi-thickness, and $v_t \lesssim c_s$, the sound speed, one may write

$$\nu = \alpha \ c_s H \qquad (2.13)$$

where α is a dimensionless parameter which measures the strength of the turbulence and we expect $\alpha \lesssim 1$. Since the disc thickness H is given roughly by $H \sim c_s/\Omega$, the formula is sometimes written $\rho\nu = \alpha P/\Omega$, where P is the pressure and ρ the density in the disc. There is no real physics in any of the above and probably the best approach to this prescription is to treat α as an observational parameter. The only measurements of ν obtained so far come from the time-dependent accretion discs involved in the outbursts of dwarf novae and the values obtained are in the range $\alpha \lesssim 0.1 - 1.0$ in the outburst and $\alpha \lesssim 10^{-2}$ in quiescence. In terms of the α-prescription, the viscous time scale now becomes

$$t_\nu \sim \alpha^{-1}(\frac{R}{H})^2\Omega^{-1} \qquad (2.14)$$

Note that $t_\nu \gg \Omega^{-1}$, the orbital timescale in the disc.

3. PROTOSTELLAR DISCS

3.1. Infall to a disc

Do we expect protostars to have discs? Most of the work done on the details of protostellar collapse has been in terms of spherically symetric collapse (see for example Appenzeller 1980). These calculations assume implicitly that rotation of the collapsing material does not affect the final outcome. There does now appear to be a general unanimity about how the final dynamical collapse in the formation of a protostar takes place in broad outline. Whatever the pre-collapse state, be it a state of hydrodynamic equilibrium or one in which magnetic forces dominate the support, and independent of why the collapse occurs, be it due to some kind of push over the self-gravitational threshold or slow removal of support through ambipolar diffusion (Shu 1983), all authors appear to agree that the final dynamical collapse entails a) the initial collapse occurring at the centre with the formation of a dense core and b) the rest of the collapsing material accreting into it at a rate given approximately by $M \sim c_s^3 /G \sim 10^{-6}$ $M_\odot y^{-1}$ (Shu 1977, Bodenheimer and Black 1976). The whole collapse occurs on the dynamical timescale for the outermost parts, which for a solar mass object is $\sim 10^5 - 10^6$ years. Because the collapse is dynamical, this implies that during the collapse there is no time for the redistribution of angular momentum through viscous or turbulent processes. This means that any angular momentum present prior to collapse is conserved during the collapse process. In other words the collapse must be to a disc-like structure and not to a central point.

How big would we expect such a disc to be? What is relevant is the specific angular momentum of the infalling material. The specific angular momentum of a disc particle at radius R in Keplerian orbit around a central mass M is

$$h_k = (GMR)^{\frac{1}{2}}$$
$$= 4.5 \times 10^{19} R_{AU}^{\frac{1}{2}} \; cm^2 s^{-1} \qquad\qquad (3.1)$$

The specific angular momenta of rotating clouds tabulated by Goldsmith and Arquilla (1985) range between 3×10^{24} and 1.2×10^{21}. If these collapsed directly to form discs the disc radii extend upwards from about 700 A.U. The dense cores in dark clouds seen in NH_3 and reported by Myers and Benson (1983) have mean size ~ 0.1 pc and mass~ 1 $M_\odot$. They are clearly not supported by rotational motions, but there is some evidence for velocity gradients across some clouds. In L63 a velocity shift of about$\Delta V \sim 0.16$ km s^{-1} is seen across the cloud of radius $R \sim 0.12$ pc and mass ~ 1.7 $M_\odot$. This gives a specific angular momentum of $\sim 6 \times 10^{21}$ cm^2 s^{-1} and so collapse would produce a disc extending out to ~ 6000 A.U. Because observational selection favours measurement of the larger velocity gradients, and because the velocity gradients are biggest in those regions of the gas (the outside of the cores) which may not ultimately collapse the above estimates should be regarded as generous. Nevertheless it does seem reasonable to expect that collapses could frequently give rise to discs whose radii are in the range $10^2 - 10^3$ A.U, and sometimes larger.

A simple picture of how such a collapse might proceed is given by Cassen and Moosman (1981). The picture is of a uniformly rotating cloud which collapses from inside out. As the collapse proceeds, shells of material arrive at the centre from larger initial radii and so with larger specific angular momentum. Since most of the mass and angular momentum is at large radii, most of the infalling mass will arrive at large disc radii. To make rough estimates of the conditions in such a disc, consider matter accreting uniformly into a disc of radius R and mass $M \sim 1M_\odot$ at a rate of 3×10^{-6} $M_\odot$ y^{-1}. The matter strikes the disc and produces a shock. As for the accretion shock in the spherically symmetric case the shock is highly radiative, so that a better estimate of the temperature in the disc is the effective temperature at which this accretion energy is radiated away

$$T_{eff} \sim \left(\frac{GMM'/R}{\sigma\pi R^2}\right)^{\frac{1}{4}} \sim 460 \ R_{AU}^{-3/4} \ K \tag{3.2}$$

Since the disc is heated locally from outside it is essentially
isothermal at temperature T_{eff} and so has thickness given by

$$H/R \sim 0.07 \ R_{AU}^{1/8} \tag{3.3}$$

Using equation (2.14) we may now calculate the viscous timescale in the
disc which is

$$t_\nu \sim 10 \ \alpha^{-1} \ R_{AU}^{5/4} \ \text{years} \tag{3.4}$$

For a disc of radius $R > 4 \times 10^3 \alpha^{4/5}$A.U. the viscous timescale is
larger than the accumulation timescale of $\sim 3 \times 10^5$ years. Thus for a
value of α around 0.1 the inner regions of the disc can accrete as
assumed by Cassen and Moosman but in the outer regions ($R \gtrsim 300$A.U.) the
accreting material begins to accumulate in the disc. This implies that
for reasonable input parameters it is to be expected that the
protostellar disc will be for the most part self-gravitating.

3.2 The self-gravitating disc

A local measure of the importance of self-gravity in a disc is given by
the Toomre Q-parameter (1984) $Q \sim c_s \Omega/G\Sigma$. If Q is less than some
critical value which is of order unity then self-gravity is of
relevance. If we write the disc mass as $M_{disc} \sim R^2\Sigma$ and the central
disc mass as M, so that for a centrally condensed disc $M_{disc} \ll M$
whereas for a uniform disc $M_{disc} \sim M$, then

$$Q \lesssim 1 \ <=> \ \frac{c_s}{R\Omega} \lesssim \frac{M_{disc}}{M} \tag{3.5}$$

This implies that even for a centrally condensed disc, it can still be
self-gravitating if it is cool enough.

What happens to a disc if it is selfgravitating? The first point
to note is that it does not necessarily fragment into self-gravitating
substructures. Any attempt to do this is resisted by the strong shear
present in a disc. Numerical simulations of a disc which is cooled
continuously in order to keep self-gravity relevant indicate that the
instabilities present give rise to clumps which are sheared out by the
rotation to give the overall appearance of transient, but continuously
regenerated, spiral arms. Such features must, on a timeaverage, give
rise to a net transfer of angular momentum outwards in the disc and
therefore to local disc heating. Sellwood and Carlberg (1984) found
that with a cooling timescale $t_{cool} \sim 10^2 - 10^3 \Omega^{-1}$ a balance between
heating and cooling could be achieved with the disc settling into a
marginally self-gravitating state with $Q \sim 1$. If the disc is strongly
self-gravitating, either due to rapid disc cooling ($t_{cool} \lesssim \Omega^{-1}$) or due
to rapid accretion of material, then any instabilities present are
likely to act more strongly. Lin and Pringle (1987) proposed that for

a strongly self-gravitating disc the net effect of such instabilities would be to act as a local viscosity

$$\nu \sim R^2 (M_{disc}/M)^2 \tag{3.6}$$

with corresponding viscous timescale

$$t_\nu \sim \Omega^{-1} (M/M_{disc})^2 \tag{3.7}$$

for the case $M_{disc} \ll M$. Numerical simulations of strongly self-gravitating discs are at present underway (Sellwood, Ruden - private communication) and the initial indications are that this proposal might not be a bad first guess but that the full situation is a great deal more complicated. In any case it seems that a self gravitating disc is able to transfer angular momentum efficiently through gravitational torques, and one may tentatively conclude that the protostellar disc evolves rapidly to a regime in which although much of the angular momentum resides in the disc, most of the accreted material finds its way rapidly to the centre.

3.3 The remnant disc

Once all the exceitement has died down and we have a star formed in the centre of the disc, we should ask how much material is likely to be left behind in this disc. Larson (1983) addressed this question by suggesting that the maximum disc mass that could be left behind is one that is marginally selfgravitating (that is Q = 1 everywhere). For a disc temperature profile $T = 300\, R_{AU}^{-\frac{3}{2}}$ (which assumes the disc is optically thin) he found the disc mass to be

$$M_{disc}\,(R) = 0.14\, R_{AU}^{\frac{1}{4}}\, M_\odot \tag{3.8}$$

In the initial stages, however, the disc is likely to be optically thick. For low accretion rates ($\dot{M} \lesssim 10^{-6}\, M_\odot y$) the dominant disc heating comes from the central protostar. If we take the star to have mass 1 $M_\odot$, radius 2×10^{11}cm, and surface temperature 4000 K then the disc temperature is approximately $T_{disc} \simeq 500\, R_{AU}^{-3/4}$ K, which for a Q=1 disc implies

$$M_{disc}\,(R) = 0.12\, R_{AU}^{1/8} M_\odot \tag{3.9}$$

How long the disc can remain there depends on the viscous timescale in the disc. We have seen that for standard discs with constant α the outer disc regions last longer. For a value of $\alpha \sim 10^{-2}$, corresponding to the quiescent, cool dwarf nova discs the timescale for a 300 A.U. disc would be over 10^6 years. Thus the expectation that such discs might be quite massive ($\sim 0.1\, M_\odot$) and quite long lasting is not an unreasonable one.

4. DISCUSSION.

It seems reasonable to expect that the final dynamical collapse to form a protostar gives rise initially to a disc of some sort, with radius probably in the range $10^2 - 10^3$ a.u. Whatever the radius is at which it forms either viscous forces or self-gravitational interactions act rapidly on the initial disc to form a massive core and a less massive disc (which contains the residual angular momentum). The evolutionary behaviour of the core is probably not too dissimilar from the spherically symmetrically accreting protostars. The implication of the expectation (Lin and Pringle 1987) that once the disc becomes self-gravitating it evolves rapidly is that once a disc is formed it does not evolve into a binary star. Thus binary stars must form through some other mechanism. Presumably the fragmentation into a multiple star system takes place at the onset of the collapse (e.g. Bodenheimer 1978, Bodenheimer, Tohline and Black 1980). This provides evidence in favour of the external triggering of the collapse, rather than the slow evolution towards a singular density distribution envisaged by Shu (1977).

The bigger the disc the longer the evolution time. However the actual evolutionary timescales for the disc are highly uncertain because little is known about the relevant viscous processes. Some discussion has been made of convection driving angular momentum transfer in such discs (Lin and Papaloizou 1985) but it is by no means evident that convective modes do transfer angular momentum, nor, even if they do, that they do so to such an extent that they can be self-sustaining. A number of the pre-main-sequence stars (in particular the T Tauri stars) show evidence that they are surrounded by discs. The initial spread in disc radius presumably reflects the spread of rotation rates present in the initial gas clouds. Thus at any point in the HR diagram one would expect a corresponding spread in accretion disc properties. As the main sequence is approached one would expect only those stars with initially large enough discs to still have a sizeable disc remnant. How the disc eventually disperses is still mostly a matter of speculation. Rings or discs of cool gas and dust have now been found around a number of stars and it is worth remembering that estimates for the mass of our own comet cloud range up to $\sim 10^{-3}$ $M_\odot$ (Bailey et al 1984) which implies an initial hydrogen and helium mass cloud (or disc) of ~ 0.1 $M_\odot$.

I have not so far discussed the fate of material which reaches the centre of the protostellar disc, except to say that it presumably forms a protostar. It should be noted however that all the matter deposited on the central object arrives with Keplerian rotation speed and so high angular momentum. It might be thought that the arrival of such high angular momentum material might present problems for the budding protostar, but that is not the case. As numerical attempts to produce fission of rapidly rotating objects have shown (Durisen and Tohline 1985), a rapidly rotating star is capable of shedding angular momentum through nonaxisymmetric dynamical instabilities on a dynamical timescale. Thus, in principle, the central object can accept all of the arising material without difficulty and will return any excess angular momentum that it cannot accommodate to the disc by tidal torques. On the other hand, if this were the only process operating we would expect premainsequence stars when they become visible to be rotating close to break up. That this is not the case (such stars are observed to be spinning well below break up) implies that some other spindown mechanism is operating: indeed not only operating, but operating far more efficiently than necessary and on a timescale of $\lesssim 10^6$ years. The obvious candidate for such a mechanism is a stellar wind tied to stellar field lines which by the magnetic lever arm effect is able to carry off material with high specific angular momentum.

ACKNOWLEDGEMENTS

This research has been conducted in part under the auspices of a special NASA astrophysics theory grant which supports a joint Center for Star Formation Studies at NASA - Ames Research Canter, U.C. Berkeley, and U.C. Santa Cruz.

REFERENCES

Adams, F.C., Lada, C.J., and Shu, F.H., 1987. Astrophys. J., in press
Appenzeller, I., 1980. In "Star Formation". Saas Fee. eds A. Maeder and L. Martinet.
Bailey, M.E., McBreen, B. and Ray, T.P., 1984. Mon. Not. R. astr. Soc., 209, 881
Bertout, C., Basri, G. and Bouvier, J., 1987. Astrophys. J., in press.
Bodenheimer, P., 1978. Astrophys. J., 224, 488.
Bodenheimer, P. and Black, D.C., 1978. In Protostars and Planets, ed. T. Gehrels. Univ. of Arizona Press, p 288.
Bodenheimer, P., Tohline, J.E. and Black, D.C., 1980. Astrophys. J., 242, 209.
Cassen, P.M. and Moosman, A., 1981. Icarus, 43, 353.
Cassen, P.M., Shu, F.H. and Tereby, S., 1985. In Protostars and Planets II, eds. D.C. Black and M.S. Matthews, Univ. of Arizona Press, p 448.
Durisen, R.H. and Tohline, J.E., 1985. Ibid., p 534.
Frank, J., King, A.R. and Raine, D.R., 1985. Accretion power in astrophysics. Cambridge Univ. Press.
Goldsmith, P.F. and Arquilla, R., 1985. In Protostars and Planets II, eds. D.C. Black and M.S. Matthews, Univ. of Arizona Press, p. 137.
Kenyon, S.J. and Hartmann, L., 1987. Astrophys. J., in press.
Larson, R.B. 1983. Rev. Mexicana Astr. Astrof., 7, 219
Lin, D.N.C. and Papaloizou, J.C.B., 1985. In Protostars and Planets II, eds. D.C. Black and M.S. Matthews, Univ. of Arizona Press, p. 981.
Lin, D.N.C. and Pringle, J.E., 1987. Mon. Not. R. astr. Soc. 225, 607.
Myers, P.C. and Benson, P.J., 1983. Astrophys. J., 266, 309.
Pringle, J.E., 1981. Ann. Rev. Astr. Astrophys., 19, 137
Sargent, A.I. and Beckwith, S., 1987. Astrophys. J., in press.
Sellwood, J.A. and Carlberg, R.E., 1984. Astrophys. J. 282, 61.

Shakura, N.I. and Sunyaev, R.A., 1973. Astr. Astrophys., 24, 337.
Shu, F.H., 1977. Astrophys. J., 214, 488.
Shu, F.H., 1983. Astrophys. J., 273, 202.
Toomre, A., 1964. Astrophys. J., 139, 1217.
Wolstencroft, R.D. and Walker, H.J., 1987. Phil. Trans. Roy. Soc.
    (London)., in press.

ACCRETION DISKS AROUND YOUNG STARS

Lee Hartmann and Scott Kenyon
Smithsonian Astrophysical Observatory
Harvard-Smithsonian Center for Astrophysics
60 Garden Street
Cambridge, MA 02138 USA

ABSTRACT. This article reviews observational constraints on disk accretion in young stellar objects, and discusses implications of this phenomenon for early stellar evolution.

I. Introduction

In recent years it has been suggested that the ubiquitous near-infrared excess emission emitted by young stellar objects arises in circumstellar disks. This interpretation of infrared spectra implies that disks are a common remnant of the star formation process. Theoretically, disks are likely to occur simply because of the angular momentum probably present in the pre-collapse interstellar material. If accretion is not perfectly radial, large quantities of infalling material could end up in the disk rather than on the central object. This possibility is reinforced by the observations of Sargent and Beckwith (1987), who estimate a disk mass ~ 0.1 $M_\odot$ for the T Tauri star HL Tau, an appreciable fraction of the mass of the central star.

The absence of near-IR excesses in older stars indicates that disks "disappear" on time scales of $\leq 10^8$ yr. Where does the material go? It may coagulate into planets or other large bodies; it may be blown out by the stellar wind; or it may fall onto the central object. The last possibility is especially interesting for its potential effect on stellar evolution.

It has become more popular to suppose that star formation involves substantial amounts of accretion from a disk onto a central core, particularly in the earliest phases (see Pringle's review). However, theoretical calculations of disk accretion are strongly dependent upon the poorly-understood physics of viscous angular momentum transfer and energy dissipation. One must turn to observations to assess the importance of disk accretion in early stellar evolution, although it is important to keep in mind that the most important phases of disk accretion, when the central object might be extremely obscured by infalling matter, may well be missed.

In this article we discuss the observational constraints on disk accretion in young stellar objects, and their implications for early stellar evolution. We concentrate on spectroscopic diagnostics, as convincing imaging information on important size scales is difficult to obtain for heavily-obscured young objects in relatively distant star forming regions. Spectroscopic indicators of disks, while providing less direct evidence than imaging, are less subject to selection effects at present. We concentrate on studies of the low-mass, pre-main sequence T Tauri stars for statistical reasons.

163

A. K. Dupree and M. T. V. T. Lago (eds.), Formation and Evolution of Low Mass Stars, 163–179.
© *1988 by Kluwer Academic Publishers.*

164

II. Disks and the Infrared Excess Emission

The energy distributions of classical T Tauri stars (TTS) show infrared excesses that cannot be explained by a single temperature blackbody, as shown in Figure 1 (Rucinski 1985; Rydgren and Zak 1987). These observations require emission over a broad range of temperatures, suggesting a spatially *extended* source of infrared radiation.

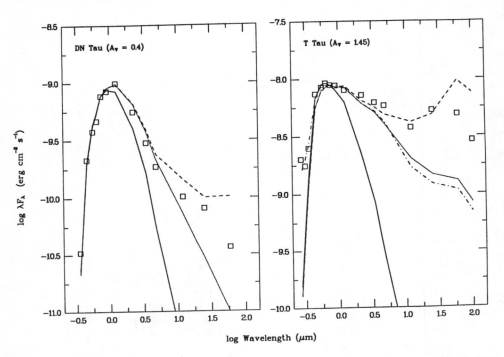

Fig. 1. Energy distributions for two typical T Tauri stars. The open squares denote the observations; the solid line indicates the energy distribution expected for a single, normal star. The broken and dotted lines show predicted spectra for circumstellar disks of various thicknesses which absorb light from the central star and reradiate as blackbodies. From Kenyon and Hartmann (1987).

Disks naturally emit at a wide range of temperatures. For both steady accretion disks and for *reprocessing* disks (disks which absorb starlight and reradiate this energy at longer wavelengths), the radial temperature distribution is $T \propto R^{-3/4}$ at large radii (Lynden-Bell and Pringle 1974 = LBP; Friedjung 1985; Adams, Lada, and Shu 1987 = ALS). Therefore a disk with a sufficiently large radial extent will exhibit emission over a region with a large temperature gradient.

The case for connecting near-infrared excesses with dust disks in young stars has been strengthened by observations of the nearby A star β Pic. Ground-based observations following IRAS detection of the infrared excess emission provided a direct optical image of scattered light from a disk extending out to about 400 A.U. (Smith and Terrile 1984).

A strong argument for the presence of disks comes from observations of broad forbidden-line emission in some TTS (Jankovics, Appenzeller, and Krautter 1983; Appenzeller, Jankovics, and Ostreicher 1984; Edwards et al. 1987). The large velocity widths of the forbidden lines ($\sim$ 200 km s^{-1}) suggest that the emission arises in stellar winds. Since the forbidden lines are collisionally de-excited at high electron densities, the emission probably comes from the low-density, distant wind, at radii $\sim$ 10 - 100 A.U. from the star (Edwards et al. 1987). Nearly all objects show blue-shifted emission. Unless TTS eject material preferentially towards Earth, there must be some large-scale occulting screen present. An extended opaque disk is the most likely explanation.

In principle, the interpretation of infrared excesses as disk emission is not unique. In practice, no one has been able to come up with a plausible alternative for T Tauri stars. Extended circumstellar dust shells also exhibit a range of temperatures, and must contribute to infrared emission, particularly at long wavelengths (see ALS). However, in some cases it is difficult to reconcile the amount of dust needed to explain the IR radiation with the modest optical extinction (Myers *et al.* 1987). Furthermore, the near-IR excess implies warm dust near the stellar surface. It is hard to see how this material can be supported in a stationary position, so it must be either falling in, flowing out, or orbiting. The near-stellar environment of most TTS is apparently dominated by hot ($\sim$ 10^4 K) material flowing outward at $\sim$ 200 km s^{-1} (cf. Hartmann 1985), which does not seem to be a favorable environment for dust. Dust in a flat disk obviously gets around this difficulty.

Chromospheric models for optical and ultraviolet line and continuum emission (e.g., Cram 1979; Calvet, Basri, and Kuhi 1984) do not produce much infrared emission. In principle a hot, gaseous envelope surrounding the central star could produce an infrared excess, as is observed in Wolf-Rayet and Be stars, but in practice envelopes with sufficient emission measure to give an IR excess produce too much optical and ultraviolet line and continuum emission.

III. Reprocessing disks

We begin our discussion by considering disks heated only by light from the central star. Circumstellar disks are quite likely to be very opaque in the visual and near-infrared regions. A disk of mass 10^{-3} M$_\odot$ and radius 100 A.U. with a uniform surface density and dust properties similar to those in the diffuse interstellar medium has a vertical visual optical depth of 10^2. The large optical depths result in an important simplification in computing the infrared spectrum. Blackbody arguments can be applied to the absorption and emission of starlight from an optically thick surface (Adams and Shu 1986; ALS), independent of the details of the dust properties.

Consider a flat, extended, opaque disk, in which all of the incident radiation is absorbed and re-emitted as a blackbody. The flux of radiation incident on a portion of the disk at distance R from the central star falls off asymptotically as R^{-3}, because the solid angle subtended by the star at the disk decreases roughly as R^{-2}, while the cosine of the angle of incidence is asymptotically proportional to R^{-1} (Adams and Shu 1986). Since the absorbed stellar flux must equal the emitted disk flux σT^4 in steady state, it follows that T$\propto$R$^{-3/4}$.

For a general temperature distribution T$\propto$R^{-n}, the asymptotic energy distribution at long wavelengths is $\lambda F_\lambda \propto \lambda^\alpha$, where the spectral index α = (2-4n)/n (Lynden-Bell and Pringle 1974). The predicted spectrum for a T$\propto$R$^{-3/4}$ disk at long wavelengths is $\lambda F_\lambda \propto \lambda^{-4/3}$. If the disk has infinite radius, the reradiated luminosity is 1/4 of the stellar luminosity (ALS).

As long as flat reprocessing disks are opaque and extend from the stellar surface out to large distances, the observed infrared energy distributions will be quite similar in shape. The effects of inclination will be small, except when the disk is seen near edge-on. The infrared excess will scale with the stellar luminosity and effective temperature. Thus, if reprocessing dominates the infrared excess emission one might expect many TTS of varying disk masses and dust properties to have similar spectra. There seems to be evidence for such a "typical" TT infrared spectrum. Rydgren and Zak (1987) showed that several TTS in Taurus have near-IR spectral indices of $\alpha \sim -2/3$, with excess "luminosities" about 50% of the "stellar luminosity"; other analyses of the ground-based IR and *IRAS* data show similar results (Kenyon and Hartmann 1987; Adams, Lada, and Shu 1988).

The average observed spectrum is not quite what the simple flat disk result predicts; the observed energy distribution is a little flatter, and is somewhat more luminous. However, it is likely that disks are not completely flat; and because the starlight falls onto a flat disk at very oblique angles, even small departures of the disk surface from complete flatness can make a substantial difference in the absorbed flux, and hence on the emergent spectrum (Kenyon and Hartmann 1987). The vertical scale height of an isothermal disk surrounding a central point mass is proportional to $H \propto (R^3T/M)^{1/2}$. Thus, if $T \propto R^{-3/4}$, then $H \propto R^{9/8}$, and the scale height increases relative to the radius. The result is an increasingly thicker disk at larger radii, with a surface tilted more toward the star than in a flat disk, increasing the radiative heating and thus the outer disk temperatures. We showed (Kenyon and Hartmann 1987) that spectra with indices of $\alpha \sim -2/3$ can be obtained with plausible amounts of disk curvature or flaring (specifically, $H \propto R^{9/8}$). The increase in disk thickness also increases the amount of starlight intercepted, and brings the total disk luminosity into reasonable agreement with observation[1].

The disk reprocessing explanation of TT IR spectra requires a disk of modest but finite thickness. In practice, it appears that the dust must be mixed with the gas to roughly three scale heights above the midplane (Kenyon and Hartmann 1987). This requirement poses a problem, because dust particles are expected to settle to the disk midplane on time scales far shorter than the age of the average T Tauri star. Modest turbulence would probably be adequate to support the dust to this extent, such as would be produced by convection (Ruden 1987). But convection requires an internal heat source, while pure reprocessing of starlight would produce temperatures which decrease toward the disk midplane, a situation which is stable against convection.

Despite the theoretical difficulties involved in suspending dust to adequate heights in the disk, it is plausible to attribute the infrared excesses of the majority of T Tauri disks to reprocessing, simply because the excess luminosity is often in reasonable agreement with the reprocessing prediction. There is no reason why another source of energy, such as accretion, should be so tied to the stellar luminosity.

Not all TTS have infrared excess emission in agreement with the simple reprocessing disk model. About 15% of the population of TTS in Taurus-Auriga have very flat infrared spectra, with spectral indices ~ 0 (Rucinski 1985; Strom et al. 1988). Using the Lynden-Bell and Pringle (1974) formula relating the temperature distribution of a disk with the spectral index of the emergent flux, this implies $T \propto R^{-1/2}$, which is the slowest temperature falloff obtainable by reprocessing of light from a central object. In addition, the excess infrared

---

[1]There has been a lot of confusion concerning disk "luminosities", which has to do with the fact that flat disks do not radiate isotropically; more flux is observed from a disk seen pole-on than equator-on. For example, even though the infinite flat disk has a luminosity of 1/4 of the stellar luminosity, the bolometric flux received from such a disk seen pole-on will be 1/2 of the stellar flux. Even at intermediate inclinations, the disk flux contribution *relative to the star* can be increased if the disk extends down to the stellar surface, occulting part of the star. Only for inclinations $\geq 70°$ is the disk contribution reduced relative to the star.

luminosity is often apparently comparable to - or even larger than - the stellar luminosity. These features are difficult to explain with a reprocessing disk unless the solid angle covered by the disk as seen from the central star is extremely large.

One way out of this difficulty is to invoke more or less spherical infall of a dusty envelope (ALS). While extended dust shells probably contribute to the far-IR excess in several objects, it is not likely that sources with flat near-infrared spectra can be explained this way. As discussed in the previous section, radially infalling dust close to the star is not consistent with the observed stellar winds and limits on optical extinction.

Another possibility is that another stellar source is present in the large IRAS beams. Roughly 1/4 of TTS have *known* visual companions within IRAS limits. An embedded star could contribute at long wavelengths, and there are objects with a second "hump" in the far-IR. However, the smoothness of the many flat-spectrum sources implies a range of temperatures are present, which may not be easy to produce. In addition, it seems unlikely that all or even most of the flat-spectrum sources can be explained in this way, given the frequency of highly-extincted sources in Taurus (Beichman et al. 1986).

A third possibility is to attribute the extra luminosity to accretion in the outer disk (ALS; Adams, Lada, and Shu 1988). We return to this possibility in the next section, noting for the moment that this mechanism requires implausibly large disk masses and/or accretion rates.

A fourth explanation, which we favor in many cases, is that the extra luminosity of flat-spectrum sources is only apparent, and is due to radiative transfer effects. The classic example of this is the Ae star Walker 90 in NGC 2264. The energy distribution of this star, shown in Figure 2, is extremely peculiar, with a negative α and much more luminosity in the IR than in the optical. The reddening of this object is only $E_{(B-V)} \sim 0.2$, according to Strom, Strom, and Yost (1971), who find from analyses of Balmer line absorption wings that W90 is an early A star somewhat above the main sequence. But this reddening, coupled with a normal extinction law, puts W90 well below the main sequence (Walker 1956).

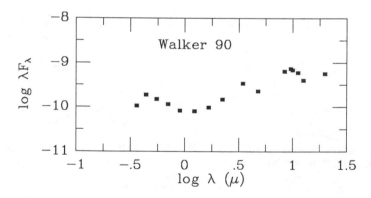

Figure 2. Energy distribution of W90; from Rydgren and Vrba (1987).

Rydgren and Vrba (1987) point out that the 10 μm flux of W90 is consistent with that of other A stars in NGC 2264. Thus, the 10 μm luminosity of W90 is not anomalous; it is the low *optical* luminosity that is the problem. The solution must be that the optical (stellar) luminosity has been underestimated by a large factor (Strom et al. 1972). Rydgren and Vrba estimate that the visual extinction is really ~ 4 magnitudes to place W90 among the other

NGC 2264 A stars; with this kind of correction the infrared excess luminosity is not larger than the stellar luminosity.

The solution to this puzzle is clearly non-standard extinction. As Strom et al. suggested, dust grains in a circumstellar shell or disk might well be larger than typical in the interstellar medium, resulting in more neutral extinction. In addition, there might be complex radiative transfer effects. Scattering into the line of sight may well be important, as suggested by the large optical polarization of W90 (~ 2 - 3%; Breger 1974), and the larger dust albedo expected at shorter wavelengths could help make the observed spectrum bluer.

TTS viewed *through* their circumstellar disks will have peculiar-looking spectra from radiative transfer effects, if not from different dust properties. At first glance it would seem that seeing a young star directly through its disk must be a very improbable event, and cannot explain very many flat-spectrum sources. We emphasize, however, that the flat spectrum sources account for only ~ 15% of the total TT population; that is, the phenomenon to be explained is not very frequent.

We expect to see a significant fraction of systems edge-on based on our flared disk model. To reproduce the *average* TTS spectrum with a reprocessing flared disk, the maximum disk thickness H relative to the radius R must be H/R ~ 0.2. As mentioned above, this corresponds to something like three scale heights, and since the disk is likely to be optically thick one would guess that stars seen through the outer disk scale height would appear like W90, while others seen more equatorially would simply be embedded IR sources. Thus, for a random distribution of disk axes we would estimate ~ 7% of the population should be like W90, and 14% should be highly extincted, low-luminosity sources. Out of a (well-studied) classical TT population of ~ 50 stars in Taurus-Auriga, approximately 8 are flat-spectrum sources. Myers et al. (1987) have found 11 embedded sources of luminosities ~ 1 $L_\odot$ in the region. A statistical comparison between observations and flared disk predictions is difficult, because luminosity selection effects must be important in finding embedded and edge-on sources. However, the above discussion suggests that a significant fraction of the flat-spectrum sources could be objects seen through the disk, based on the flared disk model.

We also note that Strom et al. (1971) found two of 42 stars studied in NGC 2264, W90 and V116, to fall well below the main sequence (see also Walker 1956), and that another two stars probably have neutral scattering "shells".

Similar effects on observed spectral energy distributions could be produced by flattened distributions of dust on scales of ~ 100 A.U. ($\leq$ 1 arc sec) which are not part of the infrared-emitting disk, but much further away. Some TTS are surrounded by so-called cometary reflection nebulae, the result of stellar scattered light observed through a hole in the interstellar medium. These nebulae are often large - several arcseconds or hundreds of A.U. at the distance of Taurus-Auriga. Depending upon the geometry and amount of dust present, it is quite possible for optical starlight *reflected* through a "hole" in the circumstallar material to be much brighter than the direct, reddened light from the central star. If the scale on which this occurs is unresolvable from the ground, the observed "stellar spectrum" will be apparently lightly reddened and have an abnormally low "luminosity". On the other hand, the infrared emission can come through with much less attenuation, and will be less sensitive to the geometry of observation; hence, the infrared luminosity will appear to be abnormally large. Many of the flat-spectrum sources - HL Tau, DG Tau, RY Tau, DO Tau - have associated optical cometary nebulae.

If scattering in a non-spherical geometry is important in producing flat spectrum sources, one would expect optical polarization to be correlated with the large infrared excesses. This is the case. We would classify T, RY, DG, HL, XZ, DO, HP, and DP Tau, and UY Aur as flat-spectrum sources in Taurus-Auriga. Bastien (1982) has reported polarizations for 7 of these 9 objects, which have an average optical polarization of 3.7%. In contrast, the average polarization of 16 other TTS in the region is 0.6%. The five objects reported by Bastien to have P > 2% are all flat-spectrum sources.

It is not easy to produce the large polarizations seen in the flat-spectrum sources HL Tau (8.3%) and DG Tau (5.5%). Strong scattering is required, and it helps to extinct the central source, enhancing the reflected light. Thus, the large optical and near-infrared polarizations seen in many flat spectrum sources are *prima facie* evidence that we have difficulties in determining the true spectrum of the central luminous object. For this reason we suspect that many, if not most, of the flat-spectrum sources are seen though flattened distributions of dust, which cause us to underestimate the stellar luminosity and produce strange-looking spectral energy distributions.

IV. Accretion Disks

Accretion is an "important" source of the observed radiation when the energy generated by infall is comparable to the stellar luminosity. Taking typical values for a T Tauri star of $M = 1\ M_\odot$, $R = 3\ R_\odot$, the accretion luminosity

$$L_{acc} = GM\dot{M}/R \qquad (1)$$

is comparable to the stellar luminosity of $\sim 3\ L_\odot$ when $\dot{M} \sim 3 \times 10^{-7}\ M_\odot\ yr^{-1}$.

We now show that such an accretion disk must be optically thick. To estimate the disk surface density requires a knowledge of the inward drift velocity, which depends upon the viscosity and is thus unknown. However, it is typical to assume that the turbulent velocity responsible for the diffusion of angular momentum is less than or equal to the sound velocity in the disk interior; and this prescription leads to drift velocities much less than the sound speed. A typical sound speed would be perhaps 5 km s^{-1}, from which we infer that the surface density of a disk accreting at $\dot{M} \sim 10^{-7}\ M_\odot\ yr^{-1}$ is $\gg 10$ g cm^{-2}. Using the argument cited above for reprocessing, one can see that if dust is the dominant opacity such a disk must be very optically thick at 1 μm. Since the disk thickness is generally < 0.1 R, the midplane gas density must be $\gg 3 \times 10^{14}$ cm^{-3}, and one can show that gas at 2000 K - 3000 K with this density and path length will also be optically thick.

Thus once again we can use optically-thick (i.e., blackbody) radiation theory to compute the observed spectrum. Further, since the disk is optically thick, it will also absorb light from the central star, so that *any accretion disk of interest must also be a reprocessing disk as well.*

Roughly speaking, the accretion disk spectrum can be derived by balancing the energy dissipation required for material to fall through radial distance ΔR with the blackbody emission from the appropriate annulus, i.e.

$$GM\dot{M}\frac{\Delta R}{R^2} = 2 \times 2\pi R\ \Delta R\ \sigma T^4 \qquad (2)$$

(see Pringle's review). Thus, in the simplest theory, wherein $\dot{M}$ is roughly constant, once again (asymptotically) $T \propto R^{-3/4}$, so one cannot distinguish easily between reprocessing and self-luminous (accreting) disks on the basis of spectral energy distributions (ALS). Since the accretion disk must also be a reprocessing disk, the infrared spectrum will be dominated by reprocessing unless the accretion luminosity is sufficiently large in comparison with the stellar luminosity.

As noted in the previous section, most TTS have IR excesses roughly consistent with pure reprocessing. Strom et al. (1988) found that about 80% of the TTS in the Taurus-Auriga molecular clouds exhibit excess infrared emission consistent with reprocessing, suggesting that accretion rates are $\leq 10^{-7}\ M_\odot\ yr^{-1}$ for most objects.

The objects with strong excess emission over and above the stellar luminosity can be sorted into two classes. One class exhibits very flat (spectral indices ~ 0) infrared spectra, so that the far-infrared luminosity is very large. As indicated above, reprocessing disks have substantial difficulties in producing such flat spectra. Accretion would seem to be a good mechanism for producing excess luminosity, but producing flat spectra out to ~ 100 μm is not easy. To see why, return to eq. (1). A flat spectrum implies $T \propto R^{-1/2}$, which in turn implies $M \dot{M} \propto R$. The problem is that the long-wavelength emission comes from R ~ 10 - 100 A.U., corresponding to thousands of R_*. A massive disk ($M \propto R$), as originally suggested by ALS, must be impossibly more massive than the central star (Kenyon and Hartmann 1987; Shu, Adams, and Lizano 1987). Alternatively, one might have non-steady accretion, with $\dot{M} \propto R$. But in this case one requires $\dot{M} \gtrsim 10^{-4}$ $M_\odot yr^{-1}$ at 100 A.U., which seems to imply uncomfortably short evolutionary times in the outer disk. Thus the apparent excess energy of the flat-spectrum sources seems unlikely to be explained by accretion. As discussed previously, we suspect that complex radiative transfer effects resulting from scattering in a disk or flattened arrangement of dust produce most peculiar spectra.

The other class of sources with strong excess emission exhibit optical and ultraviolet emission at levels which overwhelm the stellar photospheric spectrum. (Some of these objects also have flat IR spectra.) We next consider the explanation of this ultraviolet and optical excess in terms of boundary layer models.

V. The Boundary Layer

The boundary layer is the turbulent region where disk material rotating at the local Keplerian velocity comes to rest on the slowly-rotating stellar surface. In the event that the stellar rotation is small, which seems to be true for most TTS (Hartmann et al. 1986; Bouvier et al. 1986), the energy dissipated per particle in the boundary layer is the full kinetic energy of the innermost Keplerian orbit; and this in turn is equal to the energy lost in moving a particle from infinity in to the stellar radius (see Pringle's review). Thus, in a steady state, one expects the luminosity of the boundary layer to be equal to the luminosity of the accretion disk. Because the boundary layer is expected to be narrow in radial extent, it must be much hotter than the disk, and radiate at shorter wavelengths.

In the original paper by Lynden-Bell and Pringle (1974), it was suggested that the ultraviolet excess emission from TTS was produced in the boundary layer between accretion disk and star. The boundary layer has become more popular as the site of the extreme excess optical and ultraviolet emission observed in some T Tauri stars (Bertout 1986; Kenyon and Hartmann 1987; Bertout, Basri, and Bouvier 1988). The alternative model for producing the observed excess emission is a more energetic version of the solar chromosphere (e.g., Calvet, Basri, and Kuhi 1984), in which the stressing and twisting of magnetic fields rooted in the stellar photosphere produce heating above the photosphere. Simple extensions of the theory of wave energy generation developed for solar-type stars (Stein and Ulmschneider 1982) suggest that this mechanism may not be able to produce excess emission comparable to or greater than the stellar luminosity (Calvet and Albarran 1984). On the other hand, the boundary layer is under no such constraint. Since we do not understand viscosity or other turbulent processes in the disk, and so cannot predict accretion rates, we can simply set them to be as large as we like. The only limit is that one does not wish to accrete more than the mass of the star; but this is no problem, as accretion rates $< 10^{-6}$ $M_\odot$ yr^{-1} do the job, and need not be invoked for more than 10^6 yr (the characteristic age of TTS). In fact, the extreme emission stars constitute only 10% - 20% of the total population of classical TTS, suggesting that the strong-emission phase may last for only 10^5 yr.

There is some semantic confusion concerning the use of the term *chromospheres*. We define a chromosphere as a layer or layers of gas hotter than the photosphere with modest to small continuum optical depths. TTS *must* have chromospheres to produce the strong emission lines observed. Whether this emission comes from a *stellar chromosphere* or a *boundary layer chromosphere* is unresolved.

In our opinion, boundary layer energy generation is the best explanation of *extreme* optical and UV TT emission. The two strongest arguments in favor of the boundary layer explanation rather than the stellar chromosphere model are (1) the large energy fluxes involved (Kenyon and Hartmann 1987) and (2) that so far all the strong-emission stars have large IR excesses, and thus posess disks (Bertout, Basri, and Bouvier 1988). The boundary layer, with its large turbulent velocities, is also an attractive site for generating waves that can propagate outward and shock to produce the required chromospheric heating of the envelope (Kenyon and Hartmann 1987).

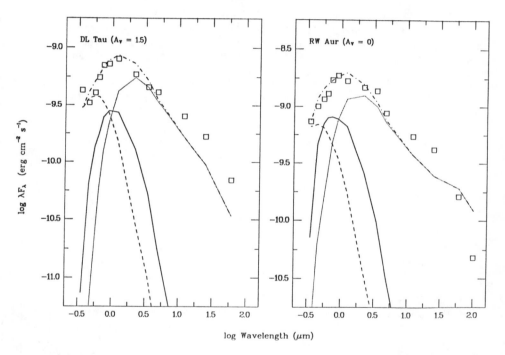

Figure 3. Schematic models (dot-dashed curve) compared with the broad-band energy distributions of extreme emission TTS (open squares). The solid line is the stellar photosphere, the dotted line is the accretion disk, and the dashed line is the boundary layer contribution. From Kenyon and Hartmann 1987.

However, it must be said that the case against stellar chromospheres is not airtight, simply because we do not really understand stellar chromospheric activity. All cool stars show some chromospheric emission, and it is known that such emission is larger in the youngest stars. It is possible that the youngest objects, with the largest fluxes, might also have the most massive disks simply because they are young. Extrapolations from other objects may be dangerous, because we do not understand magnetic dynamos very well, and very young objects might still have strong primordial fields. The theoretical estimates of stellar chromospheric energy fluxes are based on essentially dimensional arguments, and are difficult to extrapolate, especially considering our ignorance of stellar magnetic field strengths and fluxes.

If we assume boundary-layers are the site of excess UV TT emission, there are still several problems in interpreting observations. Without a theory of viscosity the boundary layer size is not determined, so even assuming single-temperature blackbody radiation, we cannot predict the boundary layer temperature. If we make the plausible assumption that the boundary layer can be no thinner than the scale of the surface it is rubbing against, and hence cannot be narrower than the stellar photospheric scale height (almost 10^{-3} R_*), we arrive at temperatures $\lesssim 2 \times 10^4$K at accretion rates $\lesssim 10^{-6}$ $M_\odot$ yr^{-1}, the latter being at the high end of what we require. It is quite probable that the boundary layer thickness is an order of magnitude larger than this, pushing temperatures down to around 8000K. Thus, it seems that the boundary layer emission should be in the *potentially* observable ultraviolet.

The adoption of a boundary layer thickness $\Delta R \sim 10^{-2}$ R_* produces temperatures around 8000 K for relevant accretion rates $\sim 3 \times 10^{-7}$ $M_\odot$ yr^{-1}, in agreement with some observational estimates (cf. Herbig and Goodrich 1986). Unfortunately, T Tauri stars are heavily reddened as a rule, and there is some occasionally disturbing evidence that reddening in star-forming regions is not always "normal". UV excesses can be quite uncertain; and for strong-emission stars, where the object temperature is unknown, no extinction correction can be reliably estimated.

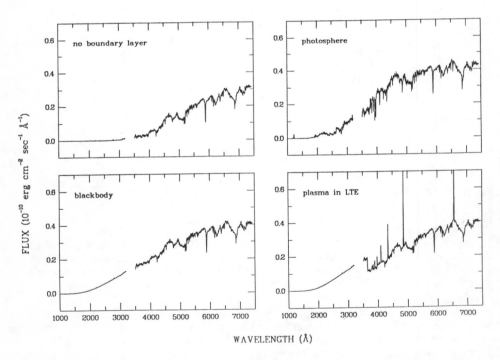

Figure 4. Models for boundary layer emission, assuming the same boundary layer temperature and emitting area, but assuming that the boundary layer emits like a photosphere, a blackbody, or a plasma in LTE. After Kenyon and Hartmann (1987).

Another problem is that the strong continuum emission required to understand the high-resolution spectra of some TTS (Kenyon and Hartmann 1987) is also accompanied by strong *line* emission (see Fig. 4), which must be explained by the boundary layer model as well. This implies that the boundary layer, or accretion, must be able to excite a *chromosphere*. The result is a complicated modelling problem, wherein optically thin plus optically thick hot gas in the boundary layer, plus a stellar contribution, add to give the optical spectrum (with an uncertain reddening correction). Given these complications, it seems premature to develop detailed models using only blackbodies for the boundary layer, in order to derive second-order effects like disk inclinations (e.g. Bertout, Basri and Bouvier). In practice, boundary-layer luminosities- and hence inferred accretion rates- must be uncertain by a factor of several.

VI. FU Ori Objects: Rapid Disk Accretion in Early Stellar Evolution

The clearest examples of pre-main sequence disk *accretion* are the FU Orionis variables (Hartmann and Kenyon 1985, 1987a,b; Kenyon, Hartmann, and Hewett 1988). Due to a fortunate combination of circumstances, we are able to make a number of observational tests of disk accretion, with the result that simple models do surprisingly well in accounting for a number of otherwise peculiar features of these objects.

The FU Ori variables are obviously in an early evolutionary phase. These heavily-reddened objects are found in regions of star formation, with a close kinematic connection with molecular gas, and exhibit reflection nebulae when not too obscured (Herbig 1977). One new object is associated with an HH object (Graham and Frogel 1985).

FU Ori variables generally exhibit ≥ 5 mag outbursts in the optical, with rise times of hundreds of days and decay times of years to > 10 yr; maximum luminosities are $> 10^2 \, L_\odot$. A pre-outburst spectrum of one object, V1057 Cyg, looks like a faint, reddened T Tauri star; at maximum light, the optical spectrum is that of a late F or early G supergiant, showing that the outburst is not simply due to removal of obscuring dust (Herbig 1977).

The light curves of the FU Ori objects are qualitatively similar to those of dwarf novae, in which eruptions are caused by a rapid increase in mass accretion, causing the disk to become very luminous. Since the rise and decay times for FU Ori outbursts are comfortably longer than orbital periods or sound speeds, disk accretion can in principle provide an explanation for the increase in luminosity, provided some mechanism for triggering a sudden increase in accretion can be found. We return to this question below.

In the accretion disk model, the large increase in optical light implies that the central star can no longer be seen in outburst; the disk dominates at all wavelengths. Thus, there is no necessity to disentangle stellar and chromospheric emission from disk emission, or worry about reprocessing of stellar radiation in the disk.

The simple steady disk model does a remarkably good job in explaining the observed energy distributions, as shown in Figure 5. Although extinction is large toward FU Ori and V1057 Cyg, the spectra are clearly broader than any single blackbody. FU Ori objects exhibit a variation of spectral type with wavelength; the optical spectra are of type G, while the strong CO and water vapor absorption are consistent with an M-type spectrum. This effect is straightforwardly predicted by the disk model with its large radial variation of temperature; at short wavelengths, one observes the inner disk, while in the infrared outer disk regions dominate.

The accretion disk model requires fairly rapid rotation if the central star has anything like a solar mass of material; and this is observed to be the case. The optical rotational velocities of the three best-studied objects FU Ori, V1057 Cyg, and V1515 Cyg are ~ 65, 45, and 20 km s^{-1}, respectively. These values of v sin i are quite large for G supergiants as suggested by the optical spectral types, but are a little small for disks; the implied < sin i > is ~ 0.4 if the central masses are 0.8 M$_\odot$ (Kenyon, Hartmann, and Hewett 1988).

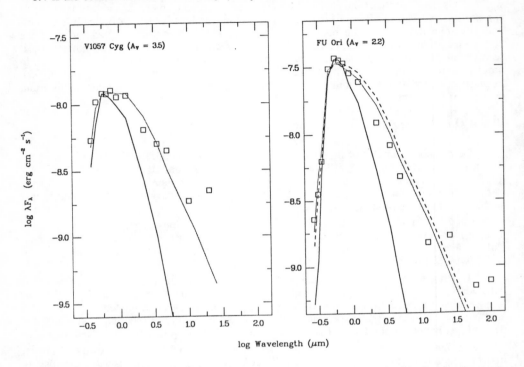

Figure 5. Energy distributions for V1057 Cyg and FU Ori, compared with the simple steady accretion disk model prediction.

Because the optical spectrum comes from the inner parts of the assumed Keplerian disk, and the infrared emission arises in cooler, outer regions, the infrared lines must be much less broadened by rotation than the optical lines. This is observed to be the case in FU Ori and V1057 Cyg (Hartmann and Kenyon 1987a,b; KHH). The agreement between observation and theory is quite good, given the uncertainties in calculating infrared CO spectra at temperatures < 2000 K (Kenyon et al. 1988).

At present there is no other model which explains the difference in rotation between the optical and infrared regions. Indeed, it is hard to imagine anything other than a disk producing this effect, and accretion provides a natural explanation of the observed outbursts.

Another potential test of the model comes from close analysis of absorption line profiles. A rotating disk produces line profiles that are slightly double-peaked; in contrast, a rotating uniform sphere exhibits roughly parabolic profiles. The doubling of line profiles is subtle, and can be overwhelmed by blueshifts due to strong winds known to emanate from these systems (Croswell, Hartmann, and Avrett 1987). There is some evidence for this line doubling in red spectra of V1057 Cyg (Figure 7). Optical spectra of FU Ori do not show this effect, but the blue-shifted, asymmetric line profiles observed suggest that mass loss effects obscure any possible disk profile doubling.

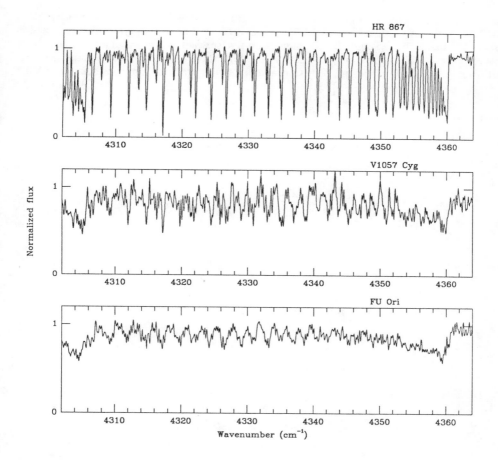

Figure 6. Infrared spectra of FU Ori and V1057 Cyg, compared with a standard M6 giant. Rotational broadening of the CO absorption lines is evident.

Limits on radial velocity variations for FU Ori and V1057 Cyg are a few km s^{-1} in *both* the optical and infrared. The upper limits to radial velocities are small fractions of the rotational velocities, and thus it is hard to invoke a companion star which can feed the accretion disk, as in close binaries. More likely, the disk is a remnant of the star formation process.

The principal problem with the disk accretion model so far is simply that the mechanism responsible for the sudden increase in accretion rate, producing the outburst, is not understood. Disk instability theories, similar in principle to ideas developed for cataclysmic variables, appear to be attractive (Lin and Papaloizou 1985), but our understanding of viscosity in disks is so poor that predictions are difficult to make. Another possibility is that a companion star on an eccentric orbit might periodically plunge through the disk, setting off an accretion event.

The accretion rates implied for FU Ori objects in outburst are quite remarkable - 10^{-4} $M_\odot\,yr^{-1}$ at maximum light. V1057 Cyg has a decay time of ~ 10 yr, implying accretion of ~ 10^{-3} $M_\odot$. FU Ori *may* be declining on a timescale of ~ 10^2 yr, although this is very uncertain; since the outburst in 1937, nearly 10^{-2} $M_\odot$ has been accreted. The statistics of FU Ori events, and the implications of these results for stellar evolution, are considered in § VII.

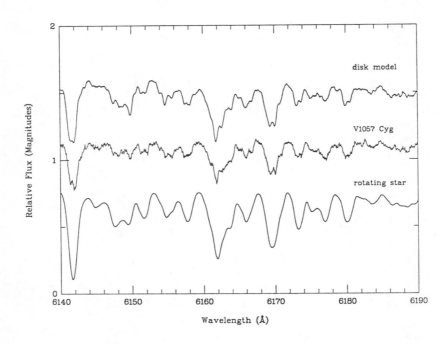

Figure 7. Comparison of the red spectrum of V1057 Cyg with a spectrum synthesis for a disk model and for a rotating uniform sphere.

Given the variety of evidence for disk accretion in FU Ori variables, it is of interest to consider the implications of the observations for TTS. The FU Ori objects, with clear spectral and *kinematic* evidence for disk accretion at very high rates, do *not* have flat energy distributions, but instead have spectra consistent with the classic steady disk model. This is another reason for our discomfort with attributing the flat-spectrum TTS to disk accretion.

FU Ori objects do not provide clear evidence for boundary-layer emission. However, the predicted boundary layer temperatures ~ 3×10^4 K push the emission far into the ultraviolet, where it is difficult to observe in these heavily-reddened objects (Hartmann and Kenyon 1985). Furthermore, at the large accretion rates typical of FU Ori objects in outburst, the stellar surface may not remain fixed, but may respond to the addition of material by expanding and spinning up; both effects will tend to drop the boundary-layer radiation by unknown amounts.

Unlike T Tauri stars, the near-infrared spectra of FU Ori objects show strong CO absorption lines. We interpret this difference as the result of different disk temperature structures in the two classes of objects. In FU Ori variables, the accretion probably heats the middle of the disk more than the outer edge, producing an outwardly decreasing temperature gradient. In TTS, if the disk is primarily heated from *above*, the outer disk regions should be hotter, not cooler, and so CO lines should be filled-in or perhaps show weakly in emission. In addition, dust may be destroyed at the hot midplane of FU Ori disks, reducing the continuum opacity and consequently increasing absorption line strengths. In contrast, TT disks will be cooler at the midplane rather than hotter, and for a given surface temperature dust is more likely to survive, providing a featureless continuum. Thus, we suggest that the difference in CO absorption properties of TT and FU Ori disks results from reprocessing dominating the disk heating in TTS, while accretion is the main energy source in FU Ori outbursts.

VII. The Importance of Disk Accretion in Stellar Evolution

The statistics of FU Ori events obviously cannot be good, since they are quite obscured and the number of observed events is so small. Nevertheless, if we assume that we have detected every FU Ori event within 1 kpc of the Sun in the last 50 years, and if all stars pass through an FU Ori phase, one estimates that a few percent of the total mass is accreted in this way (Hartmann and Kenyon 1985). Similarly, if we take the strong-emission stars to be 10% of the total TT population, and consequently assume that all TTS pass through this stage for 10% of their $\sim 10^6$ yr lifetimes, with accretion rates of a few x 10^{-7} $M_\odot yr^{-1}$, a couple of percent of an $M_\odot$ is accreted. Thus, we seem to be finding accretion of $\geq 5\%$ of the total stellar mass from disk accretion *in phases of evolution where we can observe the central object at optical or near-infrared wavelengths*. Considering the likelihood that early, embedded phases of evolution with massive disks occur, it is hard to escape the conclusion that disk accretion is a very important process in building up stars, as Cameron (1978) suggested.

This view makes it all the more surprising that young stars are so slowly-rotating (Vogel and Kuhi 1981; Bouvier et al. 1986; Hartmann et al. 1986). Clearly some efficient mechanism of angular momentum loss, probably a stellar wind magnetically coupled to the surface, must be operating at very early times. For the optically-visible phases of evolution, accretion of 5% of the stellar mass could result in rotation at ~ 50 km s^{-1} ($\sim 1/4$ breakup) assuming no angular momentum loss. This is somewhat faster than the average TT rotational velocity $\sim 10 - 20$ km s^{-1}. A wind at 10^{-8} $M_\odot yr^{-1}$ lasting for 10^6 yr, with an Alfven radius of $3 - 10$ R_* would be sufficient to remove this excess spin angular momentum; and such winds are not implausible (Hartmann, Edwards, and Avrett 1982).

The rotational velocity distributions of young cluster stars show an interesting peak at very low velocities, with a tail extending to large v sin i (Stauffer and Hartmann 1986). The spread in angular momenta among these stars appear to be larger than present among TTS (Stauffer and Hartmann 1987). Earlier arguments suggested that TTS lose very little angular momentum on their way to the main sequence in order to provide the observed rapid cluster rotators (Stauffer and Hartmann 1987), but an absence of wind braking makes it difficult to explain the slow cluster rotators. In light of the evidence for disk accretion in TTS, it perhaps makes more sense to assume that a substantial amount of angular momentum is carried off by stellar winds, but that some fraction of cluster stars have had a recent, final episode of disk accretion, which spins up the star and produces the high-velocity tail of the observed v sin i distribution.

178

This explanation of stellar rotation in clusters implicitly depends upon the idea that accretion is episodic and highly variable. Time-dependent accretion is suggested by observations of the FU Ori objects and also by the historical variability of strong emission TTS. This extreme variability of accretion will pose difficulties for theoretical attempts to understand the processes driving angular momentum transfer and energy dissipation in circumstellar disks.

References

Adams, F.C., and Shu, F.H. 1986, *Ap. J.*, **308**, 836.

Adams, F.C., Lada, C., and Shu, F.H. 1987, *Ap. J.*, **312**, 788 (= ALS).

Adams, F.C., Lada, C., and Shu, F.H. 1988, *Ap. J.*, in press.

Appenzeller, I., Jankovics, I., and Ostreicher, R. 1984, *Astr. Ap.*, **141**, 108.

Bastien, P. 1982, *Astron. Ap. (Suppl.)*, **48**, 153.

Beichman, C.A., Myers, P.C., Emerson, J.P., Harris, S., Mathieu, R., Benson, P.J., and Jennings, R.E. 1986, *Ap. J.*, **307**, 337.

Bertout, C. 1987, in *IAU Symposium 122, Interstellar Matter*, ed. I. Appenzeller and C. Jordan (Dordrecht: Reidel).

Bertout, C., Basri, G., and Bouvier, J. 1988 *Ap. J.*, in press.

Bouvier, J., Bertout, C., Benz, W., and Mayor, M. 1986, *Astron. Ap.*, **165**, 110.

Breger, M. 1974, Ap. J., **188**, 53.

Calvet, N., and Albarran, J. 1984, *Rev. Mex. Astr. Ap.*, **9**, 35.

Calvet, N., Basri, G., and Kuhi, L.V. 1984, *Ap. J.*, **277**, 725.

Cameron, A.G.W. 1978, in *Protostars and Planets*, ed. T. Gehrels and M.S. Matthews (Tucson: University of Arizona Press),

Cram, L.E. 1979, *Ap. J.*, **234**, 949.

Edwards, S., Cabrit, S., Strom, S.E., Heyer, I., and Strom, K.M. 1987, *Ap. J.*, in press.

Friedjung, M. 1985, *Astr. Ap.*, **146**, 366.

Graham, J.A., and Frogel, J.A. 1985, *Ap. J.*, **289**, 331.

Hartmann, L. 1986, *Fundamentals of Cosmic Physics*, **11**, 279.

Hartmann, L., Edwards, S., and Avrett, E.H. 1982, *Ap. J.*, **261**, 279.

Hartmann, L., and Kenyon, S.J. 1985, *Ap. J.*, **299**, 462.

Hartmann, L. and Kenyon, S.J. 1987a, *Ap. J.*, **312**, 243.

Hartmann, L. and Kenyon, S.J. 1987b, *Ap. J.*, **322**, 393.

Hartmann, L., Hewett, R., Stahler, S.E., and Mathieu, R.D. 1986, *Ap. J.*, **309**, 275.

Herbig, G.H. 1977, *Ap. J.*, **217**, 693.

Herbig, G.H. and Goodrich, R.W. 1986, *Ap. J.*, **309**, 294.

Jankovics, I., Appenzeller, I., and Krautter, J. 1983, *Pub. Astr. Soc. Pacific*, **95**, 883.

Kenyon, S.J., and Hartmann, L. 1987, *Ap. J.*, **323**, 714.

Kenyon, S.J., Hartmann, L., and Hewett, R. 1988, *Ap. J.*, in press.

Lin, D.N.C., and Papaloizou, J. 1985, *Protostars and Planets II*, eds. T. Gehrels and M.S. Matthews (Tucson: University of Arizona Press), p.981.

Lynden-Bell, D. and Pringle, J.E. 1974, *M.N.R.A.S.*, **168**, 603 (= LBP).

Myers, P.C., Fuller, G.A., Mathieu, R.D., Beichman, C.A., Benson, P.J., and Schild, R.E. 1987, *Ap. J.*, in press.

Pringle, J.E. 1988, this volume.

Rucinski, S.M. 1985, *A.J.*, **90**, 2321.

Ruden, S. 1987, private communication.

Rydgren, A.E., and Vrba, F.J. 1987, *Pub. Astr. Soc. Pacific*, **99**, 482.

Rydgren, A.E., and Zak, D. 1987, *A. J.*, **99**, 141.

Sargent, A.I., and Beckwith, S. 1987, *Ap. J.*, **323**, 294.

Shakura, N.I., and Sunyaev, R.A. 1973, *Astron. Ap.*, **24**, 337 (SS).
Shu, F.H., Adams, F.C., and Lizano, S. 1987, *Ann. Rev. Astron. Ap.*, **25**, 23.
Smith, B.A., and Terrile, R.J. 1984, Science, **226**, 1421.
Stauffer, J.R., and Hartmann, L. 1986, *Pub. Astr. Soc. Pacific*, **98**, 1233.
Stauffer, J.R., and Hartmann, L. 1987, *Ap. J.*, **318**, 337.
Ulmschneider, P., and Stein, R.F. 1982, *Astron. Ap.*, **106**, 109.
Strom, K.M., Strom, S.E., and Yost, J. 1971, *Ap. J.*, **165**, 479.
Strom, S.E., Strom, K.M., Bregman, J., and Yost, Y. 1972, *Ap. J.*, **171**, 267.
Strom, K.M., Strom, S.E., Kenyon, S.J., and Hartmann, L. 1988, *A.J.*, in press.
Ulmschneider, P. and Stein, R.F. 1982, *Astr. Ap.*, **106**, 9.
Vogel, S.N., and Kuhi, L.V. 1981, *Ap. J.*, **245**, 960.
Walker, M.F. 1956, *Ap. J. (Suppl.)*, **23**, 365.

PRE - PLANETARY DISKS AND PLANET FORMATION

W. J. Markiewicz and H. J. Völk
Max - Planck - Institut für Kernphysik
Kosmophysik
Postfach 10 39 80
D-69 Heidelberg 1
FRG

ABSTRACT. The behaviour of dust grains in a simple model of cool pre-planetary disks is discussed. Obviously grains are a most important ingredient for planetary formation, considering the fact that the terrestial planets and the presumed rocky cores of the giant planets are made of solid material. Grains drift, diffuse, melt and recondense in a turbulent disk. In particular they also coagulate to form larger, mm - to cm - size conglomerates. These processes occur while a central body forms on account of the inward mass and the outward angular momentum transport. If the turbulence in a pre - planetary disk is due to thermal convection, then grain growth can stabilize the disk by reducing its opacity. Subsequent sedimentation of the bigger conglomerates leads to a dense subdisk of grains which can fragment gravitationally to form km - size solid planetesimals. These later accumulate into solid planets or, in the massive presence of cold gas, they first form rocky cores which can acquire a big gaseous component.

1. Introduction

The present scenario for the formation of low mass stars, and possibly our Sun, has four stages (Shu and Adams 1987, Shu this volume). They are; (i) formation of a molecular cloud core within an interstellar cloud, (ii) collapse of the core to a protostar accompanied by the formation of an accretion disk and a surrounding envelope, (iii) onset of a bipolar outflow, and (iv) clearing of the disk by the stellar wind.

Our discussion will focus on the evolution of the accretion disk and surrounding envelope. This evolution will determine whether the above scenario finishes with the emergence of a planetary system.

Theories for the formation of any planetary system, and cosmogony in particular are still speculative in their global character. Available observations, interpretations of which are presently debated, can at best identify accretion disks around young stars (Bertout 1987, Rodriguez 1987). Detectors with resolution required for planetary size bodies within such disks, are only now being developed (Barberri and Nota 1987). We simply

A. K. Dupree and M. T. V. T. Lago (eds.), Formation and Evolution of Low Mass Stars, 181–192.

Table 1

The Two Sided Cosmogony	
Sciences -Astronomical -Astrophysical	-Planetary -Meteoritic -Astrochemical
Observations -Interstellar medium -Star forming regions -Pre-planetary disks	-Interplanetary medium -Atmospheres, surfaces and interiors of planets -Comets and meteorites

Theory
 -Fragmentation and collapse of
   an interstellar cloud
 -Magnetic and turbulent
   angular momentum transport
 -Ambipolar and turbulent
   diffusion of magnetic field
 -Dust transport in the disk
 -Coagulation of grains
	-Detailed Structure of
	   the Solar system
	 -Position, angular momenta,
	   composition and structure
	   of planets and satellites
	 -Chemistry and morphology
	   of meteorites

remind the reader that, just as the biological evolution theorists, so also cosmogonists and the planetary scientists have on the whole only one natural laboratory (our solar system), to look for evidence in support of and against their theories for planetary system formation. More specifically, however, there are many branches of the physical sciences that contribute towards cosmogony, just as there are many branches of the natural sciences, for example genetics and molecular biology, that contribute towards the theories of biological evolution. It is this interdisciplinary nature, that makes our subject so interesting, as well as provides the challenge of integrating the available knowledge.

In Table 1 we have tried to summarize the physical sciences, and list some of the topics of discussion, relevant to the problem of the formation of a planetary system. This table is of course incomplete, and biased by our taste and specialization. We wish to make clear the dividing line in the middle of Table 1. Problems that cross this line are rare (but see for example Cameron 1978, Morfill and Völk 1984, Lin 1985, Wood and Morfill 1987). One such problem is to find a consistent scenario for the formation of planetesimals, recently discussed by Mizuno, Markiewicz and Völk (1987) (subsequently referred to as MMV). We will have more to say on this specific topic in the following sections.

Let us finish this brief introduction with some historical remarks. Besides distinctions implied in Table 1, the various cosmogonies can be classified according to their assumption, of where the mass which eventually made up the planets and their satellites came from. In the Kant-Laplace cosmogony the formation of planets accompanies the formation of the central object. The scenario mentioned at the beginning is in accord with this cosmogony. Many of the more modern hypotheses belong to this class as well. One alternative is for matter to accrete to planets and satellites, after the formation of the respective central object. Alfvén's cosmogony (Alfvén 1954, Alfvén and Arrhenius 1976) belongs to this second class. Although many astronomical facts point to the joint formation of the star and planets (see for example Völk 1982), we should keep in mind the work of Alfvén and Arrhenius for its detailed description of possible plasma processes relevant within the context of a pre-solar nebula. The dynamic importance of the magnetic fields on the evolution of interstellar clouds to pre main-sequence stars is continually debated, both from the observational and theoretical points of view (Taylor 1987, Crutcher, Kazés, and Troland, 1987). We will now look at one cosmogony in some detail.

2. Planetesimal Cosmogony and Intermittent Turbulence

The extensive craterisation of the solar system bodies, the uniformity of their orbital speed, the existence of asteroids, the morphology of primitive meteorites, are just some of the observable facts in support of planetesimal cosmogony.

This is a hypothesis that the terrestial planets, the satellites, and the cores of the gaseous planets, were formed by the accumulation of smaller solid bodies. To justify this hypothesis, which could also be applicable to the formation of any other planetary system, we need to formulate a consistent scenario for the coagulation of the micron size and smaller interstellar grains, up to centimeter sizes. These subsequently evolve to kilometer size planetesimals (Safranov 1972, Goldreich and Ward 1973), which may finally accumulate to bodies of sizes presently observed in the solar system. The final stage of the interaction of planetesimal swarms, has been studied by many authors. See for example Safranov (1972), Cameron (1973), Hayashi (1981), Wetherill (1985) and references therein, or Spaute, Lago and Cazenave (1985). For some recent analytical estimates of planetesimal growth to O(10 cm), see Nakagawa, Sekiya, and Hayashi (1986). Methods used, such as three dimensional Monte Carlo and other simulations, help us understand the final distribution of mass and angular momenta between the solar system objects. Fragmentation of grains upon impact, although parametrically included by the above authors is still one of the problems of the planetesimal cosmogony. More work, especially laboratory experiments on grain collisions is clearly needed. Another possible difficulty is that present observations of the interplanetary matter may sometimes carry little information on the early history of the solar system. Extensive bombardment by energetic protons, if it occurred, of complex organic material, which possibly evolved from carbon rich volatiles, could have totally transformed the original matrix (Strazzulla, 1986).

Let us consider the initial phase of the planetesimal cosmogony, that is the formation of the planetesimals themselves. The environment in which the interstellar grains coagulate is the protostellar accretion disk. Such disks around T-Tauri stars, as well as in a more general context are discussed by many of the authors in this volume (see for example Pringle, this volume, and also Lin and Papaloizou 1980, Pringle 1981). More physico - chemical discussions of the pre - planetary disks, can be found in Cameron and Fegley (1982), Morfill (1983), Morfill and Völk (1984), or articles in *"Protostars and Planets II"*, edited by D. Black (1985). These accretion disks will, at least during part of their evolution, and at least in isolated regions be turbulent. More on this specific problem will follow in the next section.

As pointed out by Völk (1982, 1983), although turbulence enhances the rate of grain coagulation, it also hinders their sedimentation. The net first order effect being, that while the disk is turbulent, the time scale for the formation of O(1cm) grain conglomerates is longer than the time scale of grain radial drift towards the central object. Hence grains will be lost into the central star, before they can reach sizes large enough to sediment to the central plane and to go through the stage of swarm interaction, as mentioned above, to form planets. A possible solution to this problem is that growing grains require a relatively short, $O(10^{-3})$ of the accretion time, laminar episode, for the sedimentation to take place. Subsequent accumulation in the central plane to larger size solids can then produce a generation of planetesimals. Irregular accretion of disk material due to an inhomogenous envelope, or other evolutionary effects may allow for intermittent turbulence. This hypothesis has been linked by Weidenshilling (1984) with the time dependence, due to the grain growth, of the dust opacity of the disk, and substantiated by the numerical calculations of MMV. This point leads us to the next section.

3. Turbulent Accretion Disk

Pre - planetary disks and accretion disks in general, have been discussed by authors from many points of view. Some of the key references have been given in the previous section. One of the central questions about such disks is whether or not they are turbulent. Clearly we cannot give a complete answer with any of the existing theories. We can provide local stability criteria, as the simple example for thermal convection below. We can justify the need of turbulence by comparing results of the numerical cloud collapse simulations with and without the turbulent viscosity, or by modelling the thermal history of grains within the turbulent gas. All these arguments, at least qualitatively, give support for the turbulent disk models. Most of these aspects have been reviewed by Lin (1985) and Morfill, Tscharnuter and Völk (1985). An accretion disk can become turbulent through development of shearing instability due to the presence of the differential rotation, or thermal instability, if the temperature gradient is anywhere superadiabatic. Possible resultant shear and convective turbulence has been discussed in the context of accretion disks around compact objects by Shakura, Sunyaev and Zilitinkevich (1978). We also refer the interested reader to Moffatt (1987), for topological theories of 2-D and 3-D turbulence, which may have particular application to spiral structures in either thin or flared disks. More discussion of thick accretion disks can also be found in Abramovich et al. (1984) and Papaloizou and Pringle (1984).

One simple model of a steady disk is that of Lin and Papaloizou (1980), Lin (1981), and Morfill and Völk (1984). The dependence of temperature T, density ρ, and disk half-thickness h on the scaled distance $\xi = r/R$ from the central object, is shown in Figure 1.

If such a disk, not necessarily with the above radial dependences, is in radiative equilibrium, its vertical structure is determined by (i) the equation of hydrostatic balance,

$$\frac{1}{\rho}\frac{dP}{dz} = -\frac{GM_c}{r^3}z, \tag{1}$$

where P, M_c, r and G are, the pressure, mass of the central object, the distance from it

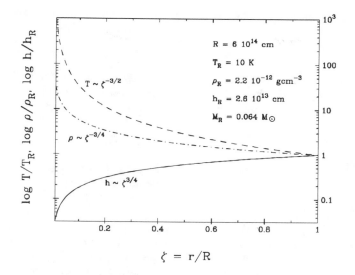

Figure 1: The radial structure of the disk. T, ρ and h are the gas temperature, density and the half-thickness of the disk normalized to their respective values at the outer radius of the disk R. ξ is the scaled radial distance from the central object.

and the gravitational constant, respectively, (ii) the radiative transfer equation,

$$-\frac{4acT^3}{3\kappa\rho}\frac{dT}{dz} = \mathcal{F}_z, \tag{2}$$

with a, c, κ and $\mathcal{F}_z$ being, the radiation constant, the speed of light, the opacity and the energy flux produced by the turbulent dissipation and (iii) the energy balance equation,

$$\frac{d\mathcal{F}_z}{dz} = \frac{3GM_c\nu}{4r^3}\rho. \tag{3}$$

where ν is the turbulent viscosity. The Reynolds number for such a disk critically depends on the prescription of the turbulent viscosity, a point most recently addressed by Lin and Pringle (1987). Values of $Re = O(10^{10})$ have been used in the calculations of MMV, but even more conservative estimates are well above the critical value of $O(10^3)$, measured for the isotropic turbulence in the laboratories. The reader should be warned however, that the astrophysical fluid flows often do not resemble those of terrestial laboratories, and the effect of compressibility on turbulence (see for example Lighthill 1955) is still very poorly understood.

Continuing our simple minded approach, we integrate equation (3), taking ν and ρ to have constant values, obtained by averaging over the disk height. Then,

$$\mathcal{F}_z = \frac{3GM_c\nu}{4r^3}\rho z. \tag{4}$$

Combining equations (1), (2) and (4) gives us the following mean radiative temperature gradient,

$$\left\langle \left(\frac{dlnT}{dlnP}\right)_{rad} \right\rangle = \frac{P}{\frac{a}{3}T^4} \frac{\kappa}{4c} \frac{3\nu\rho}{4}. \tag{5}$$

If the right hand side of equation (5) becomes larger than the adiabatic temperature gradient $(1 - \gamma^{-1})$, where γ is the ratio of specific heats, the system will be unstable. Therefore an approximate condition for the thermal convection in terms of the opacity is

$$\kappa_{crit} \simeq \frac{16}{3} \frac{c}{\nu} \frac{\frac{1}{3}aT^4}{P\rho} \left(1 - \frac{1}{\gamma}\right). \tag{6}$$

For values above κ_{crit}, we expect convection leading to turbulence. For values below κ_{crit}, laminar flow should persist. The above condition must be somewhat modified, if the stabilizing effect of magnetic fields is included. With the disk parameter values shown in Figure 1, $\kappa_{crit} = 10^{-2} \text{cm}^2\text{g}^{-1}$.

We will at this point restate the intermittent turbulence hypothesis in terms of the opacity variations. 1) The initial pre - planetary accretion disk, resultant from the stage two of star formation, is turbulent, or the accretion of interstellar grains raises the opacity above the critical value, making it turbulent. 2) Grains grow by coagulation in the turbulent disk, lowering the opacity below the critical value. 3) This stops the turbulence allowing for the sedimentation of the larger size grains to the central plane, through the laminar gas. 4) Continuing accretion of the interstellar grains increases the opacity once more, starting the turbulence and repeating the above process. We have already discussed how stage three can subsequently be followed by the formation of planets. In the next section we look at some details of stage two.

4. Coagulation of Grains in the Turbulent Gas

Let us define $n(m,t)dm$ to be the number of grains, per unit volume, within the mass range $(m, m + dm)$. Such a number density can, to first order, be thought of as an average over the pre - planetary disk. Neglecting spatial dependencies such as introduced by turbulent diffusion, sedimentation towards the central plane and drift towards the central star, the time evolution of $n(m,t)$ is given by the solution of the coagulation equation,

$$
\begin{aligned}
\frac{\partial}{\partial t}n(m,t) = & -\int_0^\infty \sigma_{rr'} < \Delta_{rr'}\delta v^2 >^{1/2} n(m,t)n(m',t)dm' + \\
& + \frac{1}{2}\int_0^m \sigma_{rr'} < \Delta_{rr'}\delta v^2 >^{1/2} n(m-m',t)n(m',t)dm' + \\
& + S(m).
\end{aligned}
\tag{7}
$$

Assuming grains to be spherical the collision cross section $\sigma_{rr'} = \pi(r+r')^2$, where r and r' are the radii of the two colliding grains. $S(m)$ is the source term due to the accretion of interstellar grains.

The turbulence induced grain - grain collision velocity $< \Delta_{rr'}\delta v^2 >^{1/2}$, was calculated using a modified version of the method of Völk et al. (1980) (subsequently VJMR). The grain - gas interaction is expressed by the Langevin equation for the randomly fluctuating component of the grain velocity $\delta\mathbf{v}$. It is

$$\frac{d}{dt}\delta\mathbf{v} = \frac{1}{\tau_f}(\delta\mathbf{v}_g - \delta\mathbf{v}), \tag{8}$$

where grain - gas coupling is defined by the grain friction time τ_f and $\delta\mathbf{v}_g$ is the random component of the gas velocity. Equation (8) has the formal solution,

$$\delta\mathbf{v}(t) = \frac{1}{\tau_f}\int_0^t dt' exp[-(t-t')/\tau_f]\delta\mathbf{v}_g(\mathbf{x}(t'),t') + \delta\mathbf{v}(0)exp(-t/\tau_f), \tag{9}$$

where, $\mathbf{x}(t)$ is the position of grain at time t. Further analysis of this solution is given in VJMR. There are however, two points missed by the above authors, worth of discussion here. The first being that the inner scale of turbulence is finite and given by the scale of molecular viscosity. VJMR assumed this inner scale to be zero. This leads to an over-estimate of the velocity between very small, $O(1\mu)$ grains, first noticed by Weidenschilling (1984). The second point has to do with the autocorrelation function for the Fourier amplitudes of the turbulent gas velocity spectrum. The exponential function used by the above authors, their equation (16), is only asymptotically valid for large times, $t \gg 1$. No analytical results are available for $t \ll 1$, but one can expect the autocorrelation function to have zero slope at $t = 0$. Hence, one simple possibility is,

$$\langle \tilde{w}(k,t), \tilde{w}(k',t') \rangle = \frac{P(k)}{4\pi k^2}\delta(k+k')\left(1 + \frac{|t-t'|}{\tau_k}\right) exp\left(-\frac{|t-t'|}{\tau_k}\right). \tag{10}$$

$P(k)$ is the one dimensional, turbulence power spectrum and τ_k is the lifetime of an eddy with wavenumber k. With this new autocorrelation function the mean square turbulence induced grain velocity $< \delta v^2 >$ is,

$$\left\langle \delta v^2 \right\rangle = \int_{k_0}^{k^*} dkP(k)\left[1 - \left(\frac{\tau_f}{\tau_k + \tau_f}\right)^2\right] + \tag{11}$$

$$+ \int_{k^*}^{k_s} dkP(k) \times \left[1 - \left(\frac{\tau_f}{\tau_k + \tau_f}\right)\right]\left[\left(\frac{tan^{-1}\chi}{\chi}\right) + \left(\frac{\tau_f}{\tau_k + \tau_f}\right)\left(\frac{1}{1+\chi^2}\right)\right],$$

where,

$$\chi = \left(\frac{\tau_f}{\tau_k + \tau_f}\right)\tau_k kV_{rel}(k),$$

k_0 and $k_s < \infty$ are the wavenumbers of the largest and the smallest eddies, k^* is the break in the eddy spectrum defined by equation (9) in VJMR and $V_{rel}(k)$ is the grain velocity relative to an eddy k. The above result should be compared with the equation (18) in VJMR. The r.m.s. induced relative velocity normalized to the r.m.s. turbulent gas velocity, between two grains with the same radius, $< \Delta\delta_{rr}v^2 >^{1/2} / < \delta v_g^2 >^{1/2}$, is plotted for the two autocorrelation functions in Figure 2, as a function of the dimensionless grain radius τ_f/τ_{k_0}. One can clearly see that the simple exponential autocorrelation (dashed-dot curve), over-estimates the induced relative velocity for the small grains, and under-estimates it for the large grains. The effect of the inner scale of turbulence can be

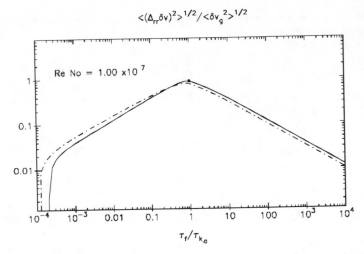

$$\langle (\Delta_{rr}\delta v)^2 \rangle^{1/2} / \langle \delta v_g{}^2 \rangle^{1/2}$$

Figure 2: The normalized r.m.s. turbulence induced relative velocity between two grains with equal dimensionless radius τ_f/τ_{k_0}.

seen in Figure 2, as the relative velocity cuts off for the very small grains. The relative velocity between any two grains, needed in the coagulation equation (7), is shown in the projection plot of Figure 3. The dotted region indicates the range of τ_f values, and hence grain sizes, for which the finite inner scale of the turbulence has an effect on $< \Delta\delta_{rr'}v^2 >^{1/2}$. In particular, along the diagonal the relative velocity is zero.

MMV have computed the evolution of the grain mass spectrum by integrating the coagulation equation (8). The description of the numerical method used is given therein. For the source term, they used the empirical $r^{-3.5}$ power law, interstellar grain size distribution of Mathis, Rumpl and Nordsieck (1977), approximately valid in the radius range $0.0005\mu < r < 1\mu$. Partial results of this calculation are shown in Figure 4. After ≈ 20 years, corresponding to a time of the order of the initial collision timescale, most of the mass is concentrated in grains with radius $O(1\mu)$. This peak in mass evolves to larger grains, roughly as $m_{peak} \propto t^6$. Approximate analysis of equation (8) predicts this time dependence. A power law for the size spectrum holds in the wide range for smaller grains, and this range extends with time. At the large mass end, the distribution drops off very quickly. The most interesting effect of the source term can be seen after $\approx 10^4$ years. In curve i, for example, a second peak in mass is clearly visible. This second generation of grains grows much faster than the first, in fact the peak mass grows approximately as t^{12}.

The final grain size distribution computed, for $t = 41000$ years, approaches a steady state, and can be approximated by a power law. The closeness of this power law to that of the source distribution, allows the interesting speculation that the interstellar grains size distribution is a result of turbulent coagulation of freshly nucleated solid particles from various astrophysical dust sources.

Finally we present the results for the opacity. At low temperatures the disk opacity is

$$< \Delta \delta_{rr'} v^2 >^{1/2} / < \delta v_g^2 >^{1/2}$$

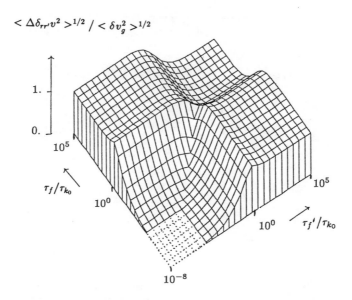

Figure 3: The projection plot of the turbulence induced collision velocity normalized to the r.m.s. turbulent gas velocity, as a function of the two grains friction times τ_f and $\tau_{f'}$. Reynolds number is 2.3×10^{10}.

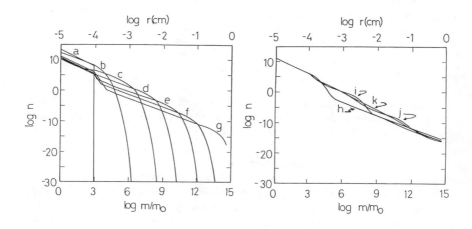

Figure 4: The grain number density $n(m)$ as a function of the grain mass m. The upper scale is logarithmic in grain radius. The epochs, in years are as follows: Curve a: 0, b: 20, c: 98, d: 222, e: 517, f: 1020, g: 2420, h: 4890, i: 10100, j: 20400, k: 41000.

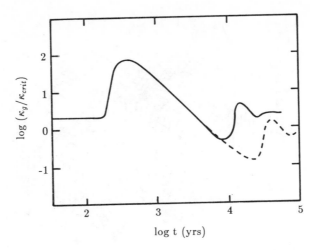

Figure 5: The grain opacity κ_g as a function of time. The solid and dashed curves are for accretion rates of 10^{-6} and 10^{-7} solar masses per year, respectively.

is dominated by the grains. The grain opacity calculated for two values of the accretion rate is shown in Figure 5. After the initial increase in the opacity to about ten times that for the solar mixture, it decreases below κ_{crit} at $t \simeq 8 \times 10^3$ years. Further evolution of the grain mass spectrum and hence also opacity is no longer consistent. Rather we should expect convection in the disk to stabilize, marking the beginning of the sedimentation phase. The evolution of the disk will now continue with gravitational fragmentation and other accumulative processes within the central plane.

5. Discussion

We have attempted to introduce the reader to the interdisciplinary field of research on the formation and evolution of a planetary system. The knowledge available, although without question of great volume, is still not well integrated and often speculative. We have also explained in some detail the dynamic grain processes in the context of the planetesimal cosmogony. However, even within this more specific problem many points have not been discussed. Is turbulence, if it persists throughout the disk, anywhere near isotropy? Grains are surely charged. How will the various possible electrical effects alter our picture? Grain fragmentation, although probably not very important since collision velocities in pre - planetary disk environment are rather low, of order ten centimeter per second, requires more empirical study.

How about the future? The possibility that the interstellar grain size distribution is due to turbulent coagulation, opens many new questions. The details of grain - turbulence interactions still have to be better understood. More laboratory work on meteorites, to possibly provide us with some crucial timescales, would be welcome. In closing, we look forward to further exploration of our solar neighbourhood, additional studies of processes leading to its formation and searches for other planetary systems.

Acknowledgments

We wish to thank the organizers and participants of this NATO Adanced Study Institute. We also like to thank H. Mizuno for release of results shown in Figures 4 and 5 prior to their publication.

References

Abramovich M.A., Piran T., Witta P.J., 1984, *Astrophys. J.*, **279**:367-383.

Alfvén H., 1954, "On the Origin of the Solar System", London: Oxford Univ. Press.

Alfvén H., 1962, *Astrophys. J.*, **136**:1005-1015.

Alfvén H., and Arrhenius G., 1975, *Structure and Evolutionary History of the Solar System*, D. Reidel.

Barberri and Nota, 1987, *Search for Planets around Nearby Stars with the HST Faint Object Camera*, preprint.

Black D., 1985, *Protostars and Planets II*, Univ. of Arizona Press.

Bertout C., 1987, in *Circumstellar Matter*, IAU Symposium No. 122, eds. I. Appenzeller and C. Jordan, pp. 23-30.

Cameron A.G.W., 1973, *Icarus*, **18**:407-450.

Cameron A.G.W., 1978, *Moon and Planets*, **18**:5-40.

Cameron A.G.W., Fegeley 1982, *Icarus*, **52**:1-13.

Crutcher R. M., Kazés I., and Troland T. H., 1987, *Astron. Astrophys.*, **181**:119-126.

Goldereich P., and Ward W.R., 1973, *Astrophys. J.*, **183**:1051-1061.

Hayashi C., 1981, in *Fundamental Problems in the theory Stellar Evolution*, IAU Symposium No. 93, eds. D. Sugimoto, D.Q. Lamb and D.N. Schramm, D. Reidel, pp. 113-128.

Lighthill M.J., 1955, in *Gas Dynamics of Cosmic Clouds*, IAU Symposium No. 2, North Holland Publishing Company, pp. 121-129.

Lin D.N.C., 1981, *Astrophys. J.*, **246**:972-984.

Lin D.N.C., and Papaloizou J., 1980, *Mon. Not. R. Astron. Soc.*, **191**:37-48.

Lin D.N.C., and Papaloizou J., 1985, in *Protostars and Planets II*, Univ. of Arizona Press, pp. 981-1072.

Lin D.N.C., and Pringle J.E., 1987, *Mon. Not. R. Astron. Soc.*, **225**:607-613.

Mathis, J.S., Rumpl, W., Nordsieck, K.H., 1977, *Astrophys. J.*, **217**:425-433.

Mizuno H., Markiewicz W. J., and Völk H. J., 1987, *Astron. Astrophys.*, submitted.

Morfill G.E., 1984, in Proc. Summer School, Les Houches, *Birth and Infancy of Stars*, eds. A. Omont and R. Lucas.

Morfill, G.E., Tscharnuter, W., Völk, H.J., 1985, in *Protostars and Planets II*, Univ. of Arizona Press, pp. 493-534.

Morfill G.E., and Völk H.J., 1984, *Astrophys. J.*, **287**:371-395.

Nakagawa Y., Sekiya M., and Hayashi C., 1986, *Icarus*, **67**:375-390.

Papaloizou J., and Pringle J.E., 1984, *Mon. Not. R. Astron. Soc.*, **208**:721-750.

Pringle J.E., 1981, *Ann. Rev. Astron. Astrophys.*, **19**:137-162.

Rodriguez L.F., 1987, *Star Forming Regions*, IAU symposium No. 115, eds. M. Peimbent and J. Juguku, pp. 239-253.

Safranov V.S., 1972, *Evolution of Protoplanetary Cloud and Formation of the Earth and the Planets*, NASA TT-F-677.

Shakura N.I., Sunyaev R.A., and Zilitinkevich S.S., 1978, *Astron. Astrophys.*, **62**:179-187.

Shu F.H., Adams F.C., 1987, *Circumstellar Matter*, IAU Symposium No. 122, eds. I. Appenzeller and C. Jordan, pp. 7-22.

Spaute D., Lago B., Cazenave A., 1985, *Icarus*, **64**:139-152.

Strazzulla G., 1986, *Icarus*, **67**:63-70.

Taylor R. J., 1987, *Mon. Not. R. Astron. Soc.*, **227**:553-561.

Völk H.J., 1982, in *Sun and Planetary System*, 6th Eur. Reg. Meet. Astron. eds W. Fricke and G. Taleki, D. Reidel, pp. 233-242.

Völk H.J., 1983, Meteoritics, **18**:412.

Völk, H.J., Jones, F.C., Morfill, G.E., Röser, S.: 1980, *Astron. Astrophys.*, **85**:316-325.

Weidenschilling S.J., 1984, *Icarus*, **60**:553-567.

Wetherill G., 1985, *Science*, **228**:877-879.

Wood J.A., and Morfill G.E., 1987, Center for Astrophysics Preprint Series No. 2492.

IRAS OBSERVATIONS

J. P. Emerson
Department of Physics
Queen Mary College
Mile End Road
London E1 4NS, England

ABSTRACT. The Infrared Astronomical Satellite (IRAS) made an all sky survey in the infrared between 7 and 135 microns in 1983. The data products already released and improved products in preparation are outlined. A simple and accurate method of calculating luminosities from IRAS flux densities is described and the IRAS colours of various classes of object are presented as an aid to locating regions of star formation. IRAS results on a sample of molecular cloud cores are discussed. These results together with ground based observations require a disk and thick envelope around the objects and are consistent with theoretical models consisting of an accreting protostar with a disk and envelope. We may well have finally located real protostars the "holy grail" of infrared astronomy. IRAS has helped determine the location and luminosity of objects driving flows but has not revealed enough extra luminosity to drive the flows by radiation pressure. The discovery, in L1551, of 60 and 100μm emission extending beyond the flow boundaries indicates the mechanical luminosity of the flow is more than ten times that previously thought. IRAS results on T Tauri stars also suggest the presence of disks many of which may be just reprocessing stellar photons but some of which are also generating their own luminosity. The disks seem to be most prominent in the T Tauri stars with the strongest Oxygen I forbidden emission lines, suggesting a connection between disks and winds.

1. INTRODUCTION

In 1983 the Infrared Astronomical Satellite (IRAS) performed an all sky infrared survey between 7 and 135μm in four broad passbands centred near wavelengths of 12, 25, 60 and 100μm and with detector sizes of $0.75' \times 4.5'$, $0.75' \times 4.6'$, $1.5' \times 4.7'$ and $3.0' \times 5.0'$ respectively (Neugebauer et al 1984), from which the IRAS Point Source Catalog (IRAS Point Source Catalog, 1985, henceforward PSC) and other products (see below) were derived (IRAS Explanatory Supplement, 1985, henceforward *Supplement*). The IRAS positions are usually good to considerably better than 15", the effective angular resolution varies from $\sim 0.5'$ at 12μm to $\sim 2.0'$ at 100μm and the catalog goes to a limiting flux density of ~ 0.5 Jy (1 Jy is 10^{-26} W m^{-2}) at 12, 25 and 60μm, and ~ 1.5 Jy at 100μm. The IRAS PSC is a uniquely useful tool because it is, to first order, a statistically complete survey of 96% of the sky. Even if objects are deeply embedded in dust and optically invisible, IRAS can

A. K. Dupree and M. T. V. T. Lago (eds.), Formation and Evolution of Low Mass Stars, 193–207.

potentially locate and characterize them by detecting the thermal emission from the surrounding dust. However one must always be cautious about interpretation of statistics of galactic objects because of the effects of confusion, shadowing and cirrus, all of which are important near the galactic plane (*Supplement*).

Three IRAS conferences have been held, one in each of the three participating nations, and their proceedings (edited by Israel 1985, Persson 1987, and Lawrence 1988) contain much information about IRAS and the results it obtained. Review articles on IRAS studies of our galaxy (Beichman 1987) and of other galaxies (Soifer, Houck, and Neugebauer 1987) appear in the 1987 *Annual Reviews of Astronomy and Astrophysics*.

Preliminary IRAS results on low mass star formation were given by Beichman et al (1984) for B5, by Emerson et al (1984) for L1551 and ESO 210-6A, and by Baud et al (1984) for Chamaeleon T1, showing that IRAS was a powerful instrument for detecting low luminosity objects in nearby dark clouds. Since then many authors have used the IRAS databases, often in conjunction with observations at other wavelengths, to interpret individual IRAS objects, or in some cases to carry out studies of a more statistical nature.

Section 2 summarises the types of IRAS data product available and forthcoming, and Section 3 describes a simple method for determination of IRAS luminosities, if distances are known, whilst in Section 4 classification of objects by their IRAS colours is discussed. Sections 5, 6 and 7 consider a few fairly extensive studies of low mass star formation, outflows and T Tauri stars, emphasising the role that disks seem to play. Conclusions are summarised in Section 8. This is intended as an illustrative rather than a complete discussion and so detailed studies of individual objects, as opposed to samples of objects, will mostly not be further discussed here.

2. IRAS DATA PRODUCTS

The first releases of IRAS data were in the form of IRAS circulars published in *Nature* and *Astronomy and Astrophysics* in 1983 and 1984. The March 1984 issue of *Astrophysical Journal Letters* described the instrument and the mission (Neugebauer et al 1984) and contained the first preliminary interpretations of data from the satellite. The major product, the PSC, was released in early 1985. An improved version 2.0 of the PSC was released in 1986, which corrected for a systematic flux density overestimation at low signal to noise levels (this only affected flux densities below 2 Jy), and used an improved calibration (for $\sim 1\%$ of the sources the flux densities changed by up to 10 %), and made other more minor improvements. No changes in position were made, although some positional uncertainties were decreased. Although version 2.0 should now be used in preference to version 1 most conclusions drawn from work with the first version should remain valid. Printed versions of the PSC and *Supplement* are currently in production.

The other most well known IRAS products are the Extended Emission Images, sometimes known as Skyflux Plates which consist of 212 overlapping $16° \times 16°$ fields with $2'$ pixel size and effective resolution $\sim 4 - 6'$ covering almost the whole sky. These come in 3 sets, 1 for each of the 3 'hours confirming' (HCON) IRAS survey coverages of the sky (*Supplement*). Version 1.0 of HCON3, for the third sky coverage, was released in early 1985 and was followed by the other 2 HCONs. Version 1.0 of HCON3 has now been superceded by version 2.0 released in mid 1986. Version 1.0 had been on a different calibration scale from the HCON1 and HCON2 plates and had patchwork artefacts,which were fixed in Version 2.0 making all 3 HCONs compatible.

A Small Scale Structure Catalog was released in early 1986 (IRAS Small Scale Structure Catalog, 1986) containing 16,740 sources with sizes $< 8'$, but does not appear to have been widely used. If a source appears in the Small Scale Structure Catalog then one must beware of its PSC flux densities, as it is probably extended.

IRAS actually spent about 40% of its time doing repeated scans of small regions of sky in various modes to enhance signal to noise ratio by coaddition of the scans (producing a gain of a factor of 3-8), and in some modes to also achieve increased spatial resolution. These data are known as Additional Observations if made with the survey instrument and as CPC data if made with the $1'$ beam 50 and 100μm Chopped Photometric Channel, and are also available. For example Young et al (1986) made use of the small edge detectors of the survey array to obtain high resolution IRAS observations of the ρ Ophiuchus cloud core, and Jennings et al (1987) made use of the CPC to obtain high resolution IRAS observations of the NGC 1333 region and its associated outflows and Herbig Haro objects.

All the point like objects seen in the additional observations were collected together in the Serendipitous Survey Catalog (IRAS Serendipitous Survey Catalog 1987) which has a sensitivity of about a factor of 4 greater than the PSC, but over only 4% of the sky. About 50% of the 43,866 Serendipitous sources are not in the PSC.

An Atlas of $8 - 22\mu$m spectra for bright IRAS sources, has been published (IRAS Low Resolution Spectra Atlas, 1986) and contains some of the brighter T Tauri stars. There are other IRAS data products released but they are of a more specialist interest and are not described here.

Not publically released (it is many hundreds of magnetic tapes) but available at processing centres in the USA, Netherlands and UK is the raw survey data. This can be coadded, either to produce, for small objects, a one dimensional scan through a specified position, or, for extended objects, an image of a small region of size about a square degree, with enhanced sensitivity and resolution over the corresponding survey products. A good example of what can be done with coadded images made in this way is the work by Edwards et al (1986) on the outflow around L1551 IRS5.

As more and more is learnt about the IRAS data, and the rich science it contains, one learns to trust it more, and wants to produce further products that were barely dreamed of before launch. A Faint Source Survey Catalog is in preparation at the IRAS Processing and Analysis Centre (IPAC) in Pasadena, and will be based on coadded, point source filtered, survey data. The intermediate product of Faint Source Survey Plates will also be released to aid in the study of objects whose position is already known from other observations. This is a large task, which is starting at high galactic latitudes because it is here that the lack of source confusion will give the greatest gains. It will work down to lower latitudes and eventually cover all the non-confused sky. It will be 95% reliable and go a factor of 3 fainter than the PSC with the first instalments scheduled for release in 1988.

There is also a corresponding plan to produce improved Extended Emission Images by coadding all the survey data into 1716 plates on the POSS ESO/SERC survey grids, using pixel size $< 2'$. The 3 HCONs images per piece of sky will thus be replaced by 1 image per piece of sky. A 'low' resolution product is scheduled for release in 1989, and there are longer term development plans for medium and high resolution images.

The existing IRAS data products, can be obtained, from either Code 633, National Space Science Data Centre (NSSDC), Goddard Space Flight Centre, Greenbelt, MD 20771, USA, or, in Europe, from the Stellar Data Centre, Centre des Donnees Stellaires (CDS), 11 Rue de l'Universite, F-67000 Strasbourg, France. The availability of the new products will be announced to the community at the appropriate time.

3. CALCULATION OF IRAS LUMINOSITY

Many workers use IRAS data to deduce the far IR luminosity of objects, but there is some confusion about the best way to do this, some of the methods in use being unnecessarily complicated and time consuming. This situation seems to arise because workers find the flux densities listed in the PSC at up to 4 wavelengths and try to fit some curve to these flux densities and integrate under it to get, for example, the 12-100μm IRAS flux, and hence the luminosity on assuming a distance. Often no simple curve (eg blackbody, power law) fits the points and the PSC fluxes are not colour corrected, and so authors spend more time trying to correct them in accordance with the source spectrum and then determine the luminosity. When one stops to consider what IRAS *actually measured*, as opposed to the way the result is quoted in the PSC, it is clear that there is a much more direct, accurate and quick way to get luminosities, and further to get them over a larger wavelength range.

IRAS carried broad band detectors and filters whose transmissions are shown and listed in the *Supplement*. Unlike the situation with a narrow band radio receiver IRAS actually determined integrated fluxes (in units of W m^{-2}) within *broad* passbands but in the PSC these measured fluxes were expressed in terms of the flux densities (in units of Janskys) at 12, 25, 60 and 100μm that, for a spectrum of the form $S(\nu) \propto \nu^{-1}$, would give the measured fluxes in the actual passbands. Thus in the process of PSC production the directly measured fluxes were divided by a passband in Hz to get flux densities in Janskies. This was done as Janskies were judged to be the form in which most astronomers would want the data presented. For deducing luminosities however one is better off with the original in band fluxes, but the equivalent, or something even more useful, may be easily recovered.

To calculate the IRAS flux components it is convenient to define four synthetic *square* wavelength passbands of 7-16, 16-30, 30-75 and 75-135μm, approximating to each of the actual IRAS passbands which are referred to as the 12, 25, 60 and 100μm bands. These synthetic passbands have the desirable properties of being contiguous, and of covering the complete wavelength range over which IRAS was sensitive to radiation. They are an idealisation to the actual (non-square and non-contiguous) IRAS bandpasses. For an input spectrum $S(\nu) \propto \nu^{-1}$ it is easy to show that the flux F_{s-l} (in W m^{-2}) between short and long wavelengths s and l is related to the flux density S_r (in W m^{-2} Hz^{-1}) at reference wavelength r by

$$F_{s-l} = S_r \times \frac{c}{r} log_e \left(\frac{l}{s} \right)$$

where c is the velocity of light which we rewrite as

$$F_{s-l} = \Delta\nu_{s-l} \times S_r$$

where $\Delta\nu_{s-l}$ are the effective bandwidths of each band. Table I shows the resulting effective bandwidths for the synthetic bands introduced above where the unit of bandwidth is correct for flux in W m^{-2} and flux density in W m^{-2} Hz^{-1}

Table I

Synthetic IRAS broadbands and corresponding effective bandwidths

Nominal λ	Lower λ	Upper λ	Effective bandwidth
r	s	l	$\Delta\nu_{s-l}$
μm	μm	μm	Hertz
12	7	16	20.653×10^{12}
25	16	30	7.538×10^{12}
60	30	75	4.578×10^{12}
100	75	135	1.762×10^{12}

As the PSC flux densities were deduced *assuming* this spectrum, $S(\nu) \propto \nu^{-1}$, these bandwidths may be used to determine the flux in these bands. For example for a source of non colour corrected flux density at 12 μm, $S_{12\mu m}$ Jy, the integrated flux between 7 and 16 μm $F_{7-16\mu m}$ in units of 10^{-14} W m^{-2} is

$$\frac{F_{7-16\mu m}}{[10^{-14}\text{W m}^{-2}]} = 20.653 \frac{S_{12\mu m}}{[\text{Jy}]}$$

If the true IRAS bandpasses were *identical* to our idealised bandpasses, these effective bandwidths would be correct for *any* spectral shape. In so far as the synthetic bandpasses are good approximations to the true bandpasses, the integrated fluxes in our synthetic passbands are good approximations to the true fluxes over these passbands for *any* smooth and not too steep spectrum. Thus this IRAS flux (in W m^{-2}) is insensitive to the actual far-infrared spectral shape. Any error involved in this procedure is likely to be less than that incurred in the alternative approach of trying to guess the true source spectrum and carrying out colour-correction and trapezoidal (or other type) integration between IRAS wavelengths. Furthermore this method of determining flux is simple and can be applied even when no information is available from the IRAS data on which to base a colour correction (e.g. a source detected in only one band). Note also that if a source is detected at all four IRAS wavelengths one may deduce the flux between 7 and 135μm which is more useful for luminosity determination than 12-100μm.

Where monochromatic flux densities are required, rather than luminosities, a colour correction should be made if the source spectrum is known. The *Supplement* describes how to do this for an assumed spectrum and tabulates colour correction factors for blackbodies and power laws, but does not list the expected non colour corrected flux density ratios for these assumed spectra, which are necessary to deduce the blackbody temperature or power law exponent to use in the colour correction. Such a table will however appear in the final printed *Supplement* (in press) and can already be found in the Small Scale Structure Explanatory Supplement (IRAS, 1986). The prescription implicit in the *Supplement* may be use to calculate similar non colour corrected ratios and correction factors for other types of spectrum (e.g. including a dust emissivity law). In practise most IRAS sources, (except for hot blackbodies), detected in more than 2 bands are not fitted by any simple spectrum, such as a blackbody or power law, so it is unclear if any gain in accuracy is achieved by colour correction in such cases. Thus if the source spectrum is not well known the appropriate colour correction is not well known either, and it is debatable if such a correction is worthwhile. When the true spectrum is known a colour correction will be worthwhile.

4. CLASSIFICATION OF SOURCES BY IRAS COLOURS

Version 2.0 of the IRAS catalog contains 245,889 objects but how do we know which of these are interesting for studies of low mass star formation? We need a classification scheme as an aid to locating low mass star forming objects.

It is useful to classify IRAS sources using their measured infrared colours [100-60], [60-25] and [25-12] defined here as the logarithm of the ratio of the IRAS catalog flux densities, S_λ in Janskies at wavelength $\lambda \mu m$ at a long and shorter pair of wavelengths. For example for the 100 to 60 μm colour

$$[100 - 60] = log_{10} \left(\frac{S_{100}}{S_{60}} \right).$$

No colour corrections (*Supplement*) have been applied in this definition.

Emerson (1987) shows [100-60] versus [60-25], and [25-12] versus [60-25] colour colour plots from the whole IRAS catalog. An empirical approach of plotting colour-colour plots of identified sources from the IRAS point source catalog shows the regions of colour-colour space occupied by various types of object. The densest clusterings of a few types of object in such colour-colour plots have been used to define the colour regions summarised in Table II, in the hope that these will be of assistance in identifying the likely nature of IRAS sources, although of course a detailed look at each object is the only method that is guaranteed.

TABLE II

Colour ranges occupied by some known types of object

Object Type		IRAS Colour indices		
		[25-12]	[60-25]	[100-60]
		$log_{10} \left(\frac{S_{25}}{S_{12}} \right)$	$log_{10} \left(\frac{S_{60}}{S_{25}} \right)$	$log_{10} \left(\frac{S_{100}}{S_{60}} \right)$
1000 K blackbody		-0.47	-0.70	-0.50
100 K blackbody		+1.64	+0.44	-0.17
50 K blackbody		+3.69	+1.79	+0.23
30 K blackbody		+6.39	+3.52	+0.70
Stars		-0.7 to -0.2	-0.9 to -0.4	-0.2 to -0.6
Bulge stars		-0.2 to +0.3	-0.8 to -0.2	-
Planetary Nebulae		+0.8 to +1.2	0.0 to +0.4	-0.4 to 0.0
T Tauri Stars		0.0 to +0.5	-0.2 to +0.4	0.0 to +0.4
Cores	Embedded	+0.4 to +1.0	+0.4 to +1.3	+0.1 to +0.7
Galaxies	Sources	0.0 to +0.4	+0.6 to +1.2	+0.1 to +0.5

The colours given for blackbodies take into account the IRAS bandpasses and so are directly comparable with the IRAS catalog fluxes, avoiding the complexities of color corrections (*Supplement*). The star's colours are based on objects associated with a stellar catalog according to the IRAS catalog. Hacking et al (1985) discuss in more detail the regions of the colour-colour diagrams occupied by K and M stars without circumstellar dust shells, by M stars with circumstellar shells and by carbon stars. Bulge stars are taken from Habing et al (1985) and are probably evolved late type M stars near the tip of the red giant branch. van der Veen and Habing (1988) discuss in detail the location of various kinds of late type stars in the 12-25-60 μm colour colour diagram for [25-12] < 0.8 and [60-25] < 0.4. Planetary nebulae colours are deduced from Pottasch et al (1984). T Tauri star colours are based on 338 IRAS sources associated with known T Tauri stars (Emerson unpublished). Cores are based on the sample of IRAS objects found associated with NH_3 cloud cores (Beichman et al 1986) discussed below. Some of these objects also populate the T Tauri region. Galaxy colours are based on objects associated with an extragalactic catalog according to the IRAS catalog. Molecular cloud hot spots, HII regions, reflection nebulae, star formation regions, which we shall collectively call embedded sources, all occupy an area similar to that of galaxies and cores. It should be remembered that these colour classifications for T Tauri sources and embedded sources are not unique, and in particular they overlap with the colours of various types of extragalactic object. For example 3C273 has colours within the T Tauri range of Table II yet it is a quasar.

5. THE CORES SAMPLE

One of the goals of IRAS was to help understand the process of star formation in our galaxy. Low mass stars have long been thought to form in the cores of molecular clouds, as is suggested by the proximity of many T Tauri stars to dust clouds (e.g. in Taurus). Prior to the launch of IRAS a survey of dark cloud cores, selected from the Palomar Observatory Sky Survey (POSS) red prints, was carried out using CO and NH_3 molecules as gas tracers (Myers Linke and Benson 1983, Myers and Benson 1983). The mean properties of these dense ammonia *cores* are listed in Table III taken from Myers (1987)

Table III

Dense Core properties

FWHM map diameter (pc)	0.05 - 0.2
Density (cm^{-3})	$10^4 - 10^5$
Kinetic Temperature (K)	9 - 12
FWHM line width (km sec^{-1})	0.2 - 0.4
Gas mass within FWHM map contour ($M_\odot$)	0.3 - 10
Free fall time ($\times 10^5$ yr)	1 - 4

Note that the line widths of these cores are close to thermal, unlike typical line widths in Giant Molecular Clouds which are much larger due to turbulence. One thus expects these clouds to collapse in about a free fall time, close to the inferred ages of the IR objects found in the cores.

IRAS data was used to search for objects associated with 95 of these cores, to a limit of 0.1 $L_\odot$ in the nearby Taurus and Ophiuchus regions, with the result that about half of the cores contained objects with dust colour temperatures corresponding to cool objects (Beichman et al 1986). The $12 - 100\mu m$ luminosities of these objects were typically 0.1 - 5 L $_\odot$, consistent with the luminosities expected of young low mass stellar objects. It was found that of the IR sources associated with these cores about half had visible counterparts on the POSS, and about half were invisible. As a rule the invisible objects and their associated molecular cores had cooler dust colour temperatures, were closer to the ammonia cores (which define the densest regions), and had greater linewidths than the visible objects. They were also associated with larger dust and CO core masses and the dust radius required to give the dust colour temperature for the observed luminosity was larger for the invisible objects (Beichman et al 1986).

Many of the visible objects were T Tauri stars and although some of the invisible objects could merely be embedded T Tauri stars the relative numbers of visible and invisible sources and the known timescales of various phases of early stellar evolution suggested that at least some of the embedded objects were very young, possibly still accreting, for if this were not so they would have already appeared out of the core either by dispersing it by their irradiation and winds, or by moving out of the core as a result of any small velocity component with respect to the core (Beichman et al 1986).

The invisible objects also had steeper IR spectra than the visible objects. A comparison of the molecular gas properties in those cores that had formed stars with those that had not showed that those with stars have a bigger line width and ratio of cloud to Jeans mass than those that had not formed stars. The larger molecular line widths in the invisible objects suggested that the embedded objects may have already added some turbulence through mass loss winds, and indeed the success rate of finding flows in these objects is higher than in other search methods (Myers et al 1988).

Myers et al (1987) extended the work of Beichman et al (1986) with further observations of 34 of these objects which extended the wavelength range of the spectra to 1.2 μm and in some cases 0.44 μm. This confirmed the identifications, and the accuracy of the IRAS positions, as well as the difference in the steepness of the spectral slope of the visible and invisible objects, and that the steeper spectrum sources were closer to the cloud core centres and also allowed a more accurate estimate of the luminosities which were mostly $\sim 1 - 2$ L$_\odot$.

Myers et al (1987) went on to model the continuum spectra of these objects. To model the near IR and optical spectra they assumed the near IR spectral shape was determined by dust extinguishing an embedded stellar blackbody source. The resulting stellar effective temperatures (~ 3000 K) were similar to those of T Tauri stars lending credence to this model, and large extinctions ($A_V \sim 30 - 90$ mag) were required to explain the steep spectrum invisible sources. This much extinction cannot be contributed by dust associated with gas at the mean core densities deduced from NH_3 observations, indicating that the material density peaks close to the embedded objects, and thus suggesting that the invisible objects have formed out of the cores. Assuming a power law variation of density n with radius r from the embedded object, $n = br^{-p}$, and using the mean density deduced from molecular line observations to set the absolute value of b, and the known core radius, they calculated the distance from the object at which dust first appears (the inner radius of the dust shell). There must be such an inner radius as if the power law continued right up to the object the extinction would be much too large to be consistent with 30-90 mag. This inner radius is of order 50 AU which is much more than the radius (~ 0.1 AU) at which dust grains would be so hot that they would sublime in the radiation field from an object of this bolometric luminosity. Therefore there must be either be a dust free cavity near the embedded objects, or any dust in this region must not contribute to the extinction.

The far IR emission was then modelled by the appropriate density gradient and column density of dust to give the fitted visual extinction, with heating by a source of the observed bolometric luminosity located at the centre of the cavity and the effective temperature of the heating source was allowed to vary until a reasonable fit to the far IR spectrum was obtained. It was found that the source effective temperature required to fit the far IR spectrum was substantially cooler than that deduced from the fits to the near IR spectrum. Also this far IR model was quite incapable of providing the observed mid IR fluxes ($5 - 30\mu$m), nor could the mid IR fluxes be produced by the sum of the far and near IR models.

These two discrepancies (conflicting effective temperature of the heating source deduced from far and near IR spectra, and lack of mid IR emission) could be qualitatively removed if the interior of the cavity surrounding the star was heated not directly by the star but by cooler material (at ~ 300 K to provide the mid IR emission) lieing between a few and 50 AU from the star, and distributed in such a way as not to greatly increase the near IR extinction. The requirement on the distribution of this dust arises because if one puts sufficient material to account for the observed $5 - 30\mu$m emission within the cavity with spherical symmetry then this same material would produce much more extinction than deduced from the near IR spectra and would totally obscure the near IR and optical flux from the star.

Myers et al (1987) were therefore driven by the observations to suggest that this cavity did actually contain dust and that it was not uniformly distributed within the cavity but rather confined to a physically thin disk around the star where it contributed to the mid IR emission (explaining the previous model deficit) but not to the extinction (unless one happened to view the star along the plane of the disk). Although these models require a disk they cannot distinguish whether its luminosity arises from reprocessing of stellar photons or is intrinsic to the disk implying that accretion onto the star, through the disk, is occurring. However it seems very plausible that at least some of the embedded objects are low mass protostars deriving more than 50% of their luminosity gravitationally from accretion.

Adams and Shu were working on theoretical models of just such objects and whereas their early models (Adams and Shu 1985) without rotation and accretion disk also failed to reproduce the observed energy distribution they found that with the inclusion of rotation (Adams and Shu 1986), which leads to formation of a disk, they could reproduce reasonably well the observed energy distributions in these steep spectrum invisible objects. This work has been further extended by Adams Lada and Shu (1987). In these models the majority of the luminosity in the invisible steep spectrum sources is accretion luminosity arising from release of gravitational energy on accretion at rates of 10^{-5} to 10^{-6} M$_\odot$ yr^{-1} onto protostars of mass 0.1 - 0.5 M$_0$. Thus it is reasonable to claim that these are, at last, *protostars* the long sought "Holy Grail" of IR astronomy (Wynn-Williams 1982). Currently observational work is underway to check predictions of these models, particularly at sub millimetre wavelengths, but whatever the details turn out to be it seems likely that the presence of disks contributes to their success, and that the presence of disks are now firmly established on observational grounds.

The coming together of the molecular line work, the IRAS observations, the ground based follow-up and the theoretical modelling seem to have led to a coherent and consistent picture in which these invisible steep spectrum objects have disks through which matter is very likely still accreting onto the stars. Although much further observational and theoretical work is required to fully check all the details it seems reasonable to conclude that IRAS succeeded in its aim of locating protostars, by finding these low mass stars in the process of formation.

6. OUTFLOWS

Various surveys for outflows around IRAS sources have been made and compared with IRAS results. For example Margulis and Lada (1987) surveyed Mon OB1 and concluded that out of 9 outflows 6 were associated with IRAS sources, and the remaining 3 could be associated with sources too faint for detection with IRAS. On the other hand only about 20% of the IRAS sources were associated with flows, indicating that for 20% of its lifetime the young sources go through an outflow phase, or perhaps that only 20% of them ever go through this phase. Searches (Myers et al 1988) around IRAS "cores" find that $\sim$ 44% have outflows, some as weak as 10^{-8} $M_\odot$ yr^{-1}. Cores with flows detected are $\sim$ 3 times as luminous as those without.

Walker et al (1986) discovered an outflow in IRAS 16293-2422 whose IRAS spectrum is well fitted by a 39 K blackbody with an emissivity law $\propto \nu$ and luminosity $\sim 23 L_\odot$. This object lies in a core and there is some evidence for infall of gas and a rotating disk (Mundy et al 1986) and it may be at an extremely young evolutionary stage.

Mozurkewich et al (1986) examined IRAS data on 45 flows to see if, with the extra luminosity revealed by IRAS, the dense neutral flows could be driven by radiation pressure. The answer in most cases was *no*, particularly for the lower luminosity flows. Over half their sources were extended and they suggested the need for two temperature components, the first at $\sim$ 165 K, and the second at $\sim$ 40 K, with the hot component permeating the whole cloud.

Edwards et al (1986) using IRAS data (see also Clark et al 1986) found 60 − 100μm emission, extending $\sim 20'$ out from L1551 IRS5 and aligned with the bipolar outflow lobes. The 60−100μm luminosity in this extended emission is $\sim$ 3 L$_\odot$ which should be compared to only 0.2 L$_\odot$ mechanical luminosity of the high velocity gas and 10 L$_\odot$ from IRS5 itself. The uniformity and magnitude of the 60-100 μm temperature is not explicable by heating from a central source (i.e. IRS5). This implies that the luminosity of the extended far IR emission must derive from the flow and that the flow luminosity is about 18% of the total stellar luminosity. This makes it even harder to countenance radiation as the driving mechanism for the outflow. The extended far IR luminosity has some 10 times the mechanical luminosity of the high velocity gas. Edwards et al (1986) suggest that the extended emission may result from energy dissipated, as photons, in shocks at the edge of the flow being absorbed and reradiated by the dust surrounding the flow. The fact that the extended far IR emission extends beyond the edge of the molecular outflow regions shows that the energy is not generated by collisions of dust with the high velocity gas.

Cohen and Schwartz (1987) searched for the exciting stars of Herbig Haro (HH) objects and determined luminosities of the HH exciting stars. They found a median luminosity of $\sim$ 9 L$_\odot$ and by comparing the IRAS flux densities with those made with a smaller beam size on the Kuiper Airborne Observatory deduced that many of the exciting stars were surrounded by extended far IR dust emission. They suggested that this extended emission was from the infalling envelopes of low mass protostars, and that the luminosities were consistent with those of young low mass stars. The IR emission is always associated with the exciting star rather than with the HH objects themselves.

7. T TAURI STARS

There is direct evidence, from imaging, of a disk around the T Tauri star HL Tau (Grasdalen et al 1984, Beckwith et al 1984) and the jets and outflows associated with some T Taus also suggest disks there. Rucinski (1985) compared the IRAS PSC with positions of T Tauri stars in Taurus-Auriga identifying 35 T Tauri stars and then showed their spectra from optical to far IR wavelengths. He noted that in many cases the spectra were of the form $\nu S(\nu) \propto \nu^1$, and suggested that this could be explained if the stars were accompanied by optically thick dusty accretion disks. Appenzeller et al (1984) and Edwards et al (1987) have suggested that the observed velocity structure in forbidden lines formed around T Taus was best explained if receding material was obscured by an opaque circumstellar disk. Edwards et al (1987) find that if the IRAS 60 μm flux density of HL Tau is from a disk at 50 K then the required disk radius is consistent with that found directly and that inferred from their model of obscuration of the forbidden line region by the disk. Adams Lada and Shu (1987 and 1988) have modelled the spectra of T Tauri stars with disks.

Cohen, Emerson and Beichman (1988) have compiled the luminosity components of T Taus in Tau-Aur using the database used by Cohen and Kuhi (1979) for optical and near IR data and IRAS for the longer wavelengths. If each of these luminosity components arose from radiation from a spherical star, or such radiation reprocessed in a spherical shell, one would expect the sum of all the components to represent the stellar bolometric luminosity which will indicate the age for stars along one of the vertical (constant log T_{eff}) convective tracks in Cohen and Kuhi's (1979) HR diagram for Tau-Aur. However no reasonable correlations of stellar properties with age along such tracks stand out from this data. This suggests, assuming the data itself is sufficiently accurate, that there is some extra luminosity component present in these objects in varying proportions in different objects. Cohen, Emerson and Beichman (1988) suggest that this additional luminosity component arises from a disk around these objects. The varying proportions that this disk component provides in different stars could be either because of different disk orientations to the observer's line of sight if the disk luminosity derives solely from the stellar luminosity, or due from the orientation effect and different intrinsic disk luminosities if the disks generate some of their own luminosity.

If a dusty optically thick disk exists around a star of luminosity L_* it will intercept, and reradiate, 0.25 of the stellar luminosity if it has infinite extent and is physically thin (Adams et al 1987, Kenyon and Hartmann 1987). Because the disk does not have spherical symmetry we must be careful in considering how this relates to the relative amounts of flux we will observe, at the earth, from the star and from the disk. If we had spherical symmetry we would expect the observed ratios to be 0.25, but this is not the case for the actual geometry (Kenyon and Hartmann 1987).

If the disk has surface area (one side) A then in terms of the radiation intensity, I, at the disk the disk luminosity $L_{disk} = \pi I 2A$ as πI is the flux density radiated into a hemisphere by a plane surface and the factor of 2 arises because the disk has two sides. If one observes this disk at an inclination i to its axis then its projected area is $Acosi$ and its solid angle at the earth, distance D away, is $Acosi/D^2$ so that the observed flux, F_{disk}, is $IAcosi/D^2$ whence $F_{disk} = 2cosiL_{disk}/4\pi D^2$. This is to be compared with $F_* = L_*/4\pi D^2$.

Thus if one looked pole on ($i = 0^o$) at an unresolved star + disk system where the disk intercepts 0.25 of the star's luminosity and measured the bolometric flux one would find

$$F_{bol} = F_{star} + F_{disk} = F_* + 2 \times 0.25F_* = 1.5F_*.$$

This is not violation of conservation of energy but is because of the non spherical geometry. The 0.25 L_* intercepted and reradiated by the disk was going in directions that would never have contributed to the flux measured by an observer looking down on the pole, but the plane disk intercepts these photons and then reradiates them towards the observer, with the geometry providing the factor of 2. On the other hand if one observed such a system along the plane of the disk one would just find $F_{bol} = 1$ F_* because none of the disk photons are reradiated parallel to the disk.

In neither of these two situations does the presence of the disk affect the flux component arising from the star as when the star is viewed from above it is not occulted by the disk, and when it is viewed from the side the disk solid angle is zero, because we have assumed a physically thin disk. At a more general viewing angle however some of the stellar radiation will be occulted by the disk and so the observed flux from both the stellar and disk components will depend on the orientation of the observers line of sight (Adams et al 1987, Kenyon and Hartmann 1987).

If the disks are not physically thin but flared (increasing in thickness with radius) they may intercept more than 0.25 L_*. The fraction of stellar luminosity intercepted depends on the adopted geometry but values of 0.5 L_* do not seem unreasonable (Kenyon and Hartmann 1987). Viewed from $i = 0^\circ$ such a system would show $F_{bol}/F_* = 2$.

The temperature distribution expected in such disks may be calculated and the resulting spectra predicted and are found to reproduce observations of a number of T Tauris reasonably well (Adams Lada and Shu, 1987 and 1988, and Kenyon and Hartmann 1987). For those objects in which a reprocessing disk does not fit the spectra Adams Lada and Shu (1988) suggest that accretion is still occurring.

Cohen Emerson and Beichman (1988) attack the question of disks for 74 T Tauri stars in Tau-Aur from the point of view of the observed fluxes (all objects are at the same distance D), rather than fitting the spectra. They estimate the actual stellar flux F_* from the observed 0.44-0.67μm fluxes derived from Cohen and Kuhi's (1979) raw data and their extinctions, A_v and the bolometric flux from the flux components including the contribution from IRAS. They then calculate the ratio F_{bol}/F_* and find that for a significant fraction of their sample it is above 2, which seems inconsistent with the idea of a simple reprocessing disk, even taking into account flaring and extinction of the star itself by the disk. In these stars they suggest that some of the disk luminosity arises from accretion.

Cohen Emerson and Beichman (1988) further find that it is precisely those stars that show most evidence of mass loss (through their Oxygen I forbidden emission lines) that have the large values of F_{bol}/F_*. Further if one looks at the fraction of the bolometric flux contained in various flux components the OI stars have a greater fraction of their bolometric flux at both 3.5-7μm and beyond 7μm (from IRAS data), as might be expected if the extra flux component arises from accretion generated luminosity in the disks. They go on to look at various other correlations in their data but conclude that the observed lack of correlations between disk indicators and many other parameters indicates that the disk properties were determined more by the formation conditions of the systems than by the masses of the central stars, or their ages.

In addition to establishing the presence of disks these results emphasise that when disks are present plotting bolometric luminosity against effective temperature and comparing with theoretical tracks in an HR diagram to deduce the age of such stars is highly dubious.

Thus it now seems, from several independent lines of evidence, that T Tauri stars are associated with disks. Many of these may be mainly reprocessing stellar photons with little intrinsic accretion luminosity, however a significant fraction, particularly those with strong OI emission indicative of mass loss, seem to have greater disk luminosity than can be explained by reprocessing and may therefore be still accreting. This suggests a connection between disks and T Tauri winds. Methods of deducing the inclination of the disks will be important in trying to determine if modest amount of accretion luminosity are being generated in even the systems with $F_{bol}/F_* < 2$.

8. CONCLUSION

The source confusion and complexity of the galactic plane make statistical studies of star formation in our galaxy difficult, the IRAS survey has however greatly increased the number of objects identified as low mass stars in early stages of evolution, including some that seem to be the long sought protostars. Various arguments all point to the importance of disks in the formation and early evolution of low mass stars. There is growing evidence that disks, winds and outflows are associated with each other. Determination of disk inclinations and further theoretical and observational studies of these objects are needed to clarify the details of these disks and their evolution as well as how much of their luminosity arises from reprocessing of stellar photons and how much from accretion. Clarification of the relationship between these disks and those around main sequence stars such as Vega and β Pictoris (Aumann 1985) may throw light on the process of planetary formation.

9. ACKNOWLEDGEMENTS

I thank the scientific organizing committee for inviting me to lecture at the ASI, and for their financial support. I thank Scott Kenyon, Lee Hartman and Gibor Basri for discussions at the ASI that helped clarify my understanding of disks. The *Infrared Astronomy Satellite* was developed and operated by the Netherlands Agency for Aerospace Programs (NIVR), the US National Aeronautics and Space Agency (NASA), and the UK Science and Engineering Research Council (SERC).

10. REFERENCES

Adams F.C. and Shu F.H., 1985, *Astrophys. J.*, **296**, 655.

Adams F.C. and Shu F.H., 1986, *Astrophys. J.*, **308**, 836.

Adams F.C. Lada C.J. and Shu F.H., 1987, *Astrophys. J.*, **312**, 788.

Adams F.C. Lada C.J. and Shu F.H., 1988, *Astrophys. J.*, in press.

Appenzeller I. Jankovics I. and Ostreicher R., 1984, *Astron. Astrophys.*, **141**, 108.

Aumann H.H., 1985, *Publ. Astro. Soc. of the Pacific.*, **97**, 885.

Baud B. Young E. Beichman C.A. Beintema D.A. Emerson J.P. Habing H.J. Harris S. Jennings R.E. Marsden P.L. and Wesselius P.R., 1984, *Astrophys. J.*, **278**, L53.

Beckwith S. Zuckerman B. Skrutskie M.F. and Dyck H.M., 1984, *Astrophys. J.*, **287**, 793.

Beichman C.A. Myers P.C. Emerson J.P. Harris S. Mathieu R. Benson P.J. and Jennings R.E., 1986, *Astrophys. J.*, **307**, 337.

Beichman C.A., 1987, *Ann. Rev. Astron. Astrophys.*, in press.

Clark F.O. Laureijs R.J. Chlewicki G. Zhang C.Y. van Osterom W. and Kester D., 1986, *Astron. Astrophys.*, **168**, L1.

Cohen M. Schwartz R.D., 1987, *Astrophys. J.*, **316**, 311.

Cohen M. and Kuhi L.V., 1979, *Astrophys. J. Suppl.*, **161**, 85.

Cohen M. Emerson J.P. and Beichman, 1988, in preparation.

Edwards S. Strom S.E. Snell R.L. Jarrett T.H. Beichman C.A. and Strom K.M., 1986, *Astrophys. J.*, **307**, L65.

Edwards S. Cabrit S. Strom S.E. Heyer I. Strom K.M. and Anderson E., 1987, *Astrophys. J*, **321**, 473.

Emerson J.P., 1987, in *Star Forming Regions, IAU Symposium 115*, editors Peimbert M. and Jugaku J., D. Reidel, 19.

Emerson J.P. Harris S. Jennings R.E. Beichman C.A. Baud B. Beintema D.A. Marsden P.L. and Wesselius P.R., 1984, *Astrophys. J.*, **278**, L49.

Grasdalen G.L. Strom S.E. Strom K.M. Capps R.W. Thompson D. and Castelatz M., 1984, *Astrophys. J.*, **283**, L57.

Habing H.J. Olnon F.M. Chester T. Gillett F.C. Rowan-Robinson M. and Neugebauer G., 1985, *Astron. Astrophys.*, **152**, L1.

Hacking P. Neugebauer G. Emerson J. Beichman C. Chester T. Gillett F. Habing H. Helou G. Houck J. Olnon F. Rowan-Robinson M. Soifer B.T. and Walker D., 1985, *Publ. Astro. Soc. of the Pacific*, **97**, 616.

IRAS Explanatory Supplement, 1985, Joint IRAS Science Working Group, US Govt Printing Office.

IRAS Low Resolution Spectral Atlas, 1986, *Astron. Astrophys. Suppl.*, **65**, 1.

IRAS Point Source Catalog, 1985, Joint IRAS Science Working Group, US Govt Printing Office.

IRAS Small Scale Structure Catalog, 1986, Joint IRAS Science Working Group, US Govt Printing Office.

IRAS Serendipitous Survey, 1987, Joint IRAS Science Working Group, US Govt Printing Office.

Israel F.P. editor, 1986, *Light on Dark Matter (1st IRAS conference)*, D. Reidel.

Jennings R.E. Cameron D.H.M. Cudlip W. and Hirst C.J., 1987, *MNRAS*, **226**, 461.

Kenyon S.J. and Hartman L., 1987, Astrophys. J., in press.

Lawrence A. editor, 1988, *From Comets to Cosmology (3rd IRAS Conference)*, Springer-Verlag.

Margulis M. and Lada C.J., 1986, *Astrophys. J.*, **309**, L87.

Mozurkewich D. Schwartz P.R. and Smith H.A., 1986, *Astrophys. J.*, **311**, 371.

Mundy L.G. Wilking B.A. and Myers S.T., 1986, *Astrophys. J.*, **311**, L75.

Myers P.C., 1987, in *Star Forming Regions, IAU Symposium 115*, editors Peimbert M. and Jugaku J., D. Reidel, 33.

Myers P.C. and Benson P.J., 1983, *Astrophys.J.*, **266**, 309.

Myers P.C. Fuller G.A. Mathieu R.D. Beichman C.A. Benson P.J. Schild R.E. and Emerson J.P., 1987, *Astrophys. J.*, **319**, 340.

Myers P.C. Heyer M.H. Snell R.L. and Goldsmith P.F., 1988, *Astropys. J.*, submitted.

Myers P.C. Linke R.A.and Benson P.J., 1983, *Astrophys. J.*, **264**, 517.

Neugebauer et al, 1984, *Astrophys. J.*, **278**, L1.

Persson C.J. editor, 1987, *Star Formation in Galaxies (2nd IRAS Conference)*, NASA CP-2466.

Pottasch S.R. Baud B. Beintema D.A. Emerson J. Habing H.J. Harris S. Houck J. Jennings R. and Marsden P., 1984, *Astron. Astrophys.*, **138**, 10.

Rucinski S.M., 1985, *Astron. J.*, **90**, 2321.

Soifer B.T. Houck J.R. and Neugebauer G., 1987, *Ann. Rev. Astron. Astrophys.*, in press.

van der Veen W.E.C.J. and Habing H.J., 1988, *Astron. Astrophys.*, in press.

Walker C.K. Lada C.J. Young E.T. Maloney P.R. and Wilking B.A., 1986, *Astrophys. J.*, **309**, L47.

Wynn-Williams C.G., 1982, *Ann. Rev. Astron. Astrophys.*, **20**, 587.

Young E.T. Lada C.J. and Wilking B.A., 1986, *Astrophys. J.*, **304**, L45.

ULTRAVIOLET OBSERVATIONS

M.T.V.T. Lago
Universidade do Porto
Grupo de Matemática Aplicada
Rua das Taipas 135
4000 Porto, Portugal

ABSTRACT. The most relevant observational results obtained with IUE in the field of T Tauri stars will be summarized. Other topics to be discussed include the application of ultraviolet observations in the study of the structure of stellar atmospheres, the origin of the continuum and molecular emission. Furthermore they provide evidence for the presence of strong stellar winds and variability. Finally a short account of the UV observational constraints to models of T Tauri stars will be presented.

1. INSTRUMENTAL CHARACTERISTICS AND THE UV DATA

By ultraviolet observations we will refer to IUE observations, the only operational ultraviolet facility at the moment.

The International Ultraviolet Explorer (IUE), a real time observatory resulting from international collaboration (NASA, ESA, SERC) was launched in 78 January 26th. Its expected lifetime (3 years) has by far been exceeded and it is still operating.
We will briefly summarize the instrumental characteristics since they are relevant for the description and understanding of the ultraviolet observations to be presented in the following sections. For a detailed description of the satellite and instrumentation see [1] and [2].

1.1. The telescope

The telescope, a Ritchey-Chrétien of 45 cm primary mirror, has a focal ratio F/15, a field of view of 16' and an image size 1" (center) to 2" (edge).
Behind the mirror (and the sun shutter which closes automatically if sunlight falls inside the telescope tube) a flat mirror mounted at 45° with the telescope tube has the dual purpose of
- reflecting the telescope field into a duplicate Fine Error Sensor (FES)

A. K. Dupree and M. T. V. T. Lago (eds.), Formation and Evolution of Low Mass Stars, 209–223.
© 1988 by Kluwer Academic Publishers.

- holding the spectrographs entrance apertures - holes drilled in
the mirror itself.

1.2. The Fine Error Sensor

The FES, limited to magnitude $V \approx 14$, operate in a dual mode
- the field camera mode for scanning the telescope field (8" steps)
an image of which (after being digitalized) is transmitted to the
ground for identification by the observer in real time
- the acquisition and tracking mode for measuring the coordinates of
the target star relative to the selected entrance aperture so that
the spacecraft may be positioned to center the star image in that
aperture; the FES is then offset to acquire a guide star for
accurate tracking ($\approx 0.3"$) during the exposure.

1.3. The Spectrographs

The two spectrographs are "echelle" type (therefore giving a compact
spectral format),
- the short wavelength (SW) covering 1180 - 2050 Å
- the long wavelength (LW) for the 1860 - 3250 Å region.

Since there is essentially no sky background in the UV range apart
from the geocorona Ly α emission line, there is no need for a slit
spectrograph. Instead, each spectrograph has two apertures which are
holes drilled in the mirror as previously referred,
- the large aperture 10" x 20" (can be closed by a shutter)
- the small aperture 3" in diameter, always open.

There are also two possible dispersion modes
- low resolution (1 single order),
$$\lambda / \Delta\lambda \approx 300 \quad \leftrightarrow \quad \Delta\lambda \approx 5 \text{ to } 6 \text{ Å},$$
- high resolution(≈ 50 orders of about 20 Å range),
$$\lambda / \Delta\lambda \approx 1 \text{ to } 1.5 \ 10^4 \quad \leftrightarrow \quad \Delta\lambda \approx 0.1 \text{ to } 0.2 \text{ Å}.$$

Because there is no seeing in space the resolution of the slitless
spectrograph is limited either by
- the pointing stability of the spacecraft (in the case of IUE the 3
axis stabilized satellite gives $\approx 1"$), or
- the point spread function, for IUE between 4" and 5", therefore
the limiting factor for the spectral resolution in the IUE data [3].

1.4. Target acquisition

The sequence of operations, a standard procedure briefly described
for reference (and the typical duration) consists of
- slew operation - to move to the next target;
the time taken by this operation depends of
course on the relative position of the two
consecutive targets, the slew velocity being
$\approx 4.5°$/minute,

≈ 5 minutes	– acquisition of the field image,
few minutes	– field and target identification,
	– acquisition of the target and its location on the aimed spectrograph aperture,
0.5 s to ≈ 20 h *	– exposure (and simultaneous acquisition of a guide star),
≈ 20 minutes	– once the exposure is finished, read out and prepare the next exposure or move to the next target and do (simultaneously) the camera preparation (i.e., uniform illumination of the camera by tungsten flood lamps in order to remove "ghost" images from previous exposures).

* the length of the exposure depends on various factors, namely,
    – the UV brightness of the target,
    – the aimed observations (strong lines, weak lines or continuum)
    – the dispersion mode used.

For T Tauri stars typical exposures range from
        few minutes for a 10th magnitude star in low resolution, to
        16 hours for the high resolution short wavelength spectrum
of that same star.
However, since these are real time observations, adjustments can be made by the observer after a preliminary analysis of the spectrum meanwhile received and displayed.

1.5 The data

The standard IUE Guest Observer tape or a tape requested from the Data Bank contains several files with the raw data, the photometric corrected image, the line by line spectrum and the flux calibrated spectrum (through the standard IUE Spectral Image Processing System – IUESIPS).
The guest observer also receives the photowrite representation of the photometric image, very useful namely in the identification of image defects.
Alternatively, one may wish to perform a non-standard data processing from the line by line spectrum. The necessary absolute flux calibration is available from the literature (for low resolution [4] and for high resolution/low resolution [5]).

2.   THE UV SPECTRA OF T TAURI STARS

2.1. General Statistics

Around 500 spectra of 90 different T Tauri stars have been obtained so far with IUE; initially, due to the magnitude limitations referred in the previous section only the brightest stars and the ones with stronger (optical) emission lines were observed and mainly in the low resolution mode.

Later on observers adventured into fainter or weaker lines stars and into high resolution spectra; however, the need for relatively long exposures due to the faintness of T Tauri stars combined with the difficulty in obtaining enough observing runs result in that only ≈ 10% of the available spectra is at high resolution. Moreover, only for 3 stars is there high resolution spectra in the short wavelength region. In fact this type of observations for T Tauri stars is at the limit of the IUE capability.

Although not all the collected data has good enough signal to noise ratio the set of spectra available at the IUE Data Bank allows interesting detailed comparative studies. There is however a strong need for carefully planned simultaneous observations at various wavelengths.

2.2. The observations

2.2.1. <u>Line identification</u>. The main interest of the UV data results from the fact that some of the strongest lines (mainly in the region 1200 to 2000 Å) arise from radiative decay from the first excited terms in ions of astrophysical interest such as, CII, CIII, CIV, SiII, SiIII, SiIV, NV. In these ions, the excited terms, 5 to 10 eV above the ground term, may be populated by collisions with electrons having energies corresponding to a Maxwellian distribution characteristic of T_e between 1.5 10^4 and 2 10^5 K. Therefore the presence of such high temperature regions may be inferred from the line identification.

TABLE I shows the most important lines usually present in the UV spectra of T Tauri stars, two of which are, for illustration, displayed in Figures 1a, 1b.

TABLE I

λ_{lab} (Å)	Identification	λ_{lab} (Å)	Identification
1239–1243	NV (1)	1808–1817	SiII (1)
1301–1306	OI (2), SiIII(4)	1859	AlIII (1)
1335–1336	CII (1)	1892	SiIII] (1)
1394–1403	SiIV (1)	1909	CIII] (0.01)
1548–1551	CIV (1)	2318–2350	CII, SiII
1657	CI(2)	2600–2630	FeII (1)
1697–1727	FeII (38)	2795–2803	MgII (1)
1663	OIII]	2930–2980	FeII (60)

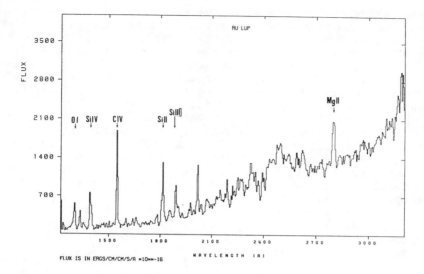

Figure 1a. IUE low dispersion spectrum of the T Tauri star RU Lupi. The spectrum is a combination of short and long wavelength spectra.

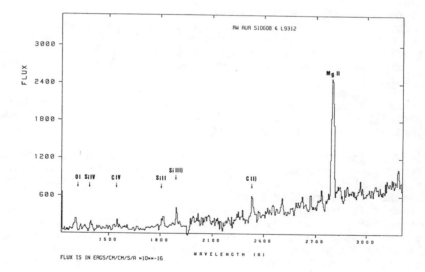

Figure 1b. IUE low dispersion spectrum of the T Tauri star RW Aur. The spectrum is a combination of short and long wavelength spectra. The position of the main emission lines is given. Note the weakness of the emission features in RW Aur when compared with RU Lupi. In the visual both stars are classified as emission intensity class 5, [7].

From the analysis of a sample of several low resolution spectra reduced in a homogeneous way one may hope to be able to single out common features characteristic of the class; the inspection of a sample of 20 T Tauri stars led to the following conclusions [6]:
    - the strongest emission lines are OI, CII, SiII, CIII, SiIII, CIV, SiIV, FeII and MgII although showing a wide range of intensity for different stars; the MgII lines are certainly the dominant feature in the spectra of all bona fide T Tauri stars; approximately 50% of the stars also show FeII(1) in emission;
    - NV is present (very weak) only in a few stars (in RU Lupi, CoD -34 7151, CoD -35 10525 and V410 Tau in the sample inspected); for the majority of the stars CIV is the highest excitation line, showing however a wide range of intensities, from a strong feature (e.g. in the first 3 stars just mentioned) to a bare detection (in the last one);
    - there seems to be no simple correlation between the intensity of the UV lines and the activity in the visual as measured by the emission intensity class [7]; for example, in the spectra of stars classified as emission class 1 there are quite strong lines of CIV and SiIV (e.g. AK Sco, SU Aur, RY Lupi) while CIV is very weak in RW Aur (emission class 5) and SiIV not detectable in AS 209 (class 4). The same is true for other lines;
    - the H_2 molecular emission, first reported in T Tau itself [8], is believed to arise from fluorescence with Ly α; these molecular lines previously observed in the UV spectrum of a sunspot, seem to be consistent with a model where the H_2 molecules are excited by collisions rather than by radiative processes; the emission probably does not come from the stellar atmosphere but from an extended region where the temperature is $\approx$ 2000 K [8]. However the H_2 emission does not seem to be a common feature in the spectra of T Tauri stars.

The general scenario just described for the T Tauri stars does not substantially change when a larger number of spectra are analysed or the sample of stars enlarged (30 stars, [9]).

2.2.2. The line fluxes. Besides the evidence for the presence of high temperature regions the UV observations are also important for the study of the structure of the atmosphere in stars through the measurement of the stellar line fluxes.
The observed fluxes can be converted into surface fluxes, F*, after correction for the distance to the star and the interstellar absorption. This may allow a direct comparison between the stellar and the solar fluxes as for example displayed in Figure 2 (Ref.10). Considering in the plots only the data points corresponding to the UV observations ($T_e \leq 2.0\ 10^5$ K) the surface fluxes from the chromospheres and transition regions in the T Tauri stars are clearly seen to be strongly enhanced relatively to the Sun. Furthermore, there are also obvious differences in the distribution of material with temperature between the three T Tauri stars.

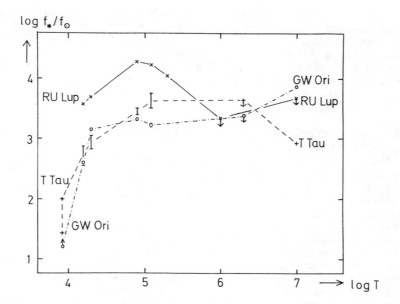

Figure 2. Ratio of stellar to solar fluxes for 3 T Tauri stars (RU Lupi, GW Ori and T Tau) as a function of temperature. The plots allow a detailed comparison with the Sun and between the stars themselves. Extra data points (relative to the coronal lines [FeX] and [Fe XIV] and to X-Ray) included in the plot are not relevant for the present discussion.

Assuming that the UV lines are formed under coronal equilibrium conditions i.e. in a region where the equilibrium is maintained by the balance between collisional ionization and excitation against effectively thin radiative decay, the emission will come from a narrow range of temperatures [11].
The surface flux will be given by

$$F^* = 5.50 \ 10^{-16} \ A \ g_{eff} \ f \ <G(T)> \int N_e \ N_H \ dh$$

where A is the abundance relative to hydrogen, g_{eff} is the effective Gaunt factor, f is the oscillator strength of the line and

$$<G(T)> = 0.7 \ G_{max} \ (T_{max})$$

is a mean value for G(T), the temperature dependent function given by

$$G(T) = T^{-1/2} \ 10^{-5040W/T} \ N_i/N_{element} \ .$$

$G_{max} \ (T_{max})$ represents the maximum value of G(T) occurring at temperature T_{max} , W is the excitation energy of the line in electron volts and $N_i/N_{element}$ the fraction of the element in ionization state i.

From the flux F* in each line formed under such conditions the
emission measure associate to that narrow temperature range, i.e.
the amount of material $\int N_e\, N_H\, dh$ required to produce all the flux
in the emission line, can be calculated.
For lines formed at $T \geq 2\ 10^4$ K we can assume the H to be full
ionized and the emission measure is

$$\int N_e^2\ dh\ .$$

Considering various lines with different values for G(T) it is
possible to obtain the distribution of emission measure with
temperature and therefore to identify the temperature ranges over
which the lines are formed ([11], [12], [13]). The atomic data
needed to perform such calculations are given in [12], [14] and [15]
and in references therein for some of the most recent calculations.
Solar abundances are usualy used.
Reference [15] also provides a very clear discussion on the use of
the emission measure diagnostic.
If N_e is known or can be calculated one could further estimate the
depth of the source region and the optical depth for each line. Such
models have been developed, for example for the star T Tau [13] but
hydrostatic equilibrium has been assumed; however this hypothesis is
certainly not supported by the observations of T Tauri stars in the
UV or other wavelengths.
Some UV lines are particularly suited for density diagnostic, namely
SiIII] λ 1892 and CIII] λ 1909 ([16], [17]); in the case of high
resolution data the electron temperature and density can also be
inferred from other line ratios such as CII] λ 2325 [18], CIII]
λ 1909 [19] and SiIII] λ 1892 [20].

2.2.3. <u>The continuum</u>. The short wavelength continuum in the spectra
of T Tauri stars is explained by optically thin hydrogenic free-free
and free-bound emission at temperatures between 10^4 and $5\ 10^4$ K
while in the long wavelength region the stellar flux is dominant
([6] and [9]).
In principle by defining in each stellar spectrum enough continuum
windows it would be possible, by fitting, to determine the blackbody temperature, the extinction and the hydrogenic emission temperature producing the observed continuum [9]; however, in the spectra
of T Tauri stars some difficulties arise because continuum windows
free from emission are hard to find on the UV spectra and consequently its number is quite small. In particular in the long wavelength region the large number of FeII emission lines may even
produce a "false continuum" as in QSO spectra [21].
In general there are no absorption lines present in the short
wavelength region since the (cool) stellar photospheric radiation is
too weak, while at $\lambda > 2000$ Å those lines are present but blended due
to the instrumental resolution.

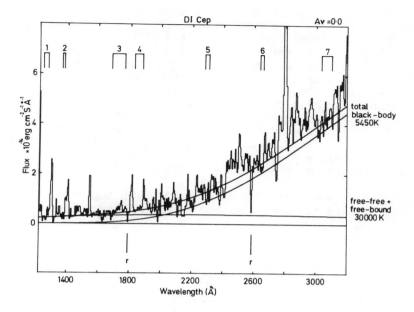

Figure 3. Low dispersion spectra of DI Cep. The spectral windows used to fit the continuum are numbered 1 to 7. The plot also displays the best fitting to the continuum: the sum of hydrogenic free-free plus free-bound emission at $3 \cdot 10^4$ K and a black body with temperature 5450 K [6].

2.2.4. <u>The line profiles</u>. Due to the instrumental constraints mentioned in the first section only with high resolution IUE data is the study of line profiles possible. Most of the high resolution observations available in the IUE Data Bank (for ≈ 15 stars) are long wavelength spectra since the strength of the MgII emission lines in T Tauri stars allows a well exposed spectrum in 1 to 2 hours. The Mg II lines are very broad, strongly asymmetric and in several stars, besides the presence of a central (narrow) interstellar absorption the lines also display a strong blue displaced absorption feature that may be attributed to a cool ($T < 5 \cdot 10^3$ K) circumstellar shell; however, the width (> 100 Km s^{-1}) and location of this feature varies from star to star [22]. The ratio of the Mg II lines is typically close to 1 thus suggesting lines optically very thick. And, as will be pointed in the next section, there is observational evidence for variability in the MgII lines at least for some stars.

It is also interesting to recall the case of RU Lupi where besides the MgII lines also the lines of FeII UV1 multiplet show classical P Cygni profiles clearly indicative of the presence of a strong stellar wind [6].

On the other hand, short wavelength high resolution spectra are very
scarce; since these observations are, for T Tauri stars, at the
limit of the IUE capabilities, of the 3 stars observed so far only
two (RU Lupi and TW Hya) have useful data. These spectra, parts of
which are displayed in Figure 4, indicate the presence of large
non-thermal motions also in the higher temperature regions. However
the higher excitation lines (SiIV, CIV, SiIII, SiII) are
considerably narrower than the chromospheric lines. These
observations give support to earlier predictions. In fact, the first
model of Alfvén driven winds proposed for T Tauri stars was
developed for the star RU Lupi based on optical data [23]; it
predicted the higher temperature lines to be narrower (FWHM = 170km
s^{-1}) than the chromospheric lines (FWHM $\approx$ 240 km s^{-1}). This agrees
well with the ultraviolet observations.
Thus the high resolution IUE data can give an important contribution
for constraining the models of the wind in T Tauri stars.

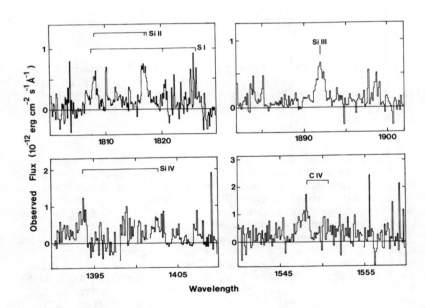

Figure 4. High resolution profiles of several lines in RU Lupi.The
spectral resolution referred in section 1 yield an instrumental
profile of FWHM $_{IUE}$ $\approx$ 25 Km s^{-1} [24] therefore much narrower than
the observed lines.

3. VARIABILITY

T Tauri stars are known to be (irregular) variables at almost all
wavelengths. The UV is no exception. The first reports of IUE data
variability refer to the star RW Aur and mention variations in the
continuum and (only) in the high ionization lines; the Mg II lines
are also said not to have changed in flux by more than 10% over a
week [25]. Much stronger variations were detected in another star
(LHα 332-21) with changes up to 60% in the Mg II lines flux when
observed 6 months apart although no variability was recorded for the
same lines in time scales of ≈ 2 1/2 hours [22].
However in none of these cases did the results involve repeated
observations at more than two different epochs, even considering the
obvious importance for the modelling of T Tauri stars of studying
the variability in the UV and the unique opportunity provided by IUE
for simultaneous sampling of the chromosphere and transition region.
More recently a sequence of observations of another star (RU Lupi)
has been carried out. Although the data is not yet fully analysed
some of the preliminary results look interesting enough to be
referred in Figures 5 to 7, [26].

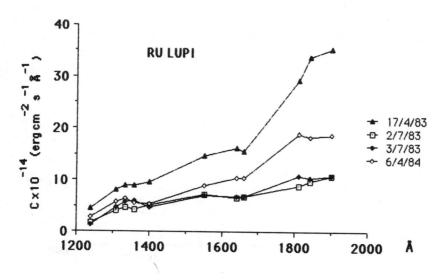

Figure 5. There is considerable variation of the intensity in the
continuum in a time scale of few months to one year; the variations
are larger at longer wavelengths therefore probably photospheric in
origin; there seems to be no variability on a time scale of one day.

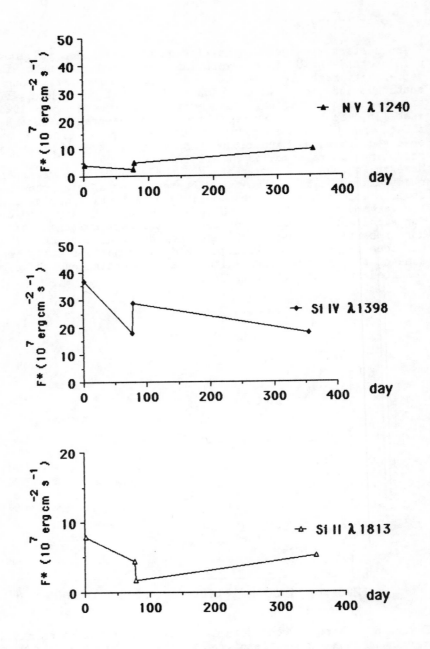

Figure 6. The plot display the stellar flux emitted in several UV lines versus the day of the observation (day zero corresponds to 17th April 83).

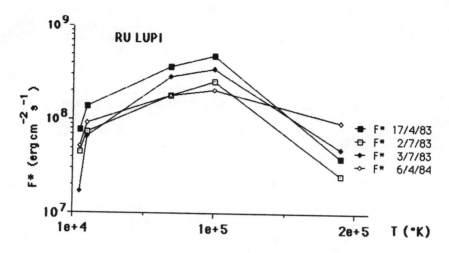

Figure 7. The points refer to stellar flux emitted in several UV lines versus the corresponding excitation temperatures; (the lines have been drawn just to help the eye); <u>the variations</u> (up to a factor of ≈ 4) <u>seem to affect all UV lines but do not seem to have a unique origin</u> in the sense that at some times all line fluxes change by approximately the same factor (for example between days 17/4 and 2/7 shown in the plot) whereas at other times, even in a time scale of one day, there seems to be no general pattern - some lines get stronger while others become weaker.

Of course these results are yet very preliminary and any conclusions must await the analysis of the (much larger) full data sample, [26].

4. CONCLUSIONS - THE IMPACT OF UV DATA ON MODELS OF T TAURI STARS

Although the ideal would be to have complete wavelength simultaneous observations, the UV data is still very critical for the modelling of T Tauri stars, as referred in previous sections. The fact is that IUE provides an unique opportunity for simultaneous sampling of the chromosphere and transition region in the stars. Therefore models are more seriously constrained in view of the evidence for:

- the existence of a temperature structure in the atmospheres of the T Tauri stars,
- the need for direct heating of the $6 \cdot 10^4$ to 10^5 K temperature region,
- outflows and strong winds even in the high temperature region,
- the presence of a large velocity gradient,
- a source of variability, most probably not unique,

- the existence of one (or more) parameters responsible for the stellar activity, much higher in these stars than in the Sun, and for a different structure in the atmosphere of T Tauri stars. These parameters might also be responsible for the differences between the various T Tauri stars.

These constraints immediately rule out hydrostatic models. Alternatively the presence of magnetic fields seem to be widely accepted while their origin and strength might still be under discussion. Still a wide range of possibilities is left open and the UV observations undoubtedly constitute a real challenge to them.

5.   REFERENCES.

[1]   Boggess, A. et al., 1978a, Nature, 275, 372.
[2]   Boggess, A. et al., 1978a, Nature, 275, 377.
[3]   Cassatella, A., Barbero, J. & Bienvenuti, P., 1985, Astr. & Astrophys., 144, 335.
[4]   Bohlin, R.C. & Holm, A.V., 1980, IUE NASA Newsletter, 10, 37.
[5]   Cassatella, A., Ponz, D. & Selvelli, P.L., 1981, IUE ESA Newsletter, 10, 31.
[6]   Lago, M.T.V.T., Penston, M.V, & Johnstone, R., 1984, Proceedings of 4th European IUE Conference, ESA SP-218, 233.
[7]   Herbig, G.H. & Rao, N.K., 1972, Astrophys. J., 174, 401.
[8]   Brown, A., Jordan, C., Millar, T.J., Gondhalekar, P. & Wilson, R., 1981, Nature, 290, 34.
[9]   Johnstone, R.M., 1985, D. Phil. thesis, University of Sussex.
[10]   Lago, M.T.V.T., Penston, M.V & Johnstone, R.M., 1985, Mon. Not. R. astr. Soc., 212, 151.
[11]   Pottasch, S.R., 1963, Astrophys. J., 137, 945.
[12]   Pottasch, S.R., 1964, Space Science Review, 3, 816.
[13]   Brown, A., Ferraz, M.C. de M. & Jordan, C., 1984, Mon. Not. R. astr. Soc., 207, 831.
[14]   Burton, W.M., Jordan, C., Ridgeley, A. & Wilson, R., 1971, Phil. Trans. R. Soc. London A., 270, 81.
[15]   Judge, P.G., 1986, Mon. Not. R. astr. Soc., 221, 119.
[16]   Doscheck, G.A., Feldman, U., Mariska, J.T. & Linsky, J. L., 1978, Astrophys. J., 226, L35.
[17]   Keenan, F.P., Dufton, P.L. & Kingston, A.E., 1987, Mon. Not. R. astr. Soc., 225, 859.
[18]   Lennon, D.J., Dufton, P.L., Hibbert, A. & Kingston, A.E., 1983, Phys. Rev., A27, 3040.
[19]   Nussbaumer, H. & Storey, P.J., 1982, Astr. & Astrophys., 115, 205.
[20]   Nussbaumer, H., 1986, Astr. & Astrophys., 155, 205.
[21]   Wills, B.J., Netzer, H & Wills, D., 1985, Astrophys. J., 288, 94.

[22]   Penston, M.V. & Lago, M.T.V.T., 1983, <u>Mon. Not. R. astr.</u>
         <u>Soc.</u>, **202**, 77.
[23]   Lago, M.T.V.T., 1979, <u>D. Phil. thesis</u>, University of
         Sussex.
[24]   Imhoff, C.L., 1983, <u>Record of the IUE 3- Agency Meeting</u>,
         pA-140, NASA, ESA, SERC.
[25]   Imhoff, C.L. & Giampapa, M.S., 1981, <u>The Universe in</u>
         <u>Ultraviolet Wavelengths: the first two years of IUE</u>, ed.
         R.D. Chapman, 185.
[26]   Lago, M.T.V.T. & Johnstone, R.M., in preparation.

X-RAYS, RADIO EMISSION, AND MAGNETISM
IN LOW-MASS YOUNG STARS

Thierry Montmerle and Philippe André

Service d'Astrophysique
Centre d'Etudes Nucléaires de Saclay
91191 Gif sur Yvette Cedex, France

ABSTRACT

In recent years, the body of X-ray and radio data on young, low-mass stars has grown steadily, thanks mainly to the "Einstein" X-ray satellite observations and archives, and to the Very Large Array radio interferometer.
The X-rays come mostly, if not predominantly, from giant, solar-like flares. The associated hot plasma is trapped in magnetic loops of sizes on the order of the stellar radius at most. The radio observations are more difficult to interpret: the detection rate is low, and the spectra are diverse. While there are definite cases of accretion, winds, or jets, it seems that the radiation is most frequently magnetic ("non-thermal") in character. Several radio flares have been detected in late-type PMS stars, and a very young B star with a dipolar field has been discovered. The associated closed magnetic structures are very large, extending up to more than 10 stellar radii. In addition, optical and/or IR data on the X-ray emitting, as well as the non-thermal radio emitting, young stars indicate that little or no circumstellar material is present around them ("naked" PMS stars).
We point out that the radio-emitting naked stars make up a new population of PMS stars, characterized by extended closed magnetic structures, and we discuss their relation with the more familiar populations of naked and classical T Tauri stars.

1. INTRODUCTION

Initially known as the "objects of Joy" after their discovery in the late forties, the T Tauri stars have long been considered as the prototypes of low-mass pre-main sequence stars. Indeed, their spectra are generally of late type, from K to M, they are overluminous ; but what made them conspicuous was the presence of frequently intense emission lines, notably Hα, in their optical spectra. The youth of these objects is attested mainly by their proximity to dark clouds and the presence of strong lithium absorption lines (lithium is subsequently

225

A. K. Dupree and M. T. V. T. Lago (eds.), Formation and Evolution of Low Mass Stars, 225–246.
© 1988 by Kluwer Academic Publishers.

destroyed in the course of stellar evolution; in the Sun, only ~ 1% of the original amount remains).

In recent years, however, the opening of new astronomical windows at IR and radio wavelengths from the ground, and in X-rays, UV and mid-IR from space, has allowed to broaden the knowledge of these objects and of pre-main sequence (PMS) stars in general to the point that we now consider the "classical" T Tauri stars (TTS), i.e., those found by optical means, as making up only a <u>minority</u> (perhaps as low as 10%) of the more general class of Young Stellar Objects (YSO).

Indeed, the IR domain has revealed deeply embedded, or dusty, optically invisible objects, of which only a fraction would be classical TTS if not shielded by obscuring material, and the X-rays have shown the existence of a new population of PMS stars, dubbed "naked" TTS because of their lack of emission lines, i.e., of dense surrounding circumstellar material.

Various aspects of classical TTS and embedded YSO are described by several authors elsewhere in this volume. Here, we will review the X-ray data (for more details, see Feigelson, 1984, 1987a), and present very recent radio results which have allowed us to isolate yet another population of low-mass PMS stars, which will turn out to be closely connected to the X-ray emitting "naked" TTS.

2. X-RAYS FROM PMS STARS

2.1. Pre-"Einstein" Predictions

Contrary to the situation holding for massive stars, for which the detection of soft X-rays with the "Einstein" satellite came as a surprise, (e.g., Rosner, Golub, and Vaiana, 1985), several predictions were made for the X-ray emission of low-mass PMS stars, mostly based on the enhanced activity one could infer from optical data, and basically linked with the existence of surface convective zones (see, e.g., Montmerle 1987).

Thermal X-rays were predicted to be associated with hot (~ 10^6 K) solar-like coronae (Bisnovatyi-Kogan and Lamzin, 1977), or from scaling with mass loss (Gahm et al., 1979, Bertout 1982), or yet associated with shocks, in the case of YY Ori stars (which display evidence for both mass-loss and accretion) and Herbig-Haro objects (Schwartz 1978). In general, such predictions led to excessive luminosities. In one case were the predictions about right (~ 10^{31} erg.s^{-1}), but they rested on a very special application of "transition radiation", a non-thermal mechanism resulting from the crossing of solids (grains) by energetic particles (Gurzadyan 1973 ; see also Yodh, Artru, and Ramaty 1973).

2.2. The "Einstein" era: Single Images

The "Einstein" satellite (1978-1981) was the first non-solar X-ray satellite to carry imaging detectors (Giacconi et al., 1981). The most widely used instrument has been the IPC (Imaging Proportional Counter), sensitive from ~ 0.2 to ~ 4 keV, and having a large field of view of $1°x1°$. Altogether, more than 10^4 IPC images have been taken during the satellite's lifetime.

Motivated by the early predictions, about 40 classical TTS were observed in single IPC pointings in the Taurus-Auriga complex, of which about 1/3 were detected (Gahm 1980, Feigelson and DeCampli 1981, Walter and Kuhi 1981, 1984). An additional 5 were found "serendipitously", i.e., fell by chance in the IPC fields, but remarkably, were found optically to be PMS stars after X-ray detection, a situation which is now more the rule than the exception (see below). Most X-ray emitting classical TTS had comparatively weak emission lines (for instance, with equivalent widths of Hα ⩽ a few 10 Å, as compared to $\sim$ 100-300 Å the extreme TTS [e.g. Cohen and Kuhi (1979)] ; the new PMS stars were of late-type (K5-M0), and had faint or absent emission lines. lines.

Searches for serendipitous X-ray emitting PMS stars are systematically conducted in large areas covering various star-forming complexes. From the examination of IPC fields from the "Einstein" archives in the Taurus-Auriga, Ophiuchus, and Corona Australis regions, Walter (1987) reports the discovery of a total of 69 X-ray) detected PMS stars, as compared to a total of 64 known classical TTS. This search is now being extended to a $\sim$ 360 square degree area, covering the whole Taurus-Auriga-Perseus complexes. In an examination of 75 IPC fields, Feigelson et al. (1988) have selected 59 serendipitous X-ray sources, 30 of which are considered as PMS candidates, to be confirmed by subsequent ground optical observations. A few additional X-ray emitting PMS stars have been found in the Chameleon, NGC 2264, λ^1Ori and L1457 region (see Feigelson 1987a).

In some cases, mosaics (i.e., adjacent, or almost adjacent, images of the same region), exist, and make-up "wide-angle" views of regions of star formation. This is the case in Orion, where Gahm (1987), and Caillault and Zoonematkermani (1987) have extended the early results (Ku and Chanan 1979, Ku, Righini-Cohen, and Simon, 1982) and found many new PMS X-ray sources ; this is also the case in the ρ Oph cloud (Montmerle et al., 1983, see below).

Altogether, about 200 X-ray emitting PMS stars are known to date, most of them of late-type, when obscuration does not prevent the spectral type to be determined. The early trend has been confirmed and the majority of the optically visible ones (but not all) have weak or absent emission lines. This has prompted Walter (1986, 1987) to define a new class of PMS stars, which he called the "naked" TTS, i.e., TTS without the active circumstellar material seen in most classical TTS, and responsible for their strong emission lines. Altogether, these "new" PMS stars turn out to be about <u>twice</u> as numerous as the classical TTS.

2.3. Time-dependent Observations

Since a raw IPC image contains photons labelled in time and energy, it is possible in particular to reconstitute the "history" of an IPC exposure (lasting typically 3000 to 5000 s), and, more precisely, of a given pixel area surrounding an interesting source.

In this way, within one observation, a rapid (factor of 5 in minutes) flarelike event was found in the very strong TTS DG Tau (Feigelson and DeCampli, 1981) ; a much slower event (factor 7 in hours) was observed in a long exposure of AS 205, a strong TTS (Walter and Kuhi, 1984). Both stars reached a peak luminosity of a few 10^{31} erg.s^{-1}. An even more powerful event has been observed in the course of <u>repeated</u> observations of the ρ Oph cloud (see below), with a peak luminosity of $\sim 10^{32}$ erg.s^{-1} and a factor of ~ 20 variability in a few hours, in the newly found X-ray source ROX-20.

In fact, when available (in Orion, e.g. Gahm 1987 ; Caillault and Zoonematkermani, 1987) and especially in ρ Oph, the IPC images available at different epochs show that <u>strong variability</u> is widespread in X-ray emitting PMS stars, making star-forming regions look like "Christmas trees" in X-rays. The most straightforward interpretation, quantitatively supported in the case of ρ Oph (see below), is that one is dealing with very strong <u>flares,</u> $\sim 10^3$-10^4 times more powerful than solar flares. From the available data, Gahm (1987) has found a $\approx 5\%$ duty cycle for the clear X-ray flares, i.e., when the intensity contrast with respect to an average value is $\gtrsim 2$.

2.4. Specific Properties of X-ray Emitting PMS Stars

Nowhere, perhaps, are the above properties of X-ray emitting PMS stars better illustrated than in the ρ Oph cloud observations of Montmerle et al. (1983 ; MKFG). The observations are still unique in that this nearby (d ~ 160 pc) star-forming region has been mapped over $\sim 2°\times2°$ in a systematic fashion at several epochs, without reference to a specific kind of X-ray source, allowing a description in time, space, and energy, of the emission characteristics of the whole region.

Briefly summarized, the results are as follows. About 50 sources were discovered in the $2°\times2°$ area investigated, the majority of which (34) lie in the central, dense core region of the cloud ($\sim 1°\times1°$; see Loren, Sandqvist, and Wootten 1983). Stellar identifications were made on the basis of coincidences with objects appearing on the Palomar Sky Survey red prints ($m_R \lesssim 18$-19), or of known optical spectra, and with IR sources and a few radio sources (see § 3). The extinction is found to be small to moderate (typical $A_v \sim 0.1$ to 5, i.e., $N_H \sim 10^{20}$-10^{22} cm^{-2}, yielding luminosities $L_x \sim 10^{30}$-10^{31} erg.s^{-1}, i.e., $\sim 10^3$-10^4 $L_{x,\odot}$.

The "ROX" (= Rho Oph X-ray) sources are associated with almost all spectral types, from early B to mid-M. Most counterparts are low-mass PMS stars, as suggested by MKFG and subsequently confirmed by optical studies of a sample of ROX sources (Bouvier and Appenzeller 1987) ; 9 of the 11 classical TTS known to be associated with the cloud have been seen at least once.

The ROX sources not associated with classical TTS stars share, in general, the property of having weak emission lines [EW(Hα) ~ few Å] (MKFG; Bouvier and Appenzeller 1987), also displayed on a more extended spatial scale by the X-ray emitting "naked" TTS of Taurus-Auriga (Walter 1987). By contrast, the classical TTS (including in ρ Oph), with stronger emission lines, show an overall tendency to be less X-ray luminous, when detected. This is illustrated in Fig. 1, after Bouvier (1987), and was already apparent in MKFG and in Walter and Kuhi (1981). This figure plots the X-ray surface flux vs. the Hα surface flux for naked TTS, ROX sources, and classical TTS. The Hα flux (or the Hα emission line profile) is in general very difficult to interpret quantitatively (see, e.g., DeCampli 1981 ; Hartmann, Edwards, and Avrett, 1982), but large values, as found in classical TTS, can be associated with dense circumstellar material (e.g., strong winds (see Felli et al., 1982)). The trend for a decrease of the X-ray flux with increasing Hα flux could thus be due, at least in part, to an absorption effect, if the X-rays are emitted close to the stellar surface. However, this interpretation is complicated by the X-ray variability, i.e., variations in the intrinsic emission: for instance, whereas in single IPC images, about 30% of the classical TTS are detected, this fraction rises to ~ 80% in the course of repeated observations, like in the ρ Oph cloud, as mentioned above.

At any rate, this situation contrasts very strongly with the steeply rising Hα-X-ray correlation, also shown on Fig. 1, for other active late-type stars, either dwarfs on the main sequence, or at later stages (in RS CVn binary systems). For these stars, the activity is solar in nature (e.g., Montmerle 1987, and refs. therein), and the Hα emission is essentially of chromospheric origin.

In Fig. 1, the transition between the PMS "region" and the main-sequence and post main-sequence correlation occurs for $F(H\alpha) \simeq 10^6$ erg. $cm^{-2}.s^{-1}$, i.e., an Hα emission line equivalent width EW(Hα) ~ 5 Å. This is also the naked TTS region, which shows that, for these stars, Hα is very likely chromospheric, whereas for the classical TTS having EW(Hα) ≫ 5 Å, Hα is very likely associated with dense ionized circumstellar material.

2.5. Links with Magnetic Structures

What neither the classical, nor the naked TTS surveys can tell us is the physical phenomenon giving rise to the X-rays, because they are made of images taken at a single epoch in most cases. But this is possible by using the ρ Oph results. Indeed, as shown by MKFG and

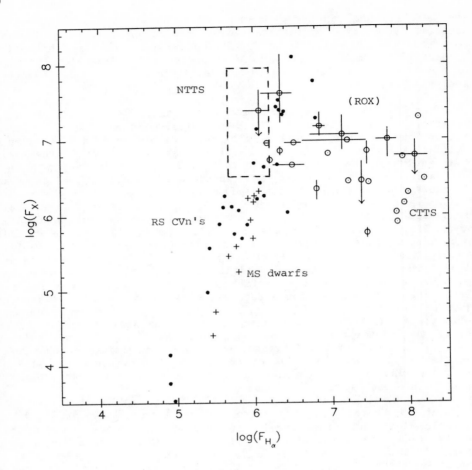

Fig. 1. Flux in the Hα line and in X-rays at the stellar surface (in erg. cm⁻².s⁻1), for main-sequence F, G, K, and M dwarfs (crosses), for RS CVn close binaries (filled circles), and for classical T Tauri stars (open circles), including some Rho Oph X-ray sources. The area where most "naked" T Tauri stars lie is also sketched (after Bouvier 1987).

Montmerle et al. (1984), the variable X-rays are produced by the thermal bremsstrahlung of a hot plasma (T $\sim 10^7$ K), predominantly associated with enhanced solar-like flares, a possible corona being too weak to be detected.

The basically solar nature of the flare events is supported by the power-law flux distribution, the tendency for the temperature to increase with the luminosity, and, indirectly, by the presence of large starspots, covering $>$ 20% of the stellar surface, on several ROX sources (Bouvier 1987). The most compelling argument, however, lies in the derivation of the physical parameters based on the observed emission measures of X-ray flares on the Sun, dMe ("flare") stars, various ROX sources, and the well-observed "superflare" in source ROX-20 (see discussion in MKFG). For an observed range in luminosities of 10^6, the temperatures are essentially the same ($\sim$ 1-3x10^7 K), and the plasma densities vary by $\sim$ 1 order of magnitude only ; the main difference lies in the size of the emitting regions, varying by 3 orders of magnitude, up to $\sim 10^{11}$ cm ($\lesssim$ 1 R_*).

Incidentally, this lends support to the interpretation of large $H\alpha$ emission in terms of dense circumstellar material, absorbing surface X-rays. For the two strong TTS DG Tau and AS 205, no X-rays were detected apart from flares (Walter and Kuhi 1984), which is consistent with the idea that large flares are produced in extended regions, sufficiently far from the stellar surface not to be absorbed, whereas the cooler smaller flares or corona would remain buried in X-ray-opaque circumstellar gas (see also the discussion below, § 4.2).

On the Sun, the X-ray emitting regions corresponds to closed magnetic loops, roughly dipolar, with measured values of the magnetic field at their feet of up to a few hundred Gauss. This value corresponds to equipartition between magnetic and gas energy densities. Since the X-ray luminosity is linked essentially with the size of the emitting regions, the X-rays from ROX sources and other PMS stars or naked TTS in fact probe the sizes of closed magnetic structures, typically 10^{10}-10^{11} cm.

In summary, two important properties characterize the X-ray-emitting PMS stars:
(i) moderate to small amount of ionized circumstellar material, traced by $H\alpha$ emission (equivalent width $\lesssim$ a few 10 Å) for weak classical TTS ; very little, if any, such material in the case of the "naked" TTS ($H\alpha$ equivalent width $\lesssim$ 5 Å, likely of chromospheric origin);
(ii) probable existence of hectogauss surface magnetic fields, associated with closed loops of typical sizes $<$ 0.1 - 1R_*, confining a hot ($\sim 10^7$ K), short-lived plasma.

3. (NON-THERMAL) RADIO EMISSION FROM PMS STARS

3.1. Available Data

The advent of the "Very Large Array" interferometer (located in New Mexico, USA, see Heeschen 1981), because of its high sensitivity (~ 0.2 - 0.3 mJy on-axis for a few hours exposure), has allowed major advances in stellar radio astronomy (see, e.g., Hjellming and Gibson, 1985).

As for the X-rays, the body of radio data now available for YSO can be conveniently broken down in two categories:
(i) observations pointed at a specific class of objects: T Tauri stars and related objects (Bieging, Cohen, and Schwartz 1984, Cohen and Bieging 1986), IR sources associated with molecular outflows (Snell and Bally 1986) ;
(ii) surveys of star-forming regions: ρ Oph cloud (André, Montmerle, and Feigelson 1987, hereafter AMF), CrA cloud (Brown 1987), Orion Trapezium region (Garay et al., 1987; Garay 1987; Churchwell et al., 1987).

In the pointed observations of 44 TTS and related objects, most (33) of the stars were not detected, down to low flux levels (~0.3 mJy). Of the 11 stars detected, 9 showed evidence for thermal (free-free) emission: all these stars are known to be otherwise associated with independent evidence of mass outflows (jets, winds, Herbig-Haro objects etc.). This association between thermal stellar radio emission and independent evidence for dense circumstellar material, as seen from IR spectra, Hα emission, or other forms of outflow, has been confirmed by the surveys of Snell and Bally (1986).

This evidence has been thoroughly reviewed by Panagia (1988) (see also Bertout, 1987), as well as by Bertout elsewhere in this volume, and will not be discussed more extensively per se here. Rather, we shall concentrate on recent surveys of wider regions of star formation, which do not focus on a specific kind of YSO, and point to the exist- ence of a new class of young stars, emitting in the radio via non thermal mechanisms (for reviews on radio emission mechanisms, ways to distinguish thermal from non-thermal, and examples, see André 1987, and Feigelson 1987b).

3.2. The ρ Oph Radio Surveys

The prime motivation of these VLA surveys was to look in the microwave range for a "radio Christmas tree", i.e., for non-thermal radio counterparts to the highly variable flaring X-ray sources found in a ~ 2°x2° area covering the dense core of the cloud (MKFG ; above § 2), based on the solar examples. The expected goal was to find other sources associated with thermal emission mechanisms (winds, accretion flows, compact HII regions etc.; see Bertout in this volume, and André 1987), thereby increasing the early sample of a few sources found previously by Brown and Zuckerman (1975), and Falgarone and Gilmore (1981).

The method has been to survey the cloud by mosaics of neighboring fields (34 in all) in an identical manner, at 4 epochs over the period 1983-1985 (to look for variability), at several frequencies (mostly 1.4, 5 and 15 GHz, to obtain spectra) and in different array configurations (to make maps of the detected sources). (For details, see AMF and Stine et al., 1988.)

Altogether, and more generally, these surveys allowed to search for radio counterparts of over 100 YSO known at various wavelengths: X-rays, optical, near IR (from the ground) and far-IR (from the IRAS satellite).

More than 200 sources were detected down to ~ 1 mJy. While the large majority of them are extragalactic, it is possible to identify 9 reliable stellar candidates on the basis of their identification with a stellar object at several wavelengths (6 of those are new radio sources). The small detection rate of stellar sources and the fact that repeated surveys essentially do not reveal new sources, contrast strongly with the X-ray results ; there is no radio Christmas tree. However, 2 sources (the well-documented object DoAr 21, see below, and the X-ray source ROX-31) have been found flaring, and are always significantly variable when not flaring. By comparison, the variability of the other sources is small (same flux density to within 50% over several years, see Stine et al., 1988). The spectra are distinctly different, when connected with variability: they are rather flat for all sources at all times ($\alpha \leq 0.4$, where $S_\nu \propto \nu^\alpha$) except when a flare occurs ($\alpha \gtrsim 1$).

Irrespective of the above characteristics, all sources share common properties at other wavelengths. Six of the 9 reliable sources are seen in X-rays (MKFG) ; the 3 remaining ones suffer high extinction (up to $A_V \sim 50$), and may therefore be undetectable X-ray emitters. A more specific property appears in the optical and/or in the IR: all have little or no Hα emission (when visible and when the spectra are known) ; all have simple blackbody spectra, i.e., no IR excess (IR class III of Lada 1987). In particular, no classical TTS is found. In other words, none of the ρ Oph stellar radio sources displays a significant amount of circumstellar material. This is reminiscent of the "naked" TTS traced by the X-rays (§2).

In order to assess the radio emission mechanisms, let us now zoom on two well-studied and very different stellar radio and X-ray sources belonging to the ρ Oph cloud: DoAr 21 and S1.

3.3. DoAr 21

This star, otherwise known under the aliases of GSS 23 and E 14 in the IR, ROX-8 in X-rays, and FG 17 in previous radio observations (see

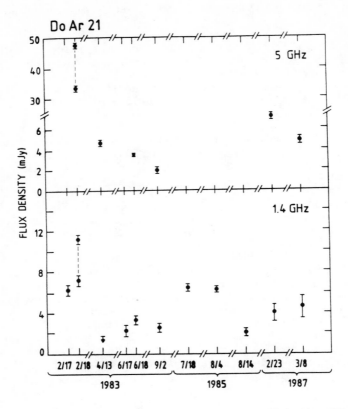

Fig. 2a. Light curve of DoAr 21, as observed with the VLA at 1.4 and 5 GHz (see Stine et al. 1988).

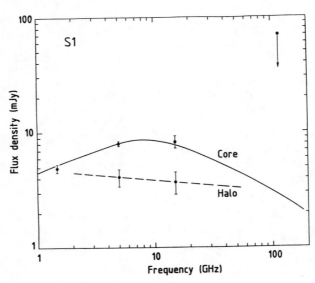

Fig. 3a.

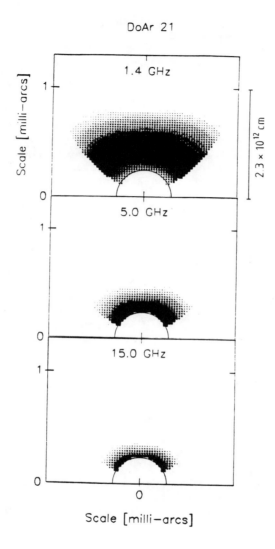

DoAr 21

1.4 GHz

5.0 GHz

15.0 GHz

Scale [milli–arcs]

Scale [milli–arcs]

2.3×10^{12} cm

Fig. 2b. Inhomogeneous model for the "quiescent" radio emission of DoAr 21; The emission takes place near the stellar surface, and results from the gyrosynchrotron mechanism of MeV electrons trapped in a dipolar loop $\sim$ 10 R_* in size, with $B_* \sim$ 150 G at the loop feet (for details, see André 1987).

Fig. 3a. Spectra of the core and halo components of S1. Dots: values from the VLA. Upper limit: 100 GHz quick-look observations with the Owens Valley interferometer (see André et al. 1987b). Continuous line: fit for the core using an inhomogeneous gyrosynchrotron emission from a complete dipole of size 10-30 R_* , with $B_* = 10^4$ G. Dotted line: fit for the halo by an HII region excited by a very young B3 star.

details in AMF), has been the seat of the first radio flare observed in a PMS star (Feigelson and Montmerle 1985). In the optical, it is a ~ K-G, R $\simeq$ 12 star, which used to display moderate Hα emission in objective-prism observations 30 years ago ; this emission has never been observed since. From its location on the H-R diagram, ~ 2 magnitudes above the main sequence, or from IR data, a radius of $R_* \simeq 4.10^{11}$ cm is inferred.

The radio properties (spectra, flux, etc.), the absence or weakness of the Hα emission or of the IR excess, as well as the late spectral type, are all shared by the few other existing flaring PMS objects: ROX-31 in ρ Oph (Bouvier and Appenzeller 1987), V 410 Tau, HP Tau/G2 and /G3 in Taurus (Cohen and Bieging 1986).

These stars have displayed rapid variability (from 7 to 11 mJy at 1.4 GHz, and from 32 to 48 mJy at 5 GHz in a few hours in the case of DoAr 21, see Fig. 2a), and/or large flux amplitudes variations (again, DoAr 21, see Fig. 2a, or V 410 Tau, which varied from 58 mJy in 1982 down to 1 mJy in 1983 at 15 GHz). Such a behaviour is a clear signature of a non-thermal radio emission mechanism. More specifically, based on recent models of radio flares on the Sun and in RS CVn systems (Klein and Trottet 1984 ; Klein and Chiuderi-Drago 1987), André (1987) has shown that the flaring and the comparatively quiescent radio emission of DoAr 21 could be explained by non-homogeneous gyrosynchrotron radiation of ~ MeV electrons trapped in a long-lived (permanent ?) large magnetic loop. In the model, the field at the loop feet is ~ 150 G and decreases according to a dipolar law ; as shown on Fig. 2b, the quiescent radio emission is mostly optically thin and takes place at ~ 1-2 R_* from the stellar surface, the loop itself being ~ 10 R_* in size. When a flare occurs, the source becomes mostly optically thick, enlarging the effective size of the emitting region and resulting in a much steeper spectrum (α ~ 1).

For future reference, it is important to notice that such a "peristellar" radio emission would be easily absorbed by a small amount of circumstellar ionized material. If in the form of a spherically symmetric stellar wind, this material is opaque if the mass-loss rate exceeds $\dot{M}_{max} \simeq 10^{-11}$ $M_\odot \cdot yr^{-1}$ (assuming a wind velocity ~ 200 km.s^{-1}, typical of PMS stellar winds).

3.4. S1

This star has long been known to be associated with a radio source (=BZ4, FG 20) ; it is a strong IR source (=GSS 35, E 25), and is associated with the brightest far IR source in the ρ Oph cloud (=FIR 1) ; it is also one of the brightest X-ray sources in the cloud (=ROX 14) (see AMF). The near-IR spectrum of Lada and Wilking (1984) is that of a pure blackbody of $T_{eff} \simeq 16\ 000$ K with a reddening of $A_v \simeq 12$. This is consistent with a B3-B5 ~ main sequence star (Lada and Wilking 1984), of radius $R_* \sim 3 \times 10^{11}$ cm and of luminosity $L_{bol} \sim 10^3$ $L_\odot$, (M $\simeq$ 4 $M_\odot$). The optical data (Bouvier and Appenzeller 1987) are still

insufficient to confirm this spectral type ; the spectrum is compatible
with a B to F type, displaying no emission lines and weak Hα absorption. However, the radio data (see below) confirm this early B spectral type. Among other remarkable features is the high L_x/L_{bol} ratio,
$\gtrsim$ 100 times the usual ratio for a main-sequence early-type star (see
Rosner et al., 1985). (This figure is uncertain because of an uncertain X-ray extinction correction.)

S1 has been observed repeatedly with the VLA, showing less than
$\sim$ 20% variability. A deep pointed exposure has been obtained in March
1987 with a C/D array, revealing unique radio features (André et al.,
1987b).

The 5 GHz map displays a resolved source, $\sim$ 30" in size. The
visibility curve (i.e., essentially the Fourier transform of the image,
the direct result of the interferometry procedure) reveals a core-halo
structure, the core itself being unresolved. This observation has
allowed to detect, for the first time in a young object, a circular
polarization ($\sim$ 7%), attributed to the core. The 5-15 GHz spectra of
both the core and halo are flat, as depicted on Fig. 3a.

By themselves, such flat spectra cannot help much in finding the
nature of the radio emission ; they merely indicate that the emission
is mostly optically thin. However, the fact that the halo is resolved
is a clear signature of the thermal nature of the emission ; it can be
shown quantitatively that it corresponds indeed to the free-free emission of a compact HII region excited by a B2-B3 star. Conversely, the
circular polarization of the core is a clear signature of a magnetical-
ly-induced (usually referred to as "non-thermal") emission mechanism.

S1 is therefore a good example of the complexity of these stellar
sources, since the global emission is mixed, i.e., thermal and non-
thermal. Other possible examples of this situation are known, also in
the ρ Oph cloud (VS 14, WL5, see André et al., 1987a).

The modest circular polarization and the detected flux in the GHz
band, along with the absence of significant variability on very different timescales (from years to hours, i.e., within one observation)
suggest, independently of any model, that the S1 core radio emission
arises in a large-scale (i.e., $\sim$ a few R_*) organized magnetic field
structure, and that the emission mechanism is gyrosynchrotron radiation
of MeV electrons in a magnetic field of order 1-100 G (see discussion
in André et al., 1987b).

Further, by drawing analogies with the radio emission from
He-rich, Bp-Ap stars recently detected by Drake et al. (1987), some of
which have measured surface magnetic fields $B_* \sim$ 10 kG, and with Jupi-
ter's magnetosphere (de Pater and Jaffe, 1984), André et al. (1987b)
suggest a model for the S1 core in the form of a rotating, mass-losing
magnetic dipole associated with a B star, as sketched on Fig. 3b. The
strong X-ray emission is explained simultaneously with the radio emis-

238

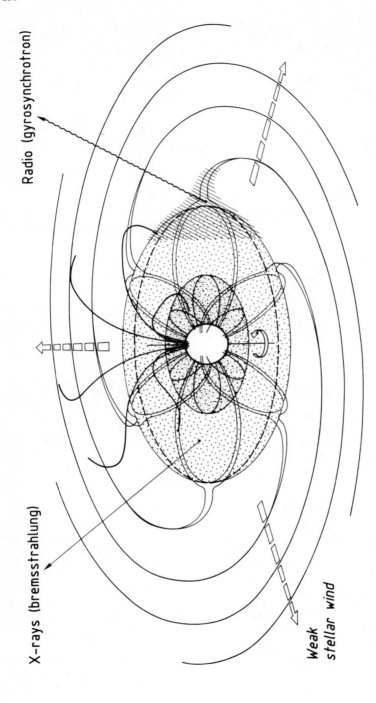

Radio (gyrosynchrotron)

X-rays (bremsstrahlung)

Weak
stellar wind

Fig. 3b. Sketch of a rotating magnetic early-type star model, as proposed for S1. The inner shaded region is a chromosphere, and the surrounding region is a magnetically trapped corona, emitting the observed X-rays. The radio emission takes place near the Alfvén radius (dashed circle), where the gas pressure from mass loss and centrifugal forces begins to exceed the magnetic pressure and opens the field lines.

sion in the framework of a model devised by Havnes and Goertz (1984) for the chemically peculiar early-type stars, featuring a magnetically trapped hot corona. The mass-loss rate assumed, $\dot{M} \sim 10^{-10}$ $M_\odot$, is consistent with that actually measured on the radio-emitting Bp star σ Ori E, and with that deduced from normal B stars (see also Drake et al., 1987). It is too small to absorb the gyrosynchrotron emission associated with the B star, in spite of the fact that it is an order of magnitude higher than the limit quoted above (§ 3.3) for the surface emission of DoAr 21. This is because the presence of an X-ray emitting, but radio-absorbing, corona, implies that the radio emission must take place just outside of the closed magnetosphere.

3.5. Other Radio Surveys of Regions of Star Formation

Only two such surveys exist to date, covering much smaller areas than the ρ Oph study ; they reveal some objects which may be similar to DꬴAr 21 or S1.

The Orion Trapezium region has been studied by Garay et al. (1987), Garay (1987), and Churchwell et al. (1987), who detected at 5 GHz 35 compact radio sources in a small 3'x3' area. This spectacular concentration of sources is certainly related to the intense ionizing environment provided by the Trapezium O stars themselves - a situation very different from ρ Oph. Several arguments suggest that a number of these sources are thermal (compact HII regions, externally ionized globules). Other sources are found to be very variable (see discussion in Feigelson 1987b) and/or associated with X-ray sources, pointing to other examples of naked non-thermal radio stars like DoAr 21. In one case, Θ^1 A Ori, a B 0.5 Vp star, the simultaneous variability and resolved extent may suggest an S1-like mixed emission.

The other region is the central part of the Cr A cloud, surveyed at 5 GHz over less than 1 square degree by Brown (1987), who found 5 stellar radio sources. This region is of the same type as the ρ Oph cloud, i.e., does not contain massive stars, but seems poor in PMS stars (only 4 classical TTS, and 4 X-ray discovered naked TTS, see Walter, 1987). All the 5 stellar radio sources reported by A. Brown (1987) are likely thermal. However, another radio source (not commented on by A. Brown) has varied from 15 mJy at 2.7 GHz in the early observations of R. Brown and B. Zuckerman (1975), to the present value of 0.7 mJy at 5 GHz. The behaviour is similar to that observed in V410 Tau, and, if stellar, this source is likely a flaring PMS stars. There is no information on the amount of circumstellar material around it, as this source has no optical or IR counterpart.

4. RADIO-EMITTING vs. RADIO-SILENT PMS STARS

4.1. Conclusions from the ρ Oph Survey

As mentioned above, the ρ Oph stellar radio sources are characterized by an absence of dense circumstellar emission. They do not show either other signs of mass ejection, which, as discussed above (§ 3.1), are usually associated with thermal stellar radio emission. Therefore, the radio properties (at varying degrees) as well as the X-ray properties, then suggest that <u>magnetized radio emission</u> is the rule, likely via the gyrosynchrotron mechanism. Out of the 9 reliable stellar candidates, this is certain in 3 cases (DoAr 21, ROX 31, and S1) probable in 5 other cases (WL 5, of class III, and VS 14, on the basis of the radio data ; VS 11, VS 22 and ROX 39 because they are associated with X-ray sources) ; the last case, YLW 15 is still open, owing to the lack of sufficient data (Stine et al. 1988).

As shown quantitatively in the cases of DoAr 21 and S1, the simultaneous absence of dense circumstellar matter and presence of a GHz radio flux at the ~ 1 mJy level then imply the existence of <u>extended closed magnetic structures</u> ($\sim 3-10$ R_*) associated with the corresponding stars.

4.2. The Low Detection Rate of Radio-emitting PMS Stars

The radio results on PMS stars seem a priori unexpected: whether we are dealing with classical TTS or with ρ Oph-type PMS objects in various regions, the detection rate is low ($\sim 25\%$ to 10%, respectively), in contrast with the X-ray situation, where the detection rate is $>$ 200% with respect to known PMS objects (taking into account X-ray extinction, i.e., for $A_V \lesssim 5-10$).

As discussed in Stine et al. (1988), simple opacity considerations may, at least in part, account for this situation. In § 3, we have seen that thermal radio emission (especially in classical TTS) is associated with independently detected outflows. For a typical (fully ionized, assumed spherical) wind velocity of ~ 200 km.s^{-1}, a $\gtrsim 0.3$ mJy flux corresponds to $\dot{M} \gtrsim 2.10^{-8}$ $M_\odot$.yr^{-1}. Since soft X-rays in the "Einstein" band are absorbed for $N_H \gtrsim 10^{22}$ cm^{-2}, only a few times 10^{-8} $M_\odot$.yr^{-1} to 10^{-7} $M_\odot$. yr^{-1} is enough to completely absorb any surface X-rays (depending strongly on the exact X-ray energy since the absorption goes as the cube of the energy). Conversely, for $\dot{M} \lesssim 2.10^{-8}$ $M_\odot$.yr^{-1}, the surface X-rays become visible, but the corresponding free-free radio flux becomes <u>undetectable</u>, given the current radio sensitivity. What about the "non-thermal" radio emission, as in ρ Oph ? We have already mentioned that the "peristellar" gyrosynchrotron emission of sources like DoAr 21 (in its quiescent state), or S1, is itself absorbed if $\dot{M} \gtrsim 10^{-11}-10^{-10}$ $M_\odot$.yr^{-1}, while such a wind remains transparent to the X-rays.

In summary, the detectability of any magnetically induced surface activity in both X-rays and radio can be schematically classified according to the associated <u>mass loss rate</u> (for $v_{wind} \simeq 200$ km.s^{-1}), as shown in the following table (which includes examples):

$\dot{M}$ ($M_\odot \cdot yr^{-1}$)	$\lesssim 10^{-11}$–10^{-10}	$\sim 10^{-10}$–10^{-8}	$\gtrsim$ few 10^{-8}
Surface X-rays	detectable (naked TTS)	detectable (TTS ROX sources, weak TTS)	absorbed (most strong Hα-emitting TTS)
peristellar radio (magnetized)	detectable (DoAr 21, S1)	absorbed (most classical TTS)	absorbed (same as above)
thermal radio (wind)	not detectable (S1)	not detectable (same as above)	detectable (same as above)

It is therefore clear that the high number of X-ray detected PMS stars, as well as the low number of radio detected ones, could be qualitatively explained if most classical TTS have a mass-loss rate $\sim 10^{-11}$–10^{-8} $M_\odot \cdot yr^{-1}$.

4.3. The Outer Magnetic Structure of PMS Stars

However, if the low amount of ionized circumstellar material is a necessary condition for the detectability of peristellar (non-thermal) radio emission, it is not sufficient for its existence. Indeed, many X-ray detected naked TTS have a very low Hα [EW(Hα) $\lesssim$ 2 Å or less], but are not detected in the radio (see the ρ Oph surveys).

This means that other conditions are necessary, and that the naked radio emitting PMS stars, do form a new, specific population of YSO. This is particularly clear in ρ Oph, where the stellar radio sources are concentrated at the edges of the dense central core, suggesting a smaller age than most ROX sources and classical TTS in the cloud (AMF). As discussed above, other members of the same population seem to exist also in the Orion, Tauris-Auriga (and perhaps Cr A) star-forming regions.

Comparing the size of their associated closed magnetic structures ($\lesssim 10 R_*$), with those characteristic of X-ray emitting PMS stars ($\lesssim R_*$, § 2), it is tempting to consider this new population as constituting the large extension tail of the magnetic field sizes of X-ray emitting objects, or even perhaps of all YSO, when associated with particularly weak stellar winds, contrary to classical TTS, as sketched on Fig. 4. This would explain, in particular, why naked radio-emitting objects are (or are presumably, if the interstellar extinction is too high) X-ray emitters, while the converse is not true. (Other illustrations of X-ray and radio emitting stars with magnetic fields large with respect to their size, are provided by the examples of the young K0 dwarf AB Dor, see Collier Cameron et al., 1987, or the RS CVn close binary systems, see Mutel et al., 1987).

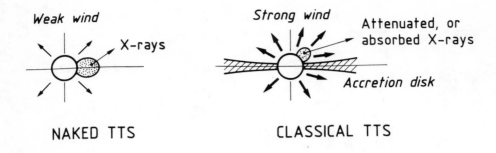

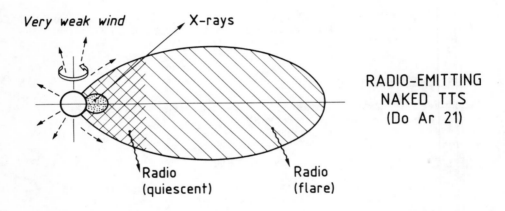

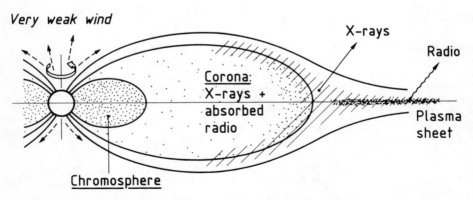

Fig. 4. Sketch of the possible relationship between various classes of PMS stars in terms of their outer magnetic structure and circumstellar environment, as can be inferred from their X-ray and radio properties.

Is there a <u>causal</u> relationship between extended closed magnetic structures and weak mass loss in the radio-emitting naked objects ? The answer to this question depends on the mass-loss mechanism. Indeed, the mechanisms usually invoked imply the dissipation of Alfvén waves above the stellar surface (Lago 1984 ; Hartmann et al., 1982), able to generate mass-loss rates up to $\sim 10^{-8}$ $M_\odot$. But the associated surface magnetic fields B_*, themselves presumed to be generated by the dynamo effect in surface convective zones, are typically of a few hectogauss, i.e., of the same order than those that the X-rays, as well as the radio emission, imply in general. (S1 is somewhat extreme, with $B_* \simeq 10$ kG, but it is an early-type star without surface convective zone, hence the magnetic field must be of a different origin ; in addition, B stars are comparatively rare in star-forming regions.) This contradiction may mean that, after all, Alfvén waves associated with stellar magnetic fields are not the right mechanism for accelerating stellar winds in late-type radio-emitting PMS stars.

Instead, one can invoke the likely presence of <u>accretion disks</u>, throught to be responsible, at least in extreme cases, for the UV and IR excesses observed in strong TTS (corresponding to IR classes I and II of Lada 1987 ; see Bertout, Basri, and Bouvier 1988, Kenyon and Hartmann 1988, and this volume). In fact, a <u>causal</u> relationship between the presence of dense winds and the existence of accretion disks, around classical TTS and related mass-losing objects, has been recently advocated by Strom, Strom and Edwards (1988) ; see also Edwards et al., (1987). In this framework, one would explain the relationship between the existence of large closed magnetic structures and the absence of dense winds, by the absence of an accretion disk around non-thermal radio-emitting objects.

However, yet another condition is required to explain the fact that, given the same (small) amount of circumstellar material (as traced, for example, by Hα), some stars have extended magnetic fields while the majority of them (i.e., the radio-silent ones) do not. In fact, we suspect that <u>fast rotation</u> is involved. From the existing limited sample of naked YSO with known v sin i, it is remarkable that all fast rotators are radio emitters (e.g. v sin i = 70 km.s^{-1} for V 410 Tau, 100 km.s^{-1} for HP Tau G2), while none of the observed slow naked rotators (v sin i $\lesssim$ 20 km.s^{-1}) has been detected in the radio. Conversely, the fast rotator SU Aur, which is surrounded by a disk (Bertout et al., 1988), is not detected.

This leads to the intriguing possibility that the specific properties of the radio-emitting non-thermal PMS stars are the result of a particular case of interplay between accretion disks, mass loss, surface magnetic fields, and rotation. This interplay may in fact be a central topic for <u>all</u> YSO, yet to be investigated, but in all likelihood closely related to early stages of evolution of newly formed stars.

ACKNOWLEDGEMENTS

We thank Jerôme Bouvier for permission to use some of his optical data before publication. TM is very grateful to Andrea Dupree and Teresa Lago for the organization of a particularly stimulating meeting held in a ...Joyful and friendly atmosphere, and kept up to high standards by huge amounts of delicious "Vinho Verde".

REFERENCES

André, Ph.: 1987, in Montmerle and Bertout (1987), p. 143.

André, Ph., Montmerle, T., Feigelson, E.D.: 1987, Astr. J., 93, 1182.

André, Ph., Montmerle, T., Stine, P.C., Feigelson, E.D., Klein, K.L.: 1987b, Ap.J., submitted.

André, Ph., Montmerle, T., Feigelson, E.D., Stine, P.: 1987a, in Circumstellar Matter, eds. I. Appenzeller and C. Jordan (Dordrecht: Reidel), p. 61.

Bertout, C.: 1982, 3rd Europ. IUE Conf., ESA, p. 89.

Bertout, C.: 1987, in Circumstellar Matter, eds. I. Appenzeller and C. Jordan (Dordrecht: Reidel), p. 23.

Bertout, C., Basri, G., Bouvier, J.: 1988, Ap. J., in press.

Bieging, J.H., Cohen, M., Schwartz, P.R.: 1984, Ap. J., 282, 699.

Bisnovatyi-Kogan, G.S., Lamzin, S.A.: 1977, Sov. Astr. 21, 720.

Bouvier, J.: 1987, Ph. D. Thesis, University Paris 7 ; also in Montmerle and Bertout (1987), p. 189.

Bouvier, J., Appenzeller, I.: 1987, in preparation.

Brown, A.: 1987, Ap. J. (Letters), in press.

Brown, R.L., Zuckerman, B.: 1975, Ap. J. (Letters) 202, L125.

Caillault, J.P., Zoonematkermani, S.: 1987, in Circumstellar Matter, ed. I. Appenzeller and C. Jordon (Dordrecht: Reidel, p. 119).

Churchwell, E., Felli, M., Wood, D.O.S., Massi, M.: 1987, Ap. J. 321, 516.

Cohen, M., Bieging, J.H.: 1986, Astr. J. 92, 1396.

Cohen, M., Kuhi, L.V.: 1979, Ap. J. Suppl. 41, 743.

Collier Cameron, A., Bedford, D.K., Rucinski, S.M., Vilhu, O., White, N.E.: 1987, M.N.R.A.S., in press.

DeCampli, W.: 1981, Ap. J. 244, 124.

De Pater, I., Jaffe, W.J.: 1984, Ap. J. Suppl. 54, 405.

Drake, S.A. et al.: 1987, Ap. J., in press.

Edwards, S., Cabrit S., Strom, S.E., Heyer, I., Strom, K.M., Anderson, E.: 1987 , Ap. J. 321, 473.

Falgarone, E., Gilmore, W.: 1981, Astr. Ap. 95, 32.

Feigelson, E.D.: 1984, in Cool Stars, Stellar Systems, and the Sun, ed. S. Baliunas and L. Hartmann (Berlin: Springer), p. 27.

Feigelson, E.D.: 1987a, in Montmerle and Bertout (1987), p. 123.

Feigelson, E.D.: 1987b, in Proc. 5th Cambridge Cool Star Workshop, eds. J.L. Linsky and R. Stencel (Berlin: Springer), in press.

Feigelson, E.D., DeCampli, W.M.: 1981, Ap. J. (Letters), 243, L89.

Feigelson, E.D., Jackson, J.M., Mathieu, R.D., Myers, P.C., Walter, F.M.: 1988, Astr. J., in press.

Feigelson, E.D., Montmerle, T.: 1985, Ap. J. (Letters) 289, L19.
Felli, M., Gahm, G.F., Harten, R.H., Liseau, R., Panagia, N.: 1982, Astr. Ap. 107, 354.
Gahm, G.F.: 1980, Ap. J. (Letters) 242, L163.
Gahm, G.F.: 1986, in Flares: Solar and Stellar, ed. P.M. Gondhalekar, in press.
Gahm, G.F., Fredga, K., Liseau, R., Dravins, D.: 1979, Astr. Ap. 73, L4.
Garay, G.: 1987, Prof. 5th IAU regional Latin-American Meeting, Merida in press.
Garay, G., Moran, J.M., Reid, M.J.: 1987, Ap.J. 314, 535.
Giacconi, R., et al.: 1981, in Telescopes for the 1980s (Palo Alto: Annual Reviews), p. 195.
Gurzadyan, G.A.: 1973, Astr. Ap. 28, 147.
Hartmann, L., Edwards, S., Avrett, E. : 1982, Ap. J., 261, 279.
Havnes, O., Goertz, C.K.: 1984, Astr. Ap. 138 , 421.
Heeschen, D.S.: 1981, in Telescopes for the 1980s (Palo Alto: Annual Reviews), p. 1.
Hjellming, R.M., Gibson, D.M. (Eds.): 1985, Radio Stars (Dordrecht: Reidel).
Kenyon, S.J., Hartmann, L.: 1988, Ap. J., in press.
Klein, K.L., Chiuderi-Drago F.: 1987, Astr. Ap. 175, 179.
Klein, K.L., Trottet, G.: 1984, Astr. Ap., 141, 67.
Ku, W.H.-M., Chanan, G.A.: 1979, Ap. J. (Letters) 228, L33.
Ku, W.H.-M., Righini-Cohen, G., Simon, M.: 1982, Science, 215, 61.
Lada, C.J.: 1987, in Star Forming Regions, eds. M. Peimbert and J. Jugaka (Dordrecht: Reidel), p. 1.
Lada, C.J., Wilking, B.A.: 1984, Ap. J. 287, 610.
Lago, M.T.V.T.: 1984, M.N.R.A.S., 210, 323.
Loren, R.B., Sandqvist, Aa., Wootten, A.: 1983, Ap. J., 270, 620.
Montmerle, T.: 1987, in Proc. 5th Eur. Solar Meeting, Solar and Stellar Physics, eds. E.H. Schröter and M. Schüssler (Berlin: Springer), in press.
Montmerle, T., Bertout, C. (Eds.): 1987, Protostars and Molecular Clouds (Saclay: CEA/Doc).
Montmerle, T., Koch-Miramond, L., Falgarone, E., Grindlay, J.E.: 1983, Ap. J., 269, 182.
Montmerle, T., Koch-Miramond, L., Falgarone, E., Grindlay, J.E.: 1984, in Very Hot Astrophysical Plasmas, Phys. Scr. T7, 59.
Mutel, R.L., Morris, D.H., Doiron, D.J., Lestrade, J.F. 1987, A.J.93, 1220
Panagia, N.: 1988, in Galactic and Extragalactic Star Formation, NATO/ ASI (Kluwer Academic Publishers).
Rosner, R., Golub, L., Vaiana, G.S.: 1985, Ann. Rev. Astr. Ap., 23, 413.
Schwartz, R.D.: 1978, Ap. J. 223, 884.
Snell, R.L., Bally, J.B.: 1986, Ap. J. 303, 683.
Stine, P.C., Feigelson, E.D., André, Ph., Montmerle, T.: 1987. Astr. J., submitted.
Strom, S.E., Strom, K.E., Edwards, S.: 1988, in Galactic and Extragalactic Star Formation, NATO/ASI (Kluwer Academic Publishers).
Walter, F.M.: 1986, Ap. J., 306, 573.
Walter, F.M.: 1987, Publ. Astr. Soc. Pac. 99, 31.

Walter, F.M., Kuhi, L.V.: 1981, Ap. J. 250, 254.
Walter, F.M., Kuhi, L.V.: 1984, Ap. J. 284, 194.
Yodh, G.B., Artru, X., Ramaty, R.: 1973, Ap. J., 181, 725.

EMISSION ACTIVITY ON T TAURI STARS

G. Basri
Astronomy Department
University of California
Berkeley, California USA

Abstract. This is a review of the efforts to understand the excess line and continuum emission in the optical and ultraviolet spectra of T Tauri stars. One of the current hypotheses about what kind of physical region gives rise to these spectral tracers is that there is extreme chromospheric activity on the surface of these very young stars. The successes and failures of this theory are discussed, as is the spectral diagnostic approach which can be used to evaluate it. While there is no doubt that strong classical chromospheric activity is present on these stars, it also appears that it is insufficient to explain the more extreme spectral anomalies. A brief discussion of the most successful alternative, boundary layer emission from continuing disk accretion, is also given. Particular attention is paid to the pressing problem of how to distinguish between the two alternatives with observations.

1. Introduction

The T Tauri stars were noticed as a class because of their spectral peculiarities. While the primary identifying characteristic is the presence of $H\alpha$ emission, as their spectra are studied in greater detail the number of puzzling features increases. They are a varied class of objects, with the least active having $H\alpha$ emission as almost the only peculiarity. The most active objects show a host of emission lines from hydrogen, helium, calcium, iron and other metals, sometimes forbidden line emission, strong continuum excesses in the ultraviolet and infrared, and weak or absent stellar photospheric absorption lines. Some objects are lightly reddened, while others are heavily extincted. We now understand them to be pre-main sequence stars which have recently emerged from their enveloping star-forming molecular cloud cores.

To the original class of T Tauri stars must now be added other known types of pre-main sequence stars, including FU Ori objects at the most active end, and the so -called "naked" T Tauri stars at the least peculiar end. Furthermore, high mass stars (greater than 2 solar masses) are really in a separate class, since they have much higher luminosities and rotations and do not undergo a fully convective phase. I ignore the Ae and Be stars in the remainder of the paper, and refer you to the article by Hartmann regarding the FU Ori objects. I adopt the acronym YSO (young stellar object/s) to refer to solar-type pre-main sequence

A. K. Dupree and M. T. V. T. Lago (eds.), Formation and Evolution of Low Mass Stars, 247–255.

stars. For these, I will refer to "moderate" (meaning a largely absorption spectrum with weak Hα emission, small or absent infrared excess, and some filling in or emission in the strongest stellar lines), "active" (meaning the classical T Tauri stars, with strong Hα emission, continuum excesses, more emission lines, but with a stellar absorption line spectrum present), and "extreme" (meaning only emission lines, strong continuum excesses, and no absorption line spectrum). For a more detailed breakdown of the observational and physical distinctions among YSO, please see Basri (1987).

2. The Chromospheric Hypothesis

The YSO were originally studied by Joy (1949), who found them on the basis of their Hα emission. A variety of theories were advanced to explain the presence of this anomalous emission, and more emission lines were discovered in some of the objects. It was noticed that many of the lines also show up in the flash spectra during solar eclipses (and Hα is very strong in active regions and flares on the Sun). Herbig (1970) and Dumont et al. (1973) proposed that one possible explanation for the emission activity is "deep chromospheres", or strong stellar activity. Such a chromosphere would be located at much larger continuum optical depths than the solar chromosphere in order to produce the observed emission. The idea is that the same causes as act to produce solar activity (magnetic and acoustic heating) would also be operating on YSO, but at a much greater levels. The atmospheric temperatures should rise significantly above their radiative equilibrium values due to the input of high levels of non-radiative heating; as the atmospheric density decreases toward the outside of the star its ability to radiate this away is reduced and the temperature rises.

This idea gained support in the last decade from work being done on the solar chromosphere (Vernazza, Avrett, Loeser 1981) and on stellar chromospheres (cf. Linsky 1980). Multi-component models for the sun that successfully predict observed spectra have the transition region and (to a lesser extent) the temperature minimum region moving to higher mass column density in more active regions, while stellar chromospheric one-component models behave similarly, with the temperature minimum located more deeply for the active main sequence stars. A rather general structure for chromospheres emerged, in which the position of the temperature minimum (or equivalently the chromospheric temperature rise) in mass column density is a function of stellar gravity and activity level. The temperature rises to 6000-7000K and levels off for some distance (the chromospheric plateau) and then abruptly rises to much higher values (the transition region) at a density also determined by gravity and activity level. These classical chromospheres will always be geometrically thin (less than R_* in extent).

Another breakthrough occurred regarding what fixes activity levels in stars of the same spectral type. It became increasingly clear from IUE and K-line studies that the fundamental stellar property determining activity levels for a given stellar mass is the stellar rotation rate. Stellar age is also a good predictor for activity, but this appears to be largely due to the evolution of rotation with age. The YSO are relatively rapid rotators for their spectral type (Hartmann et al. 1986; Bouvier et al. 1986), and so would be expected to have strong chromospheres. They most resemble the RS CVn stars, which are also cool subgiants but are evolving back off the main sequence. Forced to rotate rapidly by a close companion, they exhibit among the strongest activity levels in older stars. Indeed, studies by Bouvier (1987) show that the activity levels are similar for the moderate YSO and RS CVn stars. Calvet *et al.* (1985) also show that a single relation between CaII and MgII emission holds from the active main sequence stars through the RS CVn stars and through the moderate YSO. The same statements can be made for the various UV transition region lines, although the active and extreme YSO depart from this.

Other evidence that there is magnetic activity on YSO comes from the regular photometric variability that some of them display. This can be interpreted in many cases (Rydgren *et al.* 1984; Bouvier 1987) as dark spots being carried in and out of sight by rotation, and magnetic activity is the only current explanation for such spots. Flares, and sometimes quiescent X-ray coronae, have also been seen on a number of YSO (cf. Montmerle, this volume) which again suggests magnetic activity. Finally, Finkenzeller and Basri (1987) show explicitly that the detailed spectra of moderate YSO show all the spectral features expected from classical stellar activity, including effects in the upper photosphere which would be hard to explain any other way. It seems safe to conclude that strong classical stellar chromospheres are a ubiquitous feature of very young low mass stars.

While the solar analogy has been successful in explaining observations of stellar activity, some stars are so active that they would have to be completely covered by solar-type active regions to explain their emission fluxes. It is hard to say whether this is reasonable. Activity on the Sun appears to be concentrated in small magnetic flux tubes which have a rather low covering factor for the Sun, so in principle one could have much stronger activity on the same surface. It is also possible that very strong activity comes in a different package; the presence of huge non-circular "spots" and giant stellar-sized flares on some active main sequence stars are a warning regarding the limitations of the solar analogy. The extreme YSO appear to stretch the solar analogy beyond reasonable limits.

3. Chromospheric Spectral Diagnostics

Cram (1979) was the first to apply the full NLTE modern radiative transfer analysis to the question of whether YSO emission lines come from strong chromospheres. He concluded that many line *fluxes* could be reproduced with such a model, including CaII, NaI, and FeI lines. The Balmer lines appeared to be more problematic, particularly the great strength of Hα relative to Hβ is often impossible to explain with surface atmospheric structures. The model used was very schematic but served to explore the possibility of forming the emission spectrum in a heated stellar surface.

The question was examined in more detail with similar methods by Calvet, Basri, and Kuhi (1984, hereafter CBK). They used data from actual stars in both the UV and optical and worried both about line fluxes and continuum distributions. They tried a large variety of possible chromospheric structures to test the limits of the chromospheric hypothesis. The conclusions were that *in principle* one could reproduce the observed Balmer and Paschen continua, the fluxes and ratios of CaII and MgII, and possibly the filling in of photospheric lines with self-consistent deep chromospheres. The Hα line was again found to be impossible to reproduce (except for moderate YSO). The UV excess was explicable but the

IR excess, was not with only this stellar model. I must emphasize that this was a proof in principle, it did not address whether the chromospheric structures required could or did actually exist. The deep chromospheres used to explain active YSO would require copious non-radiative heating rates deep in the stellar atmosphere; a substantial fraction of the stellar bolometric luminosity would have to participate.

I now cover the essentials of how a deep chromosphere can produce the observed line and continuum fluxes. While the formation of spectral features is a complex subject, if we are willing to overlook a lot of important details it can be reduced to a few simple concepts. The first is that whether one sees absorption or emission is purely a matter of contrast: whether the flux in the core of the line greater or less than the flux in the nearby line wings or continuum. The second is that for a given emitting surface area the flux seen at any particular wavelength is approximately the value of the source function at that wavelength at the point in the atmosphere where it just becomes optically thick (the optical surface; $\tau = 1$). This latter fact contains two further important concepts: 1) the location of the optical surface depends on the integrated opacity in the column to a given depth, which can be a strong function of wavelength in a spectral line, and 2) the source function may well have its LTE value (the Planck function for the temperature at the optical surface at the given wavelength), but might be significantly less than that if NLTE effects are important.

The source function, optical depth, and emergent flux are related by an equation of the form

$$F \propto R^2 \int S(\tau)e^{-\tau}d\tau.$$

From this expression it is clear that the emergent flux is weighted to the place where $\tau = 1$; the exponential rapidly cuts off at larger optical depths, while the $d\tau$ factor drops off linearly at smaller optical depths. These points are embodied in the concept of the contribution function, which is just the integrand above. The form of the contribution function implies that the flux does not arise from a unique point in the atmosphere, there is an interplay between gradients in the source function and changes in optical depth. Where there is a strongly increasing source function, particularly if this occurs above the optical surface, the flux may contain contributions from a wide range of optical depths (and temperatures). It is often assumed that the contribution function large over a relatively narrow range of optical depths. In that case an emission line will be formed at the surface of the star when the line core source function is greater than the continuum source function at their respective optical surfaces.

Note the above mechanism is different than emission line formation in typical hot star cases (P Cygni profiles) where the apparent emission is due to the line optical surface having much greater geometrical area than the continuum optical surface (with the value of the local source functions playing a much reduced role). This latter mechanism may well play a role in the Hα line in YSO if it is formed many stellar radii above the photosphere, which could be the reason that chromospheric models are unsuccessful for that line.

If we want to explain why YSO have strong emission lines compared with other active stars, the contrast between line and continuum must be increased in YSO. This can be accomplished in a variety of ways: 1) S_{line} could simply be relatively larger at the line optical surface, 2) if the temperature rise occurs above the optical surface but closer to it, 3) the source function gradient in the right part of the atmosphere could be steeper. Other effects which tend to cause or enhance the formation of emission features are 1) S_{cont} could be smaller (as happens in cooler stars), 2) looking at shorter wavelengths increases the flux contrast between two different temperatures shortward of the Planck function peak wavelength, 3) if the source function rise occurs at higher densities, the likelihood of possible NLTE reductions of S_{line} will be mitigated or eliminated.

Suppose then that compared to an active main sequence star a YSO has a shallower upper photospheric temperature gradient , leading to a hotter temperature minimum, which occurs at greater mass column density and is followed by a steep rise to the chromospheric temperature plateau, which also terminates in a transition region at higher mass column density. The main sequence star is represented by Model Q in Figure 1 and the deep chromosphere with Model A. Consider first the continuum, whose optical surface is at Point 1 in both models. Suppose however, that by Point 3 the continuum optical depth has fallen to one hundredth. In Model Q the source function is also less there, so there is very little contribution to the continuum flux from that point. In Model A, however, the source function is actually larger than on the optical surface; if we chose a short wavelength, the source function might be one hundred times larger. In that case, the contribution function will have equal peaks at Points 1 and 3 and the emergent flux will be twice as bright. This is essentially the effect found by CBK which can substantially enhance the UV continuum from YSO. This is particularly effective in producing a Balmer continuum jump because the head of the Balmer continuum is much thicker than the tail of the Paschen continuum and so the chromosphere has a greater effect on it. Note this effect would also be present in all the cores of photospheric lines, which will have thicker optical depth scales and so respond to the temperature rise even more strongly.

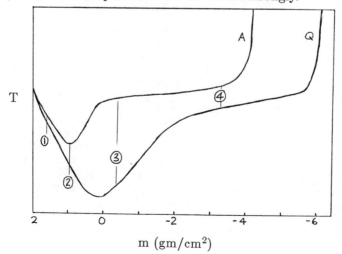

Even if the temperature rise is not deep enough to affect the continuum, the shallower upper photospheric gradient will affect the depth of photospheric absorption lines. If a line core is formed at Point 2, it is clear that the difference between the source function there and for the continuum is less in Model A than Model Q. This means that the lines will have less contrast; ie. they will appear filled in, or diluted. An equivalent way to look at this is that the core source function is larger in Model A, so the line flux will be brighter in the core for that model (and we would say the *absorption* was not as strong). As the temperature rise is moved outward to smaller optical depths, the effect will disappear first for the (initially) weakest lines, because their optical surfaces are too removed from it. From the other direction, as a star is made more active, the filling in should occur in the initially strongest lines first (they were strong because the core surface was closest to the original temperature minimum). This is presumably the effect observed by Finkenzeller and Basri (1987).

To get an emission reversal in the core of the line, the core source function just has to be brighter than the source function in the inner wings. Since optical depth decreases rapidly as one moves away from the core, the optical surface moves rapidly deeper. Thus the core source function has to get a contribution from the chromosphere while the wings get little or no such contribution to produce an emission feature in the line. If the core source function rises above the continuum value, we will see true emission. This will obviously occur for cores formed at Points 3 or 4 in Model A. At Point 4 there will be a further enhancement of the core flux by a contribution from the base of the transition region. In principle the emission might have almost the same level in Models Q and A because the temperature is similar at Point 4. In practice this part of the atmosphere is usually subject to NLTE effects because the density is low, and the coupling to the base of the transition region will be better in Model A than Model Q likely leading to a brighter line. This mode of formation can explain the CaII K-line or MgII k-line in moderate YSO. Because the densities are still rather substantial one would not expect strong hypersonic flows or turbulence to dominate in this part of the atmosphere, so the line profile should be relatively narrow ($<$ 100 km/s FWHM) and symmetric.

I digress on the subject of profile broadening, because there are actually two different ways in which a profile can become broad. The simple mechanism is just Doppler broadening, whether due to stellar rotation, organized flows, or turbulence. The basal half-width of a line dominated by Doppler broadening represents the maximum velocities present; the profile will reflect the distributions of velocities present only if the line is effectively optically thin. The other means by which a line may appear broad is opacity broadening; in essence this arises because the line will be in emission for all wavelengths where the contribution function is sufficiently large compared to the continuum. Since the optical surface moves into the star as we move away from line center (to smaller opacities), the extent of the broadening will depend on the structure of the atmosphere and the behavior of the optical depth. If the optical surface 2Å from line center is still in the chromosphere; we would expect the line to be bright at this wavelength (giving it a breadth of at least 2Å). The question of whether the broad Balmer lines actually observed in YSO can actually be produced in this way can be answered only by detailed modelling. CBK concluded that there is no arrangement which can broaden Balmer lines by the required amount with opacity and be consistent with other observed spectral features.

To summarize, the effect of pushing the chromosphere to higher mass column densities is to increase the flux in the optically thickest spectral features first; filling them in then sending them into emission. The deeper the chromosphere, the more these effects are also seen in weaker lines. The continuum finally begins to brighten, first at the Balmer jump and eventually at longer wavelengths. This will have the effect of veiling all the lines in addition to the differential veiling mentioned above. Finally we expect a strong Balmer emission jump, substantial line veiling, and many strong emission lines. Because this is a qualitative description of an active YSO spectrum, the chromospheric hypothesis is attractive.

There are a number of problems with it, however. Firstly, one expects line effects to be stronger than continuum effects. This is not always observed; sometimes there can be substantial veiling without a copious emission line spectrum. Secondly, the kinds of line profiles observed are not particularly consistent with this model. There is a tendency for all emission lines, whether strong or weak, to be very broad (hundreds of km/s) in active and extreme objects. Both this, and the asymmetries often observed are difficult to explain as surface phenomena. The Balmer decrement is also inexplicable. This model completely ignores the infrared excesses with shallow slopes that are common among active and extreme YSO. Finally, the physical mechanism which could produce a deep chromosphere is left out. It does not appear likely that the same mechanism which operates in active main sequence stars could produce such extreme atmospheres. In this regard, I must point out that some completely different mechanism might end up producing similar atmospheric structures, but I wouldn't call them classical chromospheres in that case. It therefore appears that deep classical chromospheres probably explain many of the observations of moderate YSO, but are not satisfactory for the active and extreme cases.

4. The Boundary Layer Hypothesis for YSO

Recently, an alternative to the chromospheric hypothesis has been studied in some detail by Bertout, Basri, and Bouvier (1988). It is that the active and extreme YSO are surrounded by circumstellar disks, some of which are still actively accreting. Because orbital velocities in the disk as it touches the star are much higher than the average observed stellar rotation velocity in YSO , a turbulent boundary layer might be present on the interface between star and disk. This must dissipate all the kinetic energy in the disk material, which provides a source of luminosity which can equal or overwhelm the stellar luminosity (depending on the accretion rate). Disks around YSO are discussed in several articles in this volume, including those by Pringle, Kenyon, Hartmann, and Bertout. While little firm theory exists on the details of the boundary layer, the general idea has several attractive points.

Since the evidence is rapidly mounting that disks are common among YSO, it is natural to consider the effects of accretion and boundary layers. They solve the luminosity problem; modest accretion rates ($\sim 10^{-7} M_\odot/yr$) can provide luminosities comparable to stellar luminosities. The disk models are successful in predicting the infrared and ultraviolet excesses together. Finally, detailed studies which attempt an independent determination of the true stellar luminosity (based on the photospheric spectrum) indicate that YSO sometimes have a strong extrinsic source of luminosity present. Figure 5 of Bertout's article (this volume) shows the effect a simple boundary layer model predicts on the spectrum of BP Tau if it is a K7 subgiant surrounded by an accretion disk which is consistent with the infrared spectrum. Since the stellar continuum in this figure includes the UV excess of a chromosphere similar to those found in moderate YSO, it is clear that something much brighter in the UV is present in this object (see Bertout, Basri, Bouvier 1988 for more details).

Leaving aside the question of how valid the simple boundary layer model actually is, how can we distinguish between this model and the classical chromospheric model? As mentioned above, one method is to get an accurate estimate of the photospheric luminosity. It is hardly likely that a comparable luminosity could also appear as UV and IR excesses in the context of the chromospheric model. To find the stellar luminosity, one must correct for spectral veiling, extinction, and effects of a possible passive disk. The first can be treated if the photospheric lines are visible; once a spectral type has been determined one can find the continuum which must be removed in order to render the lines similar to a standard star of that spectral type. Presumably the strongest lines would be ignored in such a procedure, as differential veiling is expected from a strong chromosphere. The best standards to use are the moderate YSO, since chromospheric effects are already present in their spectra. Similarly one can attempt to remove extinction from the unveiled spectrum using the same standards. Here one requires accurate spectrophotometry , particularly in the middle of the optical range where veiling was minimal and extinction is still appreciable. The dereddening which makes the continuum distributions match best can be applied to the object under analysis.

Correcting for a possible non-accreting disk is trickier. The disk has two effects: it occults part of the star and it reprocesses some of the optical light into the infrared. These points are covered by other authors here and by Basri (1987). In order to properly account for these effects to convert an observed flux to a corrected stellar luminosity, one must know or model the inclination of the disk. Because the infrared excess depends only on the disk, it is possible to disentangle disk effects from the stellar luminosity under the assumption of no accretion. Once one has the proper stellar luminosity, it can be compared with the observed bolometric luminosity of the system. If the bolometric luminosity is greater than 1.5 times the corrected stellar luminosity; accretion is guaranteed to be present.

Of course, even given the presence of accretion one might question the presence of a boundary layer. Given a UV excess, how much of it really comes from a chromosphere and how much from a boundary layer. The resolution of this issue requires resolution of the spectrum, both absorption and emission. In principle (as discussed above) the type of line profile expected from boundary layer or chromosphere is different; the chromospheric line should be relatively narrow and symmetric. The effect on photospheric lines of a chromosphere deep enough to produce the observed continuum excess is also calculable and can be checked. Observations of BP Tau (Basri 1987) show both a narrow and broad component in the K-line, for example. The amount of K-line emission in the narrow component is compatible with the idea that the chromosphere on BP Tau is similar to those on moderate YSO; lending credence to the idea that the UV excess mostly has other origins. Furthermore, none of the photospheric lines show a narrow emission core; indicating that the chromosphere is not as deep as required by the models for this star of CBK. They are 50% veiled, however, as if by an extrinsic continuum source. Indeed, the wavelength dependence of the veiling is predicted by either the chromospheric or boundary layer models, and can be checked against them.

Unfortunately, neither type of model is firm enough in detail that it can be unambiguously eliminated at this time. Indeed, the situation is very likely that both models are correct in that both sources of emission are present in the active YSO. The chromospheric models must face the possibility that the stellar surface is quite inhomogeneous and the physical conditions required by a one-component model may not exist anywhere on the star. The boundary layer model is still struggling with such fundamental questions as the size of the layer, the extent to which the excess energy appears as radiation, and whether the emitting region is optically thick, thin, or both. Each theory can best be checked using simultaneous spectrophotometry and echelle spectroscopy from the UV to the IR, a formidable observational task! Furthermore, a difficult and lengthy modeling effort is required to fully interpret such observations. Many of the basic physical parameters in either model remain indeterminate. Thus, it will probably be some time before these questions are settled with high confidence. The effort required will undoubtedly provide interesting times for workers in the field for the next several years.

References:

Basri, G. 1987, *Fifth Cambridge Workshop on Cool Stars, Stellar Systems, and the Sun*, Linsky and Stencel (eds.), Springer-Verlag, in press.
Bouvier, J. 1987, *Thèse de Doctorat* (Paris: Université Paris VII)
Bouvier, J., Bertout, C., Benz, W., Mayor, M. 1986, Ap. J. **265**, 110.
Bertout, C., Basri, G., Bouvier, J., 1988, Ap. J. (accepted)
Dumont, S., Heidmann, N., Kuhi, L.V., Thomas, R.N. 1973, Astr. Ap. **29**, 199.
Calvet, N., Basri, G., Kuhi, L.V. 1984, Ap. J. **277**, 725.
Calvet, N., Basri, G., Imhoff, C.L., Giampapa, M.S. 1985, Ap. J. **293**, 575.
Cram, L.E. 1979, Ap. J. **234**, 949.
Edwards, S., Cabrit, S., Strom, S., Heyer, I., Strom, K. 1987, Ap. J. **321**, 473.
Finkenzeller, U. and Basri. G., 1987, Ap. J. **318**, 823.
Hartmann, L., Hewett, R., Stahler, S., Mathieu, R. 1986, Ap. J. **309** , 275.
Herbig, G.H. 1970, Mem. Soc. R. Sci. Liège **19**, 13.
Joy, A.H. 1949, Ap. J. **110**, 424.
Linsky, J.L. 1980, *Ann. Rev. Astr. Ap.* **18**, 439.
Rydgren, A.E., Zak, D.S., Vrba, F.J., Chugainov, P.F., Zajtseva, G.V. 1984, Astr. J. **89**, 1015.
Vernazza, J.E., Avrett, E.H., Loeser, R. 1981, Ap. J. Supp. **45**, 635.

FLOWS AND JETS FROM YOUNG STARS

Reinhard Mundt
Max-Planck-Institut für Astronomie
Königstuhl 17
6900 Heidelberg
Fed. Rep. of Germany

ABSTRACT. In this paper the morphological and kinematical properties of jets and collimated "optical" flows from young stellar objects are reviewed. Numerous recently obtained high-quality CCD images are shown to illustrate the morphological properties of these flows and their associated bow-shock like structures. The currently available data on the jet sources are also reviewed. In addition to a few other topics (e.g. jet's working surface and radiative bow shocks) the kinematic properties of the driving gas of molecular flows from low-mass stars are discussed. It is suggested that it consists mainly of neutral, atomic gas with velocities in the order of 50 km/s.

1. INTRODUCTION

For almost thirty years, it has been known that at least some young stars undergo strong mass loss, with rates in the order of 10^{-7}-10^{-8} $M_\odot$/yr and wind velocities of several hundred km/s (Herbig 1960, Kuhi 1964). These values were estimated from the analysis of P Cygni profiles observed in the optical spectra of several T Tauri and Herbig Ae/Be stars. Since then it has been recognized that energetic and bipolar mass outflows are an important phase in early stellar evolution for all types of young stars. Evidence for such energetic mass outflows is provided by a variety of phenomena (see e.g. Lada 1985, Mundt 1985a, Strom and Strom 1987, Staude 1988). Many of these phenomena result from the interaction of the outflowing matter with the surrounding molecular cloud.

The two most detailed studied phenomena are: (1) high-velocity molecular gas (HVMG) and bipolar molecular flows, (2) Herbig-Haro (HH) objects and optical jets. In the former case one studies the molecular gas components near young stellar objects (YSOs) with significantly higher velocities than the typical turbulent velocities of molecular clouds (~ 1 km/s). In the vicinity of stars of low to moderate luminosity (1-100 $L_\odot$) the characteristic velocity of the molecular flows is 5-20 km/s. The HVMG itself is probably molecular cloud material which has been somehow accelerated by the stellar wind. It could for example be located in the shell of a wind-blown cavity. The necessary mass loss rates to accelerate the HVMG are rather high. Assuming momentum driven

A. K. Dupree and M. T. V. T. Lago (eds.), Formation and Evolution of Low Mass Stars, 257–279.
© 1988 by Kluwer Academic Publishers.

flows and wind velocities of about 200 km/s, estimated values range from 10^{-8} $M_\odot$/yr to several 10^{-6} $M_\odot$/yr for stars with L = 1-100 $L_\odot$. About 50% of molecular outflows are clearly bipolar. Their degree of collimation is small, and typical length-to-diameter ratios are only 2-3. The bipolar lobes extend over distances of about 0.1-1 pc. The properties of molecular outflows has extensively been discussed in a number of recent reviews (e.g. Lada 1985, Snell 1987). Therefore only one aspect, the driving agent of the molecular flows, will be discussed briefly here.

Optical emission-line jets and HH objects associated with young stars trace flow components with a much higher degree of collimation and velocity than the HVMG. The measured radial velocities can reach values of up to 450 km/s. For some jets the opening angle is less than 1 degree and their length-to-diameter ratio can have values of up to 30. The lengths of these jets range from about 0.01 to 1 pc. For details the reader is referred to Mundt, Brugel, and Bührke 1987 (hereafter MBB87). The sources of most known HH objects and jets are stars of low to moderate luminosity (1-100 $L_\odot$), i.e. low-mass stars. Whether highly collimated jets are a phenomena predominantly associated with low-mass stars is not clear yet (see also section 4).

HH objects and jets are intimately related phenomena. Both show the same emission line spectrum, which is formed behind radiative shock waves with velocities of 40-100 km/s. Furthermore, several long known HH objects form the brightest knots of a jet. On the basis of such observations it has been suggested (e.g. Mundt 1985b) that many HH objects simply represent the locations of the most brightest radiative shocks in a jet from a young star. This does not imply a unique model for all HH objects, since there is a large variety of ways to excite shocks in a collimated outflow (e.g. MBB87). For example, one expects strong shocks in the working surface of the jet, but it also possible to excite shocks along the jet (e.g. through fluid dynamical instabilities). The former aspect will be discussed in more detail in section 6.

The currently available data on jets, HH objects, and collimated optical flows have been discussed extensively in several recently published review articles (Schwartz 1983; Mundt 1985a, 1985b, 1986, 1987; MBB87; Cantó 1986, Dyson 1987, Strom and Strom 1987). Since so many recent reviews are available, only some topics and new developments will be discussed here.

2. A FEW COMMENTS ON THE DRIVING AGENT OF MOLECULAR FLOWS

It has been argued by various authors that the HVMG is formed in a swept up shell of molecular cloud material (e.g. Snell 1987). What are the properties of the wind that drives this shell? What is its velocity and momentum? Is it a molecular, neutral atomic, or an ionized wind? The wind velocity, for example, will determine to a large degree whether the wind is energy or momentum driven (e.g. Dyson 1984). Interesting in this respect are the IRAS data of L1551-IRS5 (Clark et al. 1986, Edwards et al. 1986). They show extended FIR emission along the axis of the bipolar CO lobes. If this luminous FIR emission ($\sim$ 20% of IRS5) is somehow provided by the wind from IRS5, the molecular outflow cannot be energy driven, but must be momentum driven. Note that a energy driven flow has by definition very little radiative losses.

Let us suppose that the HVMG is located in a shell of swept up molecular gas and that this shell is momentum driven by a wind of say 200 km/s. Why do we not observe HH emission at the inner edge of that shell, from the inward facing shock? Consider the case of L1551-IRS5. The estimated mass loss rate of this source is about 10^{-6} $M_\odot$/yr (Edwards and Snell 1984) in the case of a momentum driven flow with v(wind) $\sim$200 km/s. We furthermore assume that on the blueshifted side the driving wind flows within the cone-like region visible in scattered light (e.g. Lenzen 1987). This is strongly supported by high spatial resolution CO observations (e.g. Moriarty-Schieven et al. 1987). In this case the wind density would be about 300 hydrogen atoms/cm^3 at 20" from the source and about 4 times smaller at 60" from the source. Such pre-shock densities are relatively high for HH objects and one should therefore observe very bright line emission (HH emission) at the inner edge of the presumed wind cavity. Why is such emission not observed? We suggest that the wind velocities are much lower and in the order of 50 km/s. In such a case one expects shock velocities of about 10-20 km/s, since the velocity component perpendicular to the shell surface determines the shock strength. Such low shock velocities would not give rise to detectable optical emission, unless the wind is ionized.

Are there observational indications for outflowing matter from YSOs with velocities of about 50 km/s which could drive the molecular gas? A few years ago Mundt (1984) published NaD line profiles of 11 strong-emission T Tauri stars which showed rather narrow (FWHM$\sim$20 km/s) blueshifted absorption features with moderate velocities (35-110 km/s) with respect to the star. An example is displayed in Fig. 1. These absorption features showed very little variation. This, together with there low velocity dispersion, indicates that these lines are formed at large distances from the source. Mundt (1984) had severe difficulties in explaining these features through a low velocity wind or a swept up shell of gas. Therefore, the hypothesis is proposed here that low mass YSOs produce two wind components. (1) A high velocity component (v$\sim$300km/s), which we see in some stars in the form of jets and which has insufficient momentum to drive the molecular flows (MBB87). (2) A low velocity component of v $\sim$ 50-80 km/s with high momentum and poor collimation, which drives the molecular flows. Unfortunately, from the NaD data, very little can be said about the mass flux of the proposed second component, since we don't know the NaD ionization and the length scale over which the absorption is produced.

3. OBSERVATIONAL PROPERTIES OF COLLIMATED OPTICAL FLOWS AND JETS

3.1 Definition and Known Cases

Most jets and collimated optical flows have been discovered within the past 5 years through CCD imaging of HH objects and YSOs in the strongest emission lines of HH objects (e.g. Hα, [SII] $\lambda\lambda$ 6716, 6731). In order to be sure that the observed structures are indeed due to in situ formed line emission (and not due to scattered light) a continuum frame

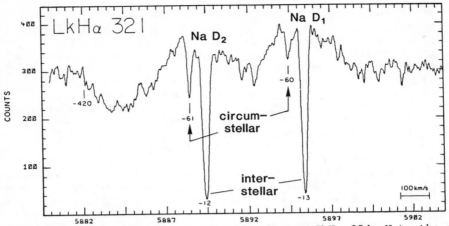

Fig. 1: NaD line profiles of the T Tauri star LkHα 321. Note the narrow absorption components at -60 km/s.

has to be taken as well.

So far no author has given any clear definition of a jet or collimated flow from a young star. The following definition of a jet on the basis of the currently available data is suggested: knotty, elongated structure with an HH-like spectrum and a length-to-diameter ratio L/D ≥ 5. In the case of a collimated flow the spectrum and morphology is the same, but the length-to-diameter ratio is smaller (2 ≤ L/D < 5). These two definitions will be used in this article. However, no attempt will be made here to classify the currently known optical outflows into these two classes.

By far the best example of a jet is the one emanating from HH34-IRS (see Fig. 2 and 3, or Reipurth et al. 1986). An example of a collimated flow is shown in Fig. 4 (DG Tau). Another example of this type is HH43 (e.g. Strom et al. 1986) or HH57 (e.g. Reipurth 1985a). So far about 20-25 jets and about 10-15 collimated optical flows have been discovered. The most extensive lists of such objects have been published by MBB87 and by Strom et al. (1986). Additional recent examples are discussed by Reipurth (1985c), Reipurth and Graham (1988), Ray (1987), Solf (1987), Hartigan and Graham 1987 and in this article.

3.2 Examples and Morphological Properties

On the following pages several examples of jets associated with YSOs are shown. Nearly all these CCD images have been taken in January 1987 at the prime focus of the 3.5 m telescope on Calar Alto, Spain. For most images displayed here a [SII]λλ6716, 6731 interference filter was used (Δλ = 70 Å). Note that in several cases it is more advantageous to image in the [SII] lines instead of Hα. The reason can be either a strong Hα background or a jet of very low excitation (or both). Most of the jets shown here have been discovered on previous CCD images (for details see

MBB87 or the corresponding figure captions). The exception is the jet emanating directly from HL Tau (Mundt, Ray, and Bührke 1987). However, all the CCD images displayed here are of superior quality than those published elsewhere.

Among the presently known jets the following morphological features seem to be common in a significant number of them:
- knots along the jet
- bright knots at the end of the jet
- bow shocks
- wiggles

(i) Knots Along the Jet

The best examples of knotty stellar jets shown here are associated with the following sources: HH34-IRS (Fig. 3), DG Tau B (Fig. 4), HH24-SSV63 (Fig. 5), and Haro 6-5 B (Fig. 7). The HH34 jet is of course the most striking example. 13 knots have been discovered by Bührke, Mundt, and Ray (1988) over a distance of 30 arcsec from the source. The typical knot separation is 1.2-2.5 arcsec (0.8-1.7 x 10^{15} cm). It should be pointed that all jets show knots on high dynamic range images. Therefore, knots are very probably a morphological property common to <u>all</u> jets.

ii) Bow Shocks and Bright Knots at the Jet's End

These morphological features are described here together, since they are physically closely related (i.e. part of the jet's working surface). This will be discussed in section 6.

One of the best examples of an association between a jet and a bow shock like structure is HH34S and HH34N (see Fig. 2). In this case a bow shock is indicated on both sides of the source. Other good examples shown here are HH1 (Fig. 10), RNO 43N, and RNO 43S (Fig. 8 and 9). A bow shock like structure is apparently associated also with the jets from DG Tau (Fig. 4) and Haro 6-5B (Fig. 7). In the case of DG Tau indications of a bow shock were also discussed by MBB87 and Bührke (1987) on the basis of long-slit spectra. Bright knots are observed at the end of several flows. The examples shown here are the collimated flow from DG Tau (Fig. 4) and the southern (redshifted) HH24 jet (Fig. 5). Other examples are the HH46/47 jet (e.g. Graham and Elias 1983) or the jet associated with AS 353A (Hartigan, Mundt and Stocke 1986).

iii) Wiggles

Nearly all jets are not perfectly straight, but show at least small changes in direction. Among the examples shown here, wiggles are most prominent in the case of Haro 6-5B, RNO 43N, and HH30. A similar case is the HH46/47 jet (Graham and Elias 1983; Raga and Mateo 1987).

3.3 Spectroscopic and Kinematic Properties

In Table 1 the typical observational properties of the jets and their sources are summarized.

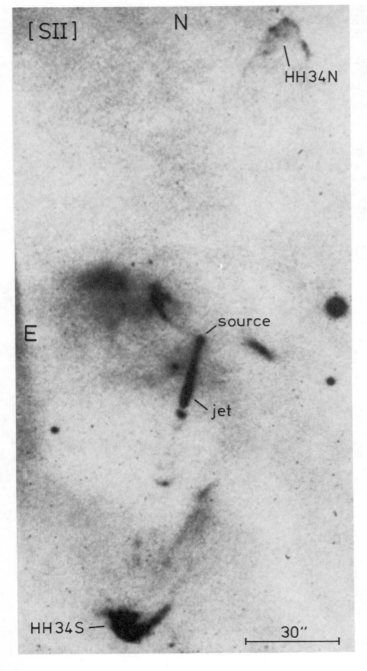

[SII]

N

HH34N

E

source

jet

HH34S

30"

Fig. 2: [SII] image of the HH34 region. A knotty jet is pointing towards HH34S (see also Fig. 2). Both, this HH object and the one on the opposite side of the source (HH34N) have the shape of a bow shock. HH34S and the jet are blue-shifted (-80--130 km/s), HH34N is redshifted (+120-+220 km/s). On all figures north is up and east to the left.

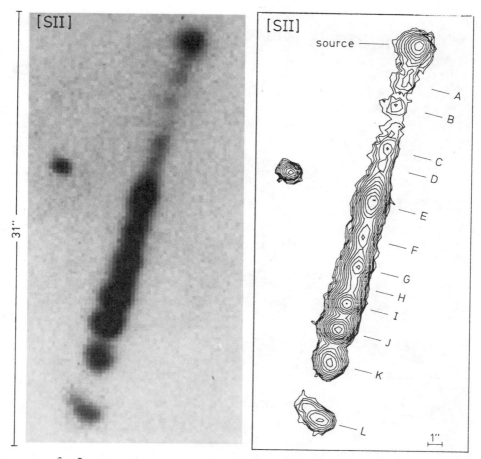

Fig. 3: [SII] image and isointensity contour plot of the HH34 jet. Twelve knots are visible in the section shown with a typical separation of 2" (from Bührke, Mundt, and Ray 1988)

As mentioned in section 1 the jets show the same spectrum as HH objects. In several cases of knotty jets and knotty collimated flows a spectrum of rather low excitation is observed (e.g. $I([SII]\lambda\lambda\,6716,\,6731)/I(H\alpha) > 1$). The best examples are the flows from HH34-IRS (Reipurth et al. 1986), SSV13 (Brugel, Böhm, and Marnery 1981), SSV63 (Solf 1987), HH1-VLA1 (Strom et al. 1985), and DG Tau B (MBB87). Their excitation can be explained by very oblique internal shocks, caused by high jet Mach numbers.

Radial velocity data are available for nearly all flows. The mean radial velocities are of the order of 100-150 km s^{-1}, with a maximum observed velocity of about 400 km s^{-1} (for details see MBB87). In several of the examples shown, the measured heliocentric radial velocities are indicated in the figures. This gives the reader some idea about the

264

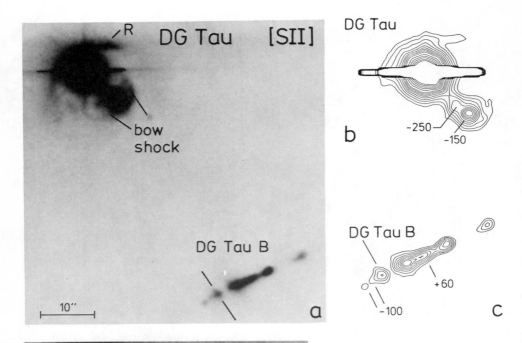

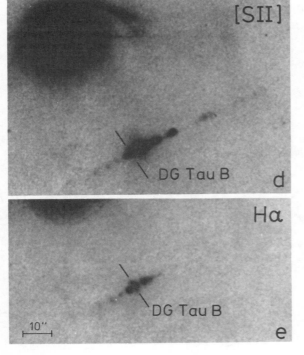

Fig. 4: DG Tau and DG Tau B region. 4a and 4d shows a [SII] image at different levels of contrast. Both stars are associated with reflection nebula; that of DG Tau has an arc-like shape and is marked by an "R". 4b and 4c shows iso-intensity contour plots in [SII] of the regions near the two stars. The indicated numbers in this and the following figures are km/s. Note the strong difference between the [SII] and Hα image of the redshifted jet of DG Tau B.

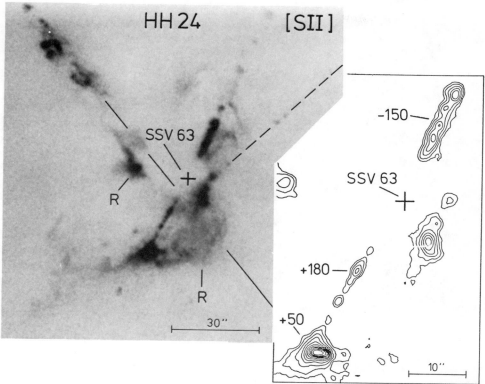

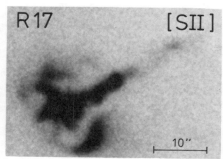

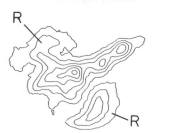

Fig. 5: HH24 region. In addition to
the bipolar jets emanating from SSV63
(Solf 1987) two aditional outflows
are suggested by this [SII] image,
which are indicated by a full and
broken line (see also Fig. 12 of
Cohen and Schwartz 1983). Note that
this region is a complicated mixture
of scattered light and in situ formed
line emission. The brightest reflection nebula are marked by an "R".

Fig. 6: Object No. 17 from Reipurth
1985b (see also Reipurth 1985c).
Bright reflection nebulae are indicated by an "R".

266

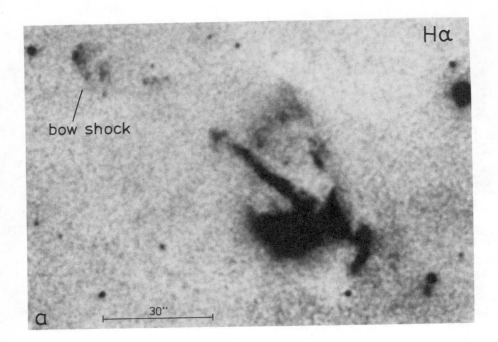

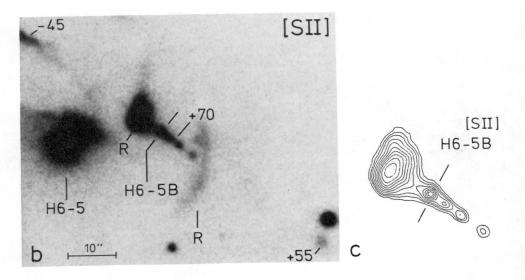

Fig. 7: Haro 6-5B region. In the Hα image in Fig. 7a the blueshifted jet is pointing towards a faint emission nebula, which has the shape of a bow shock. Note the knotty redshifted jet emanating from Haro 6-5B (see also MBB87). An "R" indicates bright reflection nebulae.

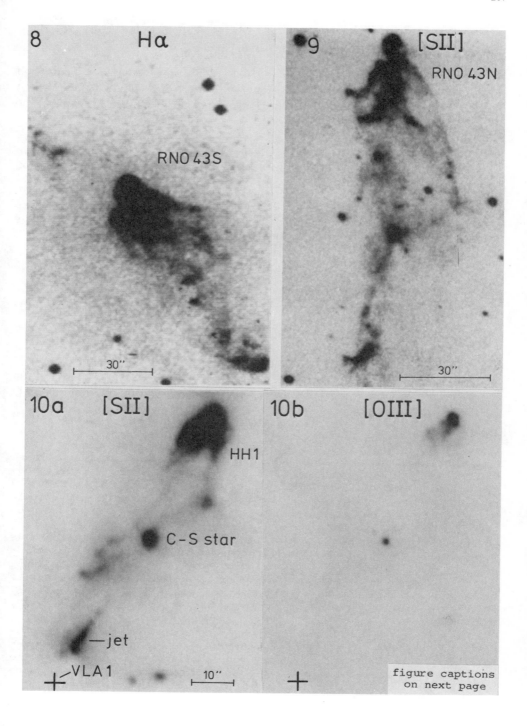

figure captions
on next page

Fig. 8 and 9: RNO 43S and RNO 43N. These are two examples of extended HH objects, of which the outer edge has the shape of a bow shock. Both objects are associated with diffuse and patchy jets.

Fig. 10: HH1 region. A similar morphology as for HH34S is indicated (e.g. knotty jet pointing towards a bow shock-like HH object). As in HH34S (see Fig. 13) the [OIII] emission originates in a relatively compact region near the apex.

kinematic properties of these flows. The relatively high radial velocities are accompanied by moderate velocity dispersions of typically 30-100 km s^{-1} (FWHM). Recent observations with the Coudé spectrograph of the 2.2 m telescope on Calar Alto (Bührke 1987) showed that the jets of HH30, HL Tau, Haro 6-5B, DG Tau B have very small velocity dispersions in the [SII] lines (FWHM < 30 km/s). Such small velocity dispersions are expected on theoretical grounds, if the emitting material is excited by oblique internal shocks (Falle, Innes and Wilson 1987).

Table 1: Typical properties of jets from young stars

projected length:	0.02 - 0.5 pc
length-to-width ratio:	10 - 20
opening angle:	5o
radial velocity:	$\lesssim$ 400 km/s
velocity dispersion (FWHM):	< 30 km/s (6 cases), $\sim$100 km/s (5 cases)
$v_{tangential}$ of bright knots:	$\lesssim$ 300 km/s
electron density:	400 - 2000 cm^{-3}
spectrum:	shock-excited emission line spectrum (HH spectrum), v_{shock} = 40-100 km/s
sources:	low luminosity IR sources (1-10 L$_\odot$) or highly extincted T Tauri like stars with reflection nebulae

Figure 11 shows the [SII] and Hα line profiles of two bright knots of the HH34 jet (from Bührke, Mundt, and Ray 1988). Two velocity components are indicated in the [SII] lines. One at -85 km/s with a FWHM < 30 km/s and an apparently resolved component at lower velocity (-53 km/s). The Hα line profiles is significantly different. However, it is shown by these authors that it can also be fitted by the same velocity components observed in the [SII] lines. Of course, the low velocity component has to be much stronger in Hα. These two components are interpreted as emission from internal shocks (high velocity component) and from the boundary layer, i.e. entrained gas (low velocity component). Similar results have been obtained for the jets associated with HH46/47 and HH24 and these have been interpreted in a similar way (Meaburn and Dyson 1987, Solf 1987).

Both blue and redshifted jets are observed. About 50% of the know flows are optically bipolar on the basis of radial velocity or proper motion data. In about half of these, there are strong differences in the morphology and spectroscopic properties between the two sides of the outflow; see e.g. DG Tau B (Fig. 4) or Haro 6-5B (Fig. 7). Proper motion data using photographic plates are available for 40% of the flows. However, mainly measurements of the brightest knots were possible, which are located normally at the end of the flow. The measured tangential velocities range from 50-350 km s^{-1}. Recently Neckel and Staude (1987) were able to measure the proper motions of the knots of the L1551-IRS5 jet with the help of CCD images. For all 4 knots in the jet they derived a tangential velocity of about 200 km/s.

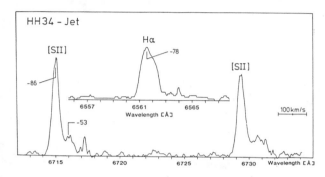

Fig. 11: [SII] and Hα line profiles of two knots of the HH34 jet. The high-velocity [SII]-component is unresolved (FWHM < 30 km/s).

For 15 of the currently known jets and collimated flows the radial velocities are high enough ($\gtrsim 50$ km/s) to study radial velocity variations along the flow axis (for details see MBB87). In general a complex velocity field is observed and sometimes several flow components are present. The only correlation in the kinematic data is the following: when a bright knot is present at the end of the jet a radial velocity decrease is observed within or near the knot (or the data are at least consistent with a decrease). Known examples are the flows associated with T Tau, DG Tau, HH24, HH33/40, HH46/47, and AS 353A. The corresponding kinematic data in the case of DG Tau and HH24-SSV63 are indicated in Figure 4 and 5. However, there is no indication that the flow velocity gradually decreases along the jet (before the bright knot at the jet's end), which means that the kinetic energy is transported rather efficiently along the jet (MBB87; Bührke, Mundt, and Ray 1988).

The electron density in the jets, based upon the [SII] 6716/6731 line ratio, has typically values of 400-2000 cm^{-3}. In 9 of the 12 flows with reasonable [SII] data N_e is decreasing along the flow direction, or at larger distances from the source the average electron density is smaller. This correlation is not surprising, since for a <u>diverging flow</u> with constant mass flux and velocity the density should decrease with increasing distance from the source. In the case of the jet from L1551-IRS and HH34-IRS it can indeed be shown that the product of jet cross section and N_e is constant along the jet within a factor of about three, although the electron density decreases by a much larger factor (MBB87; Bührke, Mundt, and Ray 1988).

4. SOURCE PROPERTIES

In all cases, where the jet source is detectable on a CCD frame (e.g. an I frame) a reflection nebula can be detected on deep exposures. In a few cases only a reflection nebula is detectable, since the source itself is so highly extincted; probably by a circumstellar disk (e.g. VLA 1-HH 1/2, Strom et al. 1985). About 70% of these nebula can be described as conical or cometary in nature, and perhaps most interestingly in 8 cases the nebular axis and jet axis is approximately aligned. One of the best examples are the jets associated with L1551-IRS5 (e.g. Mundt and Fried, 1983) and 1548C27 (Mundt et al. 1984). A similar morphology is observed in some of the cases shown here (see e.g. Fig. 2).

For the nebula associated with Haro 6-5B, L1551-IRS5 and R Mon detailed polarization maps have been obtained (see respectively Gledhill, Warren-Smith, and Scarott 1986; Lenzen 1987; Aspin, McLean, and Coyne 1985). These and other observations (e.g. Hodapp 1984 and references therein) show that SSV13, Haro 6-5B, DG Tau, L1551-IRS5, R Mon, and 1548C27 are highly polarized objects with polarization ranging from 5-22%. A interesting fact is that for all these highly polarized objects (except 1548C27; see Craine, Boeshaar, and Byard 1981) the electric polarization vector is approximately perpendicular to the jet (measured angles 72-89°). These observations are attributed to the presence of a circumstellar disk) the axis of which defines the outflow axis. In such a situation the high polarization can in principle be explained by scattering within the polar lobes of the disk (Elsässer and Staude 1978). For a very detailed discussion of the available polarization data of outflow sources and YSOs the reader is also referred to Bastien (1987).

Most sources are (or appear) highly reddened ($A_V \gtrsim 3-5$). The high A_V values and the associated reflection nebula argue for a considerably smaller average age compared to typical T Tauri stars. Very probably the jet sources belong to the youngest optically selected YSOs. The flows from the sources with large optical brightnesses have all high radial velocities of $\gtrsim 150$ km s^{-1}. This suggests that their low A_V-values result from a relatively large angle (greater than say 30°) between the line of sight and the plane of the presumed cirumstellar disk.

Nearly all known jet sources have luminosities of 0.5-60 $L_\odot$. Only R Mon has a much higher luminosity (~ 740 $L_\odot$). This does not mean that highly collimated flows are predominantly associated with low luminosity stars, since the currently available data are heavily biased towards these YSOs. In addition to R Mon a few other cases of well collimated flows from medium luminosity objects ($L \sim 10^3$ $L_\odot$) are known (Sanders and Willner 1985, Lane and Bally 1986). Nevertheless it should be seriously investigated whether jet phenomena are more common among low-mass YSOs.

About 14 of the sources have been observed spectroscopically at Hα and all show active emission, mostly typical for strong emission T Tauri stars. All sources which have been spectroscopically studied in detail show P Cygni line profiles in at least one line. At least seven sources show clearly the characteristics of T Tauri stars (e.g. FeII lines). In several cases the emission line spectrum of the star has not been observed directly but through their scattered light (see e.g. Cohen, Dopita, and Schwartz 1986).

The same observing technique was used by Mundt et al. (1985) for L1551-IRS5. Their observations showed that this star is very probably a FU Orionis star. This was confirmed by near IR spectroscopy carried out by Carr, Harvey, and Lester (1987). They showed that the first overtone bands of CO are in absorption, as in other FU Orionis stars. In SSV13 (HH7-11), DG Tau, AS 353A, and 1548C27 the CO bands were observed in emission (Carr 1988; see also Hamann, Simon, and Ridgway 1987). HH57-IRS is a further collimated outflow source, which has been recognized as a FU Orionis object (Graham and Frogel 1985, Reipurth 1985a).

5. ACCRETION DISKS AND COLLIMATED OUTFLOWS

Recently obtained data on FU Orionis stars strongly suggest that these objects are surrounded by luminous accretion disks (for details see Hartmann, this volume; Hartmann and Kenyon 1985, 1987). Furthermore, the CO band emission described in section 4 is probably formed in an accretion disk (Carr 1988). In this case the disk is optically thin in the continuum, which is not the case for the FU Orionis stars, since there the same lines appear in absorption.

It has been proposed by Edwards and Strom (1987) that the energetic mass outflows of YSOs are probably due to viscous accretion disks. In the case of T Tauri stars it has been shown by Bertout, Basri, and Bouvier (1987) that several of their properties can be explained by an accretion disk model (see also Kenyon and Hartmann 1987).

The wind properties of FU Orionis may give us a clue to the formation of collimated ouftlows. According to Crosswell, Hartmann, and Avrett (1987) the observed line profiles can be interpreted by a wind emanating directly from an accretion disk. This means, the wind would be intrinsically bipolar and already collimated. Whether the degree of collimation is sufficient to explain highly collimated jets is unclear. Maybe this is not necessary, since the subsequent collimation could take place on much large scales (e.g. $\sim$10 AU) through ambient pressure gradients.

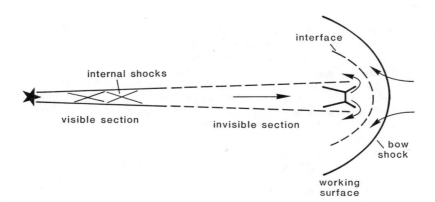

Fig. 12: Schematic structure of the working surface of a jet. The sketched "streamlines" refer to a reference frame moving with the working surface.

6. BRIGHT KNOTS AND BOW SHOCKS ASSOCIATED WITH THE JET'S WORKING SURFACE

The region, where a jet collides with the ambient medium has been called the "working surface" by Blandford and Rees (1974). The schematic structure of this region is sketched in Figure 12. A large fraction of the jet's kinetic energy may be thermalized here and therefore this region can be rather bright. Indeed, for about 70% of the currently known jets a bright knot or bow shock-like region is observed at the end of the flow.

Whether a detectable, radiative bow shock is produced by the working surface depends on a number of factors, besides ambient gas density and orientation of the flow. It has been shown by Raga (1986) that one should observe an extended radiative bow shocks if (1) the bow shock velocity is high ($\gtrsim 150$ km/s) and/or if (2) the ambient medium is (at least partially) ionized. To fullfill the first requirement the working surface velocity v_{ws} should be $\gtrsim 150$ km/s. For a jet velocity v_j of about 300-400 km/s this is the case if the ratio between jet density and ambient gas density $\rho_j/\rho_a \sim 1$ (for details see e.g. Mundt 1986). The following relation is approximately valid between these quantities:

$$\rho_j/\rho_a = v_{ws}^2/(v_j - v_{ws})^2$$

A v_{ws}-value of $\gtrsim 150$ km/s means that the HH knots embedded in the extended bow shock structures discussed here should have correspondingly high proper motion and/or radial velocities. For most cases with known proper motion or radial velocity data this requirement is fullfilled (e.g. tangential velocities of about 300 km/s have been measured for HH1 and RNO 43S). For the remaining cases where the space velocities are apparently lower (e.g. RNO 43N, v(tangential)~120 km/s) it can be argued that the ambient medium may be partially ionized; e.g. one observes on long-slit spectra background emission typical for a low excitation HII region.

The fact that the jet cannot be traced in all cases from the source towards the bow shock apex (e.g. HH1, HH34S, Haro 6-5B) has been explained by Mundt (1986) by a section of free expansion (invisible jet section in Fig. 12). In such a situation no internal shocks form, which make the gas visible. An alternative explanation would be a jet of very high Mach number (low temperature), of which any internal shocks are so oblique that the shock velocities (i.e. the normal velocity components to the shock surface) are too low to give rise to optical line emission.

In addition to the observed bow shock structures near the end of the jet, there are also kinematical arguments that the bright knots or regions observed there are associated with the working surface. In the working surface a flow deceleration should occur. As discussed in section 3.3 such a deceleration is indeed observed for all those flows, which have radial velocities high enough to measure such an effect by medium resolution spectroscopy.

7. PHYSICAL STRUCTURE OF RADIATIVE BOW SHOCKS

Objects like HH34S are rather attractive for testing our current understanding of the physical structure of radiative bow shocks. Models of such bow shocks have been calculated by Rózyczka and Tenorio-Tagle (1985), Raga and Böhm (1985), Raga (1986) and by Hartigan, Raymond, and Hartmann (1987). In addition Raga and Böhm (1987) have calculated time dependent bow-shock models. They showed that the knotty structure observed in all bow-shock like HH objects can in principle be explained by thermal instabilities (see e.g. Fig. 13 for HH34S). Unfortunately, all these models assume that the bow shock is formed by a rapidly moving solid sphere and not by a jet. Therefore, if one also observes shock heated jet gas in the working surface, and not only external gas heated in the bow shock, these models may be a rather poor approximation to objects like HH34S.

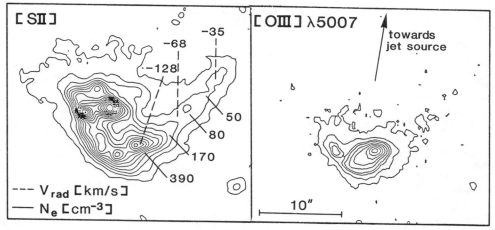

Fig. 13: Isointensity contour plots of HH34S in [SII] and [OIII]. Note that [OIII] emission is detected mainly near the apex of HH34S (from Bührke, Mundt, and Ray 1988).

Recently Bührke, Mundt and Ray (1988) have carried out a detailed spectroscopic and imaging study of HH34S. There study indicates at least in part agreement between observations and theory. Some results are illustrated in Fig. 13. According to this figure the highest excitation region, which is traced by the [OIII] emission, is restricted to the apex of the bow shock. As indicated by Fig. 10 a similar result was obtained for HH1. Furthermore, Fig. 13 illustrates that both the measured electron density and radial velocity is decreasing along the western wing of the bow shock. Such a behaviour is indeed expected on theoretical grounds. However, the agreement between the observed and predicted velocity structure in other parts of HH34S is less satisfactory. This may suggest that one is observing cooling jet gas as well and not just cooling ambient gas.

8. SMALL GROUPS OF OUTFLOW SOURCES

It is not unusual to observe several optical outflow sources on one CCD frame, e.g. on an area of about 2'x2' (or within 0.1x0.1 pc at a distance of 150 pc). Examples are DG Tau and DG Tau B (projected separation 0.04 pc), the HH24 region with apparently 3 sources within about 0.07 pc (Mundt 1987b), or the Reipurth 20 region with 3 outflows within 0.3 pc (Reipurth 1985b, Mundt 1987b). Another example is the NGC 1333/HH7-11 region. There Liseau, Sandell and Knee (1987) argue that a cluster of sources is responsible for the high velocity molecular gas north-west of HH7-11.

By far the best example is probably the HL/XZ Tau region in the L1551 dark cloud (Mundt, Ray, and Bührke 1987). The CCD frames shown in Figures 14 and 15 indicate that there are 4 jet sources within an area of about 0.05x0.2 pc. For details the reader is referred to the corresponding figure captions. The sources in this area are HL Tau, VLA1-HL Tau (a VLA radio source), the HH30-star, and the Hα jet source. The latter source has not been detected yet, but it must be located SE of XZ Tau (see also arrow in Fig. 14c). In total 6 young stars are known in this area of the L1551 cloud (the 4 outflow sources, XZ Tau, and LkHα 358). The bipolar jet from HL Tau (p.a. = 48°) is approximately perpendicular to the disk-like structure (p.a. = 120°) found by Sargent and Beckwith (1987) around that star through high spatial resolution CO observations.

The average duration τ_{OF} of the optical outflow phase of a low-mass YSO has been estimated by MBB87 to about $2x10^4$ yrs (τ_{OF} is the time over which HH emission can be found on deep CCD frames near a YSO). Note that τ_{OF} is the sum over all active phases, not counting any dormant phases in between. If these stars are located near the "birthline" of low-mass stars, they should have an age of about 10^5 yrs (Stahler 1983). This means τ_{OF} is probably relatively small compared to the expected ages of the outflow sources. Why are 4 stars out of a small association of 6 stars active at the same time? If this "simultaneous" activity is not accidental, they must be at the same evolutionary phase (within about $2x10^4$ yrs). If this is true, the formation of these 4 stars may have somehow been triggered by an external process. However, even if one triggers their formation simultaneously, it is still surprising, that they appear at the same evolutionary phase after about 10^5 yrs. Note that different initial conditions (e.g. in density) should lead to different phases. Can phases of high mass accretion be triggered in 4 closely spaced YSOs, e.g. through simultaneous gravitational disturbances of the outer parts of their infalling clouds? If so, we can understand the above observation, providing jets result from accretion disks and are observable only during phases of high mass accretion.

According to Strom and Strom (1987) and Ray (1987) the outflow directions of different sources in the same dark cloud tend to be aligned among each other and with respect to the magnetic field. In the L1551 cloud the same correlation is suggested for most outflows, including the one from IRS5. The clear exception is the Hα-jet south of XZ Tau. Furthermore, the relatively good alignment of the other 4 flows within an angle of 37° is in part only apparent. Note that the different

Fig. 14: The low contrast [SII]
image in the lower right hand
corner gives an overview of the
HL Tau/HH30 region. Two independent jets, emanating from the
VLA source and from HL Tau are
indicated on the [SII] frame
shown in 14a and 14b. These two
jets apparently criss-cross each
other. This is also illustrated

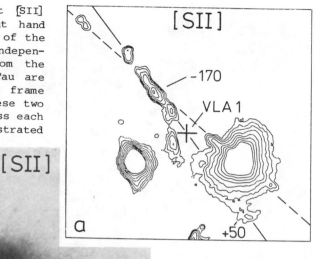

in the contour plot in
14a, where the two
outflow directions are
indicated by a broken
and full line. Note
also the Hα-jet south
of XZ Tau. The arrow
marks an emission knot
which may be associa-
ted with the source of
that jet.

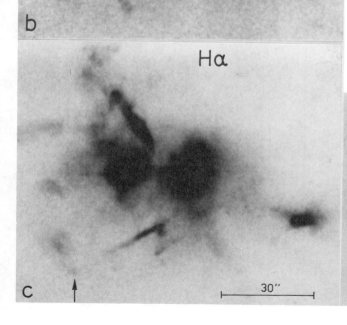

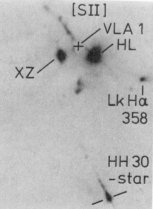

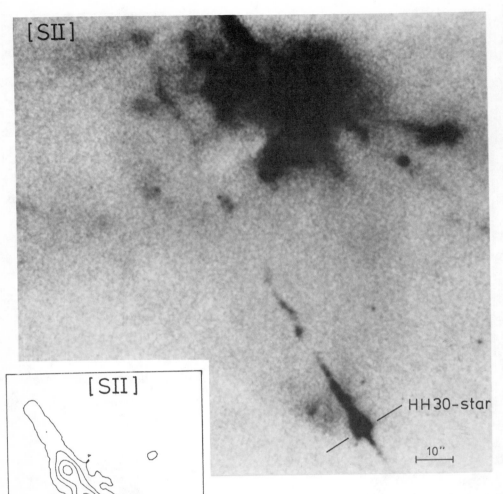

Fig. 15: High contrast [SII] image, showing the 2 arcmin (0.1 pc) long jet emanating from the HH30-star. A corresponding contour plot of this jet near the source is also shown. (from Mundt, Ray, and Bührke 1987)

flow orientation with respect to the line of sight have to be taken into account. For example, on the basis of the observed radial velocities it can be shown that the HH30 jet is practically located within the plane of sky (MBB87), while the one from HL Tau is probably inclined by about 20-30° with respect to that plane.

The author thanks Dr. T. Ray for valuable comments and critically reading the manuscript.

References

Aspin, C., McLean, I.S., and Coyne, G.V. 1985, Astr.Ap. **149,** 158
Bastien, P. 1987, Ap.J. **317,** 231
Bertout, C., Basri, G., and Bouvier, J. 1987, preprint
Blandford, R.D., and Rees, M.J. 1974, M.N.R.A.S. **169,** 395
Brugel, E.W., Böhm, K.-H., Mannery, E., 1981, Ap.J.Suppl. **47,** 117
Bührke, T. 1986, Ph.D. Thesis, University of Heidelberg
Bührke, T. 1987, priv. comm.
Bührke, T., Mundt, R., and Ray, T. 1988, Astr. Ap.,submitted
Cantó, J. 1986, in "Cosmical Gas Dynamics", ed. F.D. Kahn, VNU Science
    Press (Utrecht)
Carr, J.S., Harvey, P.M., and Lester, D.F. 1987, Ap.J.(Letters)
    321, L71
Carr, J.S. 1988, this volume
Clark, F., Laureijis, R., Chlewieki, G. Zhang, C., Oosterom, W., and
    Kester 1986, Astr.Ap. **168,** L1
Cohen, M., Schwartz, D.R. 1983, Ap.J. **265,** 877
Cohen, M., Dopita, M.A., and Schwartz, R.D., 1986 Ap.J. (Letters)
    307, L21
Craine, E.R., Boeshaar, G.O., and Byard, P.L. 1981, A.J. **86,** 751
Crosswell, K., Hartmann, L., and Avrett, G. 1987, Ap.J. **312,** 227
Dyson, J.E. 1984, Astrophys.Space Sci. **106** 181
Dyson, J.E. 1987, in "Circumstellar Matter", eds. I. Appenzeller and C.
    Jordan, Reidel (Dordrecht), p. 159
Edwards, S., and Snell, R. 1984, Ap.J. **281,** 237
Edwards, S., Strom, S., Snell, R., Jarett, T., and Strom, K. 1986,
    Ap.J. **307,** L65
Edwards, S., and Strom, S.E. 1987, in "Fifth Cambridge Cool Star
    Workshop", eds. J.L. Linsky and R. Stencel, Springer Verlag
Elsässer, H., and Staude, H.J. 1978, Astr.Ap. **70,** L3
Falle, S.A.E.G., Innes, D., and Wilson, M.J. 1987, M.N.R.A.S. **225,** 741
Gledhill, T.M., Warren-Smith, R.F., and Scarott, S.M. 1986,
    M.N.R.A.S. **223,** 867
Graham, J.A., and Elias, J.H. 1983, Ap.J. **272,** 615
Graham, J.A., and Frogel, J.A. 1985, Ap.J. **289,** 331
Hamann, F., Simon, M., Ridgway, S.T. 1987, NOAO-preprint, No. 105
Hartigan, P., and Graham, J.A. 1987, A.J. **93,** 913
Hartigan, P., Raymond, J., Hartmann, L. 1987, Ap.J. **316,** 323
Hartmann, L., and Kenyon, S.J. 1985, Ap.J. **299,** 462
Hartmann, L., and Kenyon, S.J. 1987, Ap.J. **312,** 243

Herbig, G.H. 1960, Ap.J. Suppl. **4,** 337
Hodapp, K.-W. 1984, Astr.Ap. **141,** 255
Kenyon, S.J., and Hartmann, L. 1987, Ap.J., in press
Kuhi, L.V. 1964, Ap.J. **140,** 1409
Lada, C.J. 1985, Ann. Rev. Astron. Astrophys. **23,** 267
Lane, A.P., and Bally, J. 1986, Ap.J. **310,** 820
Lenzen, R. 1987, Astr.Ap. **173,** 124
Liseau, R., Sandell, G., and Knee, L.B.G., Astr.Ap., in press
Meaburn, J., and Dyson, J.E. 1987, M.N.R.A.S. **225,** 863
Moriarty-Schieven, G.H., Snell, R.L., Strom, S.E., Schloerp, F.P., Strom, K.M., and Grasdalen, G.L. 1987, Ap.J. **319,** 742
Mundt, R. 1984, Ap.J. **280,** 749
Mundt, R. 1985a in "Nearby Molecular Clouds", ed. G. Serra, Lecture Notes in Physics, **217,** Springer Verlag (Heidelberg), p. 160
Mundt, R. 1985b in "Protostars and Planets II", eds. D. Black and M. Matthews, University of Arizona Press (Tucson), p. 414
Mundt, R. 1986, in "Jets from Stars and Galaxies", Can. J. Phys. **64,** 407
Mundt, R. 1987a, in "Circumstellar Matter", eds. I. Appenzeller and C. Jordan, Reidel (Dordrecht), p. 147
Mundt, R. 1987b, unpublished obs.
Mundt, R., Brugel, E.W., and Bührke, T. 1987, Ap.J. **319,** 275 (MBB87)
Mundt, R., Bührke, T., Fried, J.W., Neckel, T., Sarcander, M., and Stocke, J. 1984, Astr.Ap. **140,** 17
Mundt, R., and Fried, J.W. 1983, Ap.J. (Letters) **274,** L83
Mundt, R., Ray, T., and Bührke, T. 1987, in preparation
Mundt, R., Stocke, J., Strom, S.E., Strom, K.M., and Anderson, E.R. 1985, Ap.J. (Letters) **297,** L41
Neckel, T., and Staude, J. 1987, Ap.J. (Letters), in press
Raga, A.C. 1986, A.J. **92,** 637
Raga, A.C., and Böhm, K.H. 1985, Ap.J. Suppl. **58,** 201
Raga, A.C., and Böhm, K.-H. 1987, A.J., in press
Raga, A.C., and Mateo, M. 1987, A.J. **94,** 684
Ray, T. 1987, Ast.Ap. **171,** 145
Reipurth, B. 1985a, Astr.Ap. **143,** 435
Reipurth, B. 1985b, Astr.Ap. Suppl. **61,** 319
Reipurth, B. 1985c, in "(Sub)-Millimeter Astronomy", ESO-IRAM-Onsala Workshop, eds. P. Shaver and K. Kjar, p. 459
Reipurth, B., Bally, J., Graham, J.A., Lane, A., and Zealey, W.J. 1986, Astr.Ap. **164,** 51
Reipurth, B., and Graham, J.A. 1988, Astr.Ap., submitted
Rózyczka, M., and Tenorio-Tagle, G. 1985, Astr.Ap. **147,** 220
Sanders, D.B., and Willner, S.P. 1985, Ap.J. (Letters) **293,** L39
Schwartz, R.D. 1983, Ann. Rev. Astron. Astrophys. **21,** 209
Snell, R.L. 1987, in "Stars Forming Regions", eds. M. Peimbert and J. Jugaku, Reidel (Dordrecht), p. 213
Sargent, A.I., and Beckwith, S. 1987, Ap.J., in press
Solf, J. 1987, Astr.Ap. **184,** 322
Stahler, S.W. 1983, Ap.J. **274,** 822
Staude, J. 1988, in "Structure of Galxies and Star Formation", proceedings of the 10th Europ. Reg. Astr. Meeting of the IAU

Strom, S.E., and Strom, K.M. 1987, in "Star Forming Regions", eds. M.
     Peimbert and J. Jugaku, Reidel (Dordrecht), p. 255
Strom, S.E., Strom, K.M., Grasdalen, G.L., Sellgren, K., Wolff, S.,
     Morgan, J., Stocke, J. and Mundt, R. 1985, A.J. **90**, 2281
Strom, K.M., Strom, S.E., Wolff, S.C., Morgan, J., and Wenz, M. 1986,
     Ap.J. Suppl. **62,** 39

Note Added to Section 2

Recent HI observations by Lizano et al. (1987, preprint) with the Arecibo radio telescope show that some YSOs have neutral atomic winds of moderate velocities. For the HH7-11 region the HI line profiles indicate velocities of up to 170 km/s. Moreover, the estimated mass loss rate of 3×10^{-6} $M_{\odot}$/yr suffices to drive the observed bipolar CO flow. Their more ambiguous data for L1551-IRS5 show smaller widths in the HI line profiles of up to 50 km/s (estimated $\dot{M} = 4 \times 10^{-6}$ $M_{\odot}$/yr). Similar HI velocities are indicated for the NGC 2071 outflow source (Bally and Stark 1983, Ap.J. (Letters) 266, L61). These HI observations support the conclusions made here on the basis of the narrow blueshifted NaD absorption features, that low-mass YSO generate neutral atomic flow components with $v \sim$ 50-80 km/s. Lizano et al. believe that for L1551-IRS5 the real wind velocities are much higher than 50 km/s, due to geometrical projection effects ($v_{w} = 50$ /cos $\Phi = 190$ km/s, $\Phi = 75^{\circ}$, $\Phi =$ angle between line of sight and flow direction). Such a Φ-value is far too large in consideration of the large radial velocities of the jet (up to -350 km/s) and $\Phi = 45^{\circ}$ is much more likely. Furthermore, if the cavity seen in scattered light (see section 2 and Lenzen 1987) is formed by the wind, it is very poorly collimated near the source (full opening angle $\sim 110^{\circ}$). If so, the observed line width at zero intensity of 50 km/s is practically identical with the true wind velocity.

Pinch Instabilities in Young Stellar Object Jets:

T.P. Ray*, T. Bührke and R. Mundt
Max-Planck-Institut für Astronomie
Königstuhl, D-6900 Heidelberg 1

Abstract. It is proposed that the quasi-periodic knot-like features seen in jets from young stars may, in some cases, be due to Kelvin-Helmholtz instabilities; in particular, non-disruptive reflection modes. Numerical simulations have shown that such modes saturate at finite amplitude through the formation of a periodic train of oblique internal shock waves. These shocks can explain the characteristic spectrum of the jet and the formation of knots. We predict that the knots should move out with a substantial fraction of the jet velocity. The case of HH34 is examined in detail.

1. Introduction.

Young stellar object (YSO) jets are produced by stars of low to intermediate luminosity ($L \sim 1$-100 $L_\odot$). For a review of this phenomenon see Mundt (this volume). YSO jets are spectroscopically identical to Herbig-Haro (HH) objects so their emission arises from shock excited gas. Modelling implies that the shock velocities are very small, $V_{shock} \sim 40$ km/s (Mundt, Brugel and Bührke, 1987), whereas a typical jet velocity $V_{jet} \sim 300$ km/s. It follows that the shocks must be very oblique.

All YSO jets contain knots and, in at least 4 cases, a series of 4-6 periodically spaced knots have been seen (Mundt, 1987). Examples include DG Tau B and HL Tau (Mundt, Brugel and Bührke, 1987). Similar structures are observed in extragalactic jets e.g. M87 (Biretta, Owen and Hardee, 1983) and NGC 6251 (Perley, Bridle and Willis, 1984). In a seminal paper, Stewart (1971) suggested that the knots in the M87 jet might be due to Kelvin-Helmholtz (K-H) instabilities. Hardee (1982) and Perley, Bridle and Willis (1984) have considered whether the large scale "wiggles" seen in some extragalactic jets are K-H helical modes. However, there is a problem. The parameter of these jets are not well known, so it is difficult to check whether a K-H instability model is applicable. YSO jets are different. Through

---

* Alexander von Humboldt Fellow on leave from School of Cosmic Physics, Dublin Institute of Advanced Studies, Dublin.

A. K. Dupree and M. T. V. T. Lago (eds.), Formation and Evolution of Low Mass Stars, 281–283.

spectroscopy and direct imaging, we can measure such important numbers as jet densities, velocities (both radial and transverse), sound speed, etc. Therefore, in principle, it should be easier to find out whether the knots and, in some cases, wiggles (Mundt, this volume) of YSO jets are due to K-H instabilities. Moreover, these investigations might help us to gain insight into extragalactic jets as well.

2. Kelvin-Helmholtz Instabilities in YSO Jets.

The importance of Kelvin-Helmholtz (K-H) instabilities in jets has been investigated by several authors (see for example: Hardee, 1979; Ferrari, Trussoni and Zaninetti, 1981; Ray, 1981; Birkinshaw, 1984). The simplest K-H mode is the fundamental pinching mode, sometimes, referred to as the "sausage" mode. This is essentially a longitudinal oscillation along the jet axis. If, however, the Mach number of the jet $M_{jet} > 1 + \eta^{\frac{1}{2}}$ where η is the ratio of the jet, to ambient medium densities, then the so called "reflection modes" dominate over fundamental modes (Payne and Cohn, 1985). This condition should be fullfilled in YSO jets as their Mach number are so high (see e.g. Mundt, Brugel, Bührke 1987). Payne and Cohn (1985) have shown that these modes produce waves at angles

$$\theta = \pm \sin^{-1} (1 + \eta^{\frac{1}{2}}/M_{jet}) \qquad (1)$$

to the jet axis. In the non-linear regime, such waves give rise to biconical internal shockwaves (Norman, Smarr and Winkler, 1984). While some energy is lost at these shocks, most is transported towards the "working-surface".

So if these modes grow in a YSO jet, we do not expect significant energy changes along its length providing the jet is steady. Moreover, given the high Mach number of YSO jets, we expect θ (equation (1)) to be small and hence the normal shocks velocity to be a small fraction of the velocity. Such shock would then give rise to the observed emission.

3. The case of the HH34 Jet.

Recent observations of the HH34 jet (Bührke, Mundt and Ray, 1987 and see Mundt, this volume) have shown that the jet consists of at least 12 distinct knots with a typical spacing of 1.2-2.5". The jet is resolved transversely for several of the outermost knots. Their diameters are approximately constant (~ 0.7") giving a knot to diameter ratio ~ 3. This constancy implies that the jet is not free, at least along some of its length, but is instead somehow confined. This could be a result of gas pressure from the surrounding medium. Observations of the HH34 jet and bowshock (Bührke, Mundt and Ray, 1987) suggest the jet is denser than the external medium. Using equation (1) we find $\theta \sim 7°$ for a jet Mach number $M_{jet} = 25$ and $\eta = 5$ (i.e. $V_{jet} \sim 350$ km/s and T_{jet}, the jet temperature, $\sim 10^4$ K). We would therefore expect oblique shock waves to form with a normal shock velocity of approximately 350 sin 7° km/s i.e. about 45 km/s. This is approximately the value calculated on the basis of spectroscopic observations (Bührke, Mundt and Ray, 1987). Furthermore, assuming the first reflection mode is dominant we expect a knot spacing to jet diameter

$$0.4\ M_{jet}/(1 + \eta^{\frac{1}{2}}) \sim 3.0 \qquad (2)$$

where we have allowed for a small reduction in knot spacing due to non-linear effects (Payne and Cohn, 1985). This value is consistent with the observations. There are several reasons why we invisage the first reflection mode will be the major instability. One reason is that in the reflection mode regime, the spatial growth rate of the most unstable modes barely increase with wavenumber (Payne and Cohn, 1985). All modes will therefore grow at about the same rate with a wave packet length given by equation (2). This conclusion appears to be justified by numerical simulations (Norman, Winkler, and Smarr, 1984). Moreover, we can see from equation (2) that higher order reflecting modes have $kD_j > 1$ where k is the wavenumber. As pointed out by Ferrari, Massaglia and Trussoni (1982) and by Ray (1982), such modes would tend to be stabilized by the presence of a shear layer. Such a layer is suggested by the HH34 jet observations (Bührke, Mundt and Ray, 1987). If the observed knots in the HH34 jet are due to either a single reflection mode, or a nonlinear superposition of reflecting modes, then the knots should move out with a velocity V_{knot} V_{jet}. This is observationally testable over a period of a few years.

References

Biretta, J.A., Owen, F.N. and Hardee, P.E., 1983, Ap.J. 274, L27
Birkinshaw, M., 1984, Mon. Not. R. astr. Soc., 208, 887
Bührke, T., Mundt, R. and Ray, T.R. 1987, submitted to Astron. Astrophys.
Ferrari, A., Massaglia, S. and Trussoni, E., 1982, Mon. Not. R. astr. Soc., 198, 1065
Ferrari, A., Trussoni, E. and Zaninetti, L., 1981, Mon. Not. R. astr. Soc., 196, 1051
Hardee, P., 1979, Ap.J. 234, 47
Hardee, P., 1982, Ap.J. 261, 457
Mundt, R., Brugel, E.W., Bührke, T. 1987, Ap.J. 319, 275
Norman, M.L., Smarr, L., and Winkler, K.-H. A., 1984, Numerical Astrophysics: A Meeting in Honor of James Wilson, eds. J. Centrella, R. Bowers and J. LeBlanc (Portola Valley: Jones and Bartlett).
Norman, M.L., Winkler, K.-H.A., and Smarr, L.L., 1984, Physics of Energy Transport in Extragalactic Radio Sources, eds. A. Bridle and J. Eilek.
Payne, D.G. and Cohn, H., Ap.J. 291, 655
Perley, R.A., Bridle, A.H. and Willis, A.G., 1984, Ap.J. Suppl., 54, 291
Ray, T.P., 1981, Mon. Not. R. astr. Soc., 196, 195
Ray, T.P., 1982, Mon. Not. R. astr. Soc., 198, 617.
Stewart, P., 1971, Astrophys. Sp. Sci., 14, 261

The NATO ASI Blues

I came to Viana
To learn from the pros
How to drink vinho verde
And say, "Obrigado"

It rained the first day
The second through twelfth
By next week Viana
Will be in the gulf

Now there's Shu and Jim Pringle
Hartmann and Calvet
And a whole lot of water
From Scott Kenyon's bidet
(big riff)

Now Baliunas has spots
And Mundt has his jets
Shu talks way too fast
And Strom's in Massachusetts

There's too much food
And too much wine
Ten people have died
From Turkish moonshine

I'm tired of viewgraphs
The screen is too small
And without Doug Gough
It won't focus at all

(riff)
The vinho is gone
The talks are all through
I'm left down and out, oh yeah!
I'm left without a cup of coffee, people!
I'm left on the street without any escudos
I'm left with just a teaspoonful, oh yeah,
A teaspoonful of those NATO ASI blues..
(big riff)

- *P. Hartigan*

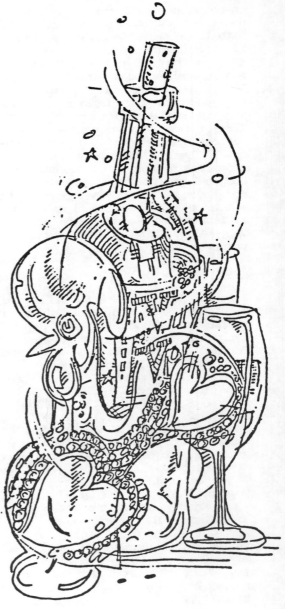

LARGE LINE WIDTHS IN HH OBJECTS – THE BOW SHOCK MODEL

Patrick Hartigan
Harvard-Smithsonian Center for Astrophysics
Mail Stop 16
60 Garden St
Cambridge MA 02138
USA

ABSTRACT. The line widths of Herbig-Haro objects sometimes exceed 400 km s^{-1} within an object only about 1500 AU in size. Many of the line profiles are double-peaked, with the stronger of the two peaks occurring near zero radial velocity. These unusual spectral characteristics together with the observed mixture of high and low excitation lines provide essential clues to the geometry and physics of the radiative shocks responsible for the emission. In this paper the predictions of a radiative bow shock model are briefly reviewed, with special emphasis placed on the explanation of broad-lined HH objects.

I. Introduction

Schwartz (1975) first noticed the similarity between HH object spectra and those of supernova remnants, and suggested that the emission lines observed arose behind a radiative shock. Direct evidence for supersonic motions in HH objects now exists from the detection of large ($\gtrsim$ 100 km s^{-1}) proper motions and radial velocities of some knots (see Schwartz 1983 for a review). These motions are supersonic since the sound speeds in the radiating gas are known to be ~ 10 km s^{-1} from studies of forbidden line ratios (Brugel, Böhm, and Mannery 1981). The mixture of high excitation lines such as C IV λ1550 and low excitation lines like [O I] seen in HH objects is expected from gas cooling behind a radiative shock (Raymond 1979, Dopita 1978), although the observed line ratios generally do not agree at all with those predicted for planar shocks.

One of the startling properties of HH objects is the presence of line widths exceeding several hundred km s^{-1} within a single knot of typical size 1500 AU. Line widths larger than 250 km s^{-1} can be found in the HH 1–2 region (Schwartz 1981), in Cep A (Hartigan et al. 1986), near M42 (Taylor et al. 1986), HH32 (Mundt, Stocke and Stockmann 1983; Hartigan, Mundt and Stocke 1986), and L1551 (Stocke et al. 1988). How can such enormous velocities exist within a small, gravitationally unbound object? The predicted observational differences between the existing models of HH emission (eg. jet model: Norman, Smarr and Winkler 1984; Meaburn and Dyson 1987; bullet model: Norman and Silk 1979; Hartigan, Raymond, and Hartmann 1987; cloudlet model: Schwartz 1981; cavity model: Cantó and Rodriguez 1980, etc.) arise from the various locations, orientations, and shapes (curved or planar) of the radiative shocks exciting the gas in each model. The large line widths and double-peaked profiles seen in several HH objects provide an important constraint for any model. In this short contribution I summarize the most important results of the simple bow shock models (i.e. the

A. K. Dupree and M. T. V. T. Lago (eds.), Formation and Evolution of Low Mass Stars, 285–294.

bullet and cloudlet models), and show how they can be used to explain the line profiles, line widths, line ratios, and position-velocity diagrams seen toward some HH objects. More detailed discussion of the bow shock models can be found in Raga and Böhm (1986), Hartigan, Raymond and Hartmann (1987), and references therein.

II. The Bow Shock Model

The shocked cloudlet and bullet models for HH objects both have the emission arising from a radiative bow shock formed around a clump of gas in the flow. In the shocked cloudlet model the bow shock forms as stellar wind flows past a dense obstacle in the wind (the shock curves in a direction away from the star), whereas in the bullet model a dense clump is accelerated by the star and rams into the ambient cloud, forming a bow shock around the obstacle (that curves in a direction toward the star).

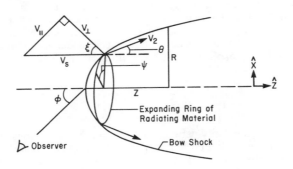

Figure 1: Bow Shock Geometry (from Hartigan, Raymond, and Hartmann 1987)

Both scenarios are modelled in the same way, with the calculations performed in the frame of reference of the bow shock and then shifted in velocity to the star's frame of reference. The so-called 'cloudlet shock' propagating into the obstacle is of low enough shock velocity that it does not contribute significantly to the observed emission for dense cloudlets and bullets. One first specifies the bow shock shape (figure 1), the preshock density η, the velocity of the bullet (or wind for the shocked cloudlet model), and the observer's orientation angle ϕ. The incident angle ξ is then determined for each point on the bow, which allows one to determine the velocity V_1. The emission lines are taken to arise behind a radiative shock of velocity V_1, which yields the relative line intensities once a suitable grid of planar shock models has been constructed (eg. Hartigan, Raymond and Hartmann 1987). To compute the line profile one needs to know the velocity V_2 and deflection angle θ of the emitting gas. These are determined by taking the parallel component to be continuous across the shock, and then computing the factor by which V_1 is reduced before the gas radiates (typically by a factor of 40 or so — a factor of 4 immediately behind the shock and an additional factor of about 10 due to radiative cooling).

The preceeding analysis assumes that the cooling distance behind the shock is small compared to the size of the HH object. If this is not valid, the gas will be able to flow to a different velocity field before it radiates. The distance required for postshock gas to cool to 10^3 K (for shock velocities greater than about 60 km s^{-1}) is given by a power law,

$$d_{c3} = 12 \left[\frac{V_\perp}{100 \text{ km s}^{-1}} \right]^{4.67} \left[\frac{100 \text{ cm}^{-3}}{\eta} \right] \text{ AU} \tag{1}$$

so the assumption of small cooling distance is valid for all but the highest velocity HH objects, and even in these fails only near the bow's apex where $V_\perp$ is large.

III. Predicted Line Ratios, Line Profiles, and Position-Velocity Diagrams

Can the bow shock model produce the large line widths seen in HH objects? It turns out that the full width near zero intensity of a line emitting behind a bow shock equals the incident shock velocity V_s *independent* of the shape of the bow shock, the orientation angle ϕ, the preshock density, elemental abundances, reddening, etc. This behavior can be understood in terms of a simple analytical calculation. For each of the radiating rings in figure 1, the maximum radial velocity occurs at the top of the ring, and the minimum radial velocity at the bottom. Neglecting the small component of $V_\perp$ that remains after the gas cools ($V_2 = V_\parallel$) we find that the maximum radial velocity MX seen from the bow shock occurs when $\xi = \pi/2 - \phi/2$, and has a value $(V_s/2)(1+\cos\phi) + \gamma$, where $\gamma = 0$ or $-V_s\cos\phi$ for the cloudlet and bullet models, respectively (the calculation was done in the frame of reference of the bow shock, and must be translated to the frame of reference of the observer in the case of a bullet). A similar calculation of the minimum velocity MN shows it occurs when $\xi = \phi/2$, so that MN = $-(V_s/2)(1-\cos\phi) + \gamma$. Hence, the FWZI = MX $-$ MN = V_s for any emission line that radiates both at $\xi = \phi/2$ and $\xi = \pi/2 - \phi/2$.

This is an extremely useful result, since it implies that the observed values of MX and MN for any line profile immediately give both the shock velocity V_s, and, choosing one of the two models, the orientation angle ϕ. This result is also independent of the slit width provided the slit is aligned with the long axis of the bow shock. A narrow slit will alter the line profile, however. Hence, large line widths from a small area emerge naturally whenever a curved shock is present. The bow shock models also make several predictions that are less intuitive. The models in figure 2 show that as the viewing angle ϕ decreases from 90 degrees, the profiles become increasingly asymmetric, with two peaks present for viewing angles less than about 45 degrees when the shock velocity exceeds about 150 km s^{-1}. The stronger of the two peaks lies near zero radial velocity for the bullet model, and at high radial velocity for the cloudlet model. Of the seven clearly double peaked profiles known only HH 32D is fit by a cloudlet model, with the rest having stronger low velocity peaks (eg. figure 3). Both models predict a spatial separation of high and low velocities in the sense that the high velocity emission occurs closer to the star than the low velocity emission when ϕ is oblique. Such spatial separations are indeed observed in both the Cepheus A (Hartigan et al. 1986) and HH 32 (Solf, Böhm, and Raga 1986) systems. The proper motions predicted from the V_s and ϕ of the bullet models agree well with those observed in the HH 1-2 region, L1551, and HH 32.

It is difficult to compare the observed and predicted line ratios for many HH objects since reliable dereddened ultraviolet and visual fluxes exist only in the HH 1-2 region. However, application of the model to HH 2A′ has resulted in a significant improvement over previous (planar) attempts (Raymond, Hartigan, and Hartmann 1988).

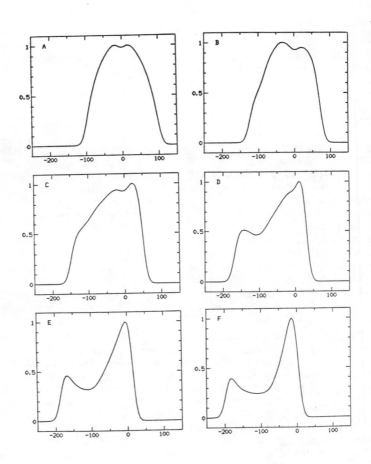

Figure 2: Predicted Hα line profiles for a 200 km s^{-1} bow shock (bullet geometry) viewed at different orientation angles φ (see figure 1). Radial velocities in km s^{-1} are plotted vs. fluxes normalized to unity at the peak. Models A − F have φ = 90, 75, 60, 45, 30, and 15 degrees, respectively. The line widths are all 200 km s^{-1} although the line profiles vary considerably in shape (from Hartigan, Raymond, and Hartmann 1987).

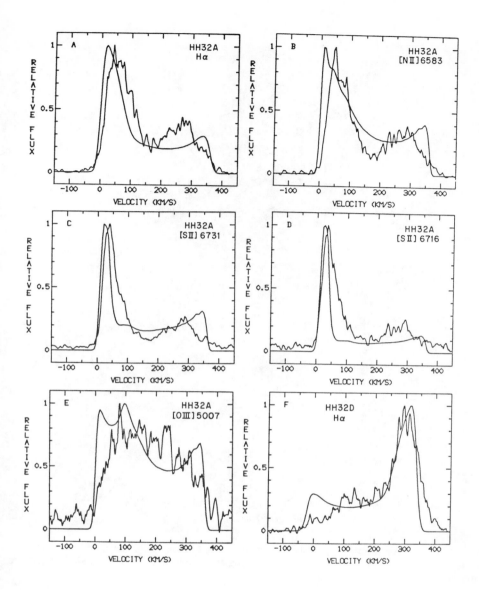

Figure 3: Model and observed profiles for HH 32A and 32D. The observations were taken from Hartigan, Mundt, and Stocke (1986). HH 32A was modeled with a 360 km s^{-1} bullet, and HH 32D with a shocked cloudlet (from Hartigan, Raymond, and Hartmann 1987).

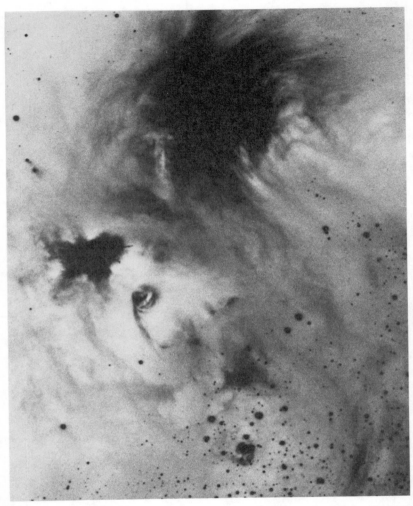

Figure 4: A negative print of a red photograph of the R CrA region taken by J. Graham (Hartigan and Graham 1987). A number of Herbig-Haro objects and young stellar objects are visible in this image (see figure 5).

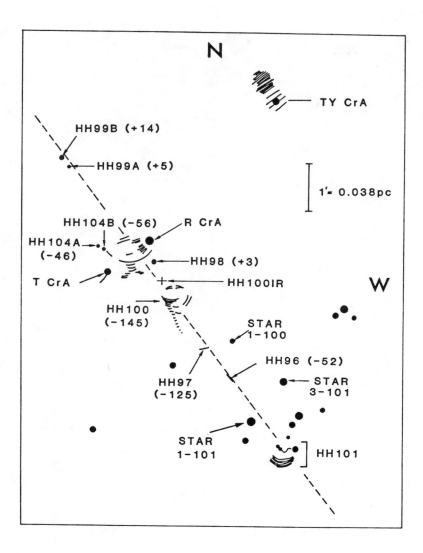

Figure 5: An identification sketch for figure 4. Velocities of the HH objects and locations of the brightest young stars in the field are indicated (from Hartigan and Graham 1987).

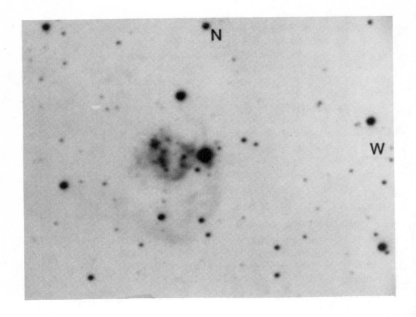

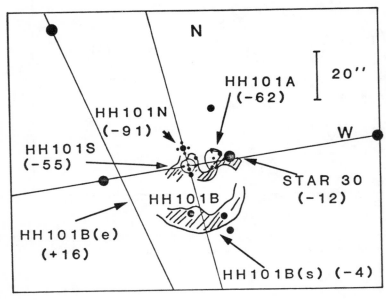

Figure 6: This photograph of HH 101 was taken by J. Graham under extraordinary seeing conditions, allowing structures as small as ~ 80 AU to be visible. At this spatial scale the emission from HH 101 breaks up into about 30 small knots. The lines on the sketch show the slit positions used to determine radial velocities (from Hartigan and Graham 1987).

IV. Concluding Remarks

A simple radiative bow shock model predicts the enigmatic double peaked line profiles and large line widths seen toward some HH objects, as well as the observed mixture of high and low excitation lines, and the spatial separation of high and low velocity peaks. In those cases where the bow shock model works well, a bullet geometry is usually indicated. Regardless of the input parameters, bow shocks should always include the systemic velocity of the system within the line profile. This is not always the case (eg. L1551 Stocke et al., 1988; HH 7-11 Solf and Böhm 1987; HH24 Solf 1987), and although the model can sometimes be modified to match observations by having a bullet ramming into a moving medium, there are clearly some cases, especially those regions with prominent jets (eg. Mundt 1988) where the shock structure seen is more complex than a single bow shock.

It is difficult to accelerate a cloudlet significantly by a wind without disrupting the cloudlet (Różyczka and Tenorio-Tagle 1985), and since most bullet models have the star accelerate the bullet initially by a stellar wind, this could pose a problem for the bullet model as well. One possibility might be to have a nonsteady jet which breaks up into blobs that subsequently move out to form bullets. Certainly at some level of clumpiness, a jet model like those favored by Mundt, Brugel and Bührke (1987), with a jet working surface, backflowing cocoon, extended bow shock, and contact discontinuity, will become indistinguishable from a series of interstellar bullets. A major difficulty with the present state of the jet model is the lack of any quantitative predictions concerning the line widths, line ratios, line profiles, position-velocity diagrams, etc. that are expected from the shocks in the working surface, cocoon, and in the jet itself (although considerable progress has been made recently in the latter area by Falle, Innes, and Giddings 1987; see also Ray 1988). We *can* say that no encompassing backflowing cocoon has ever been seen around a jet, and that large extended bow shocks are presently visible in only a few objects, so it is not clear that the continuous jet model has the correct shock placements.

At a level of about 100 AU or so, there are indications that the structure of some HH objects becomes very complex. The photographs of HH 101 in Figures 4 – 6 of the R CrA region taken by J. Graham (Hartigan and Graham 1987) under extraordinary seeing conditions reveal that HH 101 breaks up into about 30 distinct knots at the subarcsecond level. To model such complexity probably requires a sort of interstellar meteorology that is likely to be specific to this region. The launch of the Hubble Space Telescope will undoubtedly reveal additional inhomogeneties in many HH objects. In summary, the simple bullet model works extremely well in predicting the rather confusing kinematics seen in some HH objects, although the model does not account for the velocity structure seen in all systems. The most natural way to get large line widths from a small area is to have a strongly curved shock, and this seems to occur in the broad-lined HH objects.

The author would like to thank J. Raymond, L. Hartmann, and S. Kenyon for their comments on a draft of this manuscript.

References

Brugel, E.W., Böhm, K.-H., and Mannery, E. (1981). *Ap. J. Supp.*, **47**, 117.

Cantó, J., and Rodriguez, L.F. (1980). *Ap. J.*, **239**, 982.

Dopita, M.A. (1978). *Ap. J. Supp.*, **37**, 117.

Falle, S.A.E.G., Innes, D.E., and Wilson, M.J. (1987). *Mon. Not. R. Astr. Soc.*, **225**, 741.

Hartigan, P., and Graham, J.A. (1987). *Astr. J.*, **93**, 913.

Hartigan, P., Mundt, R., and Stocke, J. (1986). *Astr. J.*, **91**, 1357.

Hartigan, P., Lada, C.J., Stocke, J., and Tapia, S. (1986). *Astr. J.*, **92**, 1155.

Hartigan, P., Raymond, J., and Hartmann, L. (1987). *Ap. J.*, **316**, 323.

Meaburn, J., and Dyson, J.E. (1987). *Mon. Not. R. Astr. Soc.*, **225**, 863.

Mundt, R. (1988). This volume.

Mundt, R., Brugel, E.W., and Bührke, T. (1987). *Ap. J.*, **319**, 275.

Mundt, R., Stocke, J., and Stockman, H.S. (1983). *Ap. J. (Letters)*, **265**, L71.

Norman, C., and Silk, J. (1979). *Ap. J.*, **228**, 197.

Norman, M.L., Smarr, L., and Winkler, K.-H.A. (1984). *Numerical Astrophysics*, eds. J. Centrella, J. LeBlanc, M. LeBlanc, and R. Bowers (Boston: Jones and Bartlett).

Raga, A.C., and Böhm, K.-H. (1986). *Ap. J.*, **308**, 829.

Ray, T.P. (1988). This volume.

Raymond, J.C. (1979). *Ap. J. Supp.*, **39**, 1.

Raymond, J.C., Hartigan, P., and Hartmann, L. (1988). To appear in *Ap. J.*, March 1.

Różyczka, M., and Tenorio-Tagle, G. (1985). *Acta. Astr.*, **35**, 213.

Schwartz, R.D. (1975). *Ap. J.*, **195**, 631.

Schwartz, R.D. (1981). *Ap. J.*, **243**, 197.

Schwartz, R.D. (1983). *Ann. Rev. Astr. Ap*, **21**, 209.

Solf, J. (1987). preprint.

Solf, J., and Böhm, K.-H. (1987). *Astr. J.*, **93**, 1172

Solf, J., Böhm, K.-H., and Raga, A.C. (1986). *Ap. J.*, **305**, 795.

Stocke, J., et al. (1988). *Ap. J.*, submitted.

Taylor, K., Dyson, J.E., Axon, D.J., and Hughes, S. (1986). *Mon. Not. R. Astr. Soc.*, **221**, 155.

SOME ASPECTS OF T TAURI VARIABILITY

Gösta F. Gahm
Stockholm Observatory
S-133 00 Saltsjöbaden
Sweden

ABSTRACT

A review is presented of light variations of T Tauri stars in ultraviolet, visible and
infrared light. Special attention is given to variability connected to stellar rotation
and pulsation and to flarelike events. Very complex patterns of variability is
observed for some stars and RY Lupi is selected as a case study.

1. INTRODUCTION

One very characteristic property of the T Tauri stars is that they vary in brigthness with
time. This subgroup of young stellar objects have optical spectra of late spectral type and are
known to represent low mass stars in early phases of contraction to the main sequence. The
variability of T Tauri itself was discovered in 1852 and several of the stars have been
followed by observers over this century. We therefore have an immense amount of published
material on the seemingly irregular optical brightness changes of these objects. A large
portion of the early works was based on photographic observations. Since several of the
classical T Tauri stars have relatively large amplitudes in the optical, there has also been a
large number of visual estimates collected by both amateurs and professional astronomers and
these estimates continue to be a source of reference with regard to what happens to individual
stars. Over the last decades, of course, photoelectric photometry in broad and narrow
filterbands provides the most important observational data basis. These photoelectric
observations have revealed the properties of the stars from the ultraviolet to the infrared
and has established the presence of small amplitude fluctuations of a very regular nature for
several of the stars.

Many of the stars are also known to vary in linear polarization, spectral appearance and
at X-ray and radio wavelengths. In the following we will concentrate on some of the findings
from mainly groundbased, broad-band observations in ultraviolet, visible and infrared light.

2. PATTERNS OF VARIABILITY

2.1. Some General Properties

The light variations of the T Tauri stars are very different from star to star both with regard

A. K. Dupree and M. T. V. T. Lago (eds.), Formation and Evolution of Low Mass Stars, 295–304.

to amplitudes and characteristic time scales of the fluctuations. It also happens that one and the same star may change its general behaviour drastically over a certain period. As described in an early review by Herbig (1962) it has been common practise to group the stars according to the appearance of the light curve. In an extension of a classification system introduced by P. Parenago, Herbig gives the following classification system for the light curves: The variable is more frequently bright (Class I), at mean brightness (Class II), faint (Class III) or shows no preference for any level in its range (Class IV). In addition, any characteristic time scale is noted for instance by an s if it is short or l if it is long. These descriptions already tell us something about the diversity of patterns of variability and can be found in certain catalogues of T Tauri stars, the most important being the one by Herbig and Rao (1972).

For many stars only few observations exist and hence published values of for instance the total range in visual magnitude can be misleading. Nevertheless, overall statistical properties have been sought over the years. One result, which seams to have been rediscovered a few times, is that stars which are *most frequently bright* have *weak* optical *emission* line spectra Plagemann, 1967 Gahm, 1970; Weaver and Frank, 1980).

The amplitudes of the variations are different at different wavelengths. *The general rule* is that the stars become *redder* with *decreasing* brightness. This is not always the case, however, and as was apparent already in the early photographic work by Badaljan (1958, 1962, 1964) there are stars that turn bluer with decreasing brightness or that show more or less "grey" variations. Very complex patterns of colour changes are seen for certain stars and we will discuss one such case in more detail in Chapter 3.

So far, rather little monitoring of T Tauri stars has been made at near-infrared or infrared wavelengths. Already Nandy and Pratt (1972) concluded that the range of variability in the I band is smaller than in the U, B and V bands. The general impression is that the stars are relatively steady longward of 1 μ (Rydgren et al. 1984b, Fridlund et al. 1984) but several stars with significant infrared variability have been noted (e.g. Cohen, 1976).

It is clear that in order to understand how and why the stars vary in so many different ways one must monitor the stars during several nights, preferably both spectroscopically and photometrically over a wide range of the electromagnetic spectrum. This is not an easy task since co-ordinated observations involving several instruments on the ground, and possibly in space, can pose extraordinary administrative and economic problems. Nevertheless, several relatively ambitious case studies have been performed over the years.

We will not give a comprehensive list of all the works in this direction but rather focus on a few topics that are, or may be, of relevance to the question of the origin(s) of T Tauri variability.

2.2. Rotational and Pulsational Modulation of Light

One of the major break-throughs in T Tauri star research during the last decade has been the discovery that the stars are relatively slow rotators (small $v \sin i$) and that many show small-amplitude periodic light variations in accordance with the expected rotational velocities. (Vogel and Kuhi, 1981; Rydgren and Vrba, 1983; Rydgren et al. 1984a, 1985; Bouvier et al. 1986; Hartmann et al. 1986).

It is clear that a number of stars vary predominantly because of rotational modulation of the light from a changing mixture of hot plages, cool spots and a "normal" stellar photosphere. The periods fall in the time interval between 1 and 10 days. Early stellar

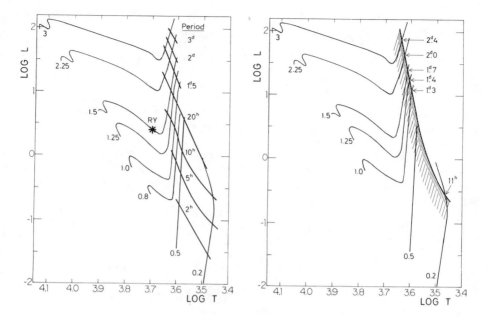

Figure 1 a: The left HR-diagram shows the expected periods of pulsations for convective contracting stars of indicated mass in solar units. The position of RY Lupi is marked.
Figure 1 b (right): The beginning of intensive deuterium burning is represented by the full-drawn curve. The corresponding pulsational periods are given.

rotation is discussed in more detail separately at this Advanced Study Institute. Here, we will draw the attention to a few of the details.

For several of the stars the observations are not carried through several periods and we must await additional observations for confirmation of periodicity. It is difficult to make any precise statement on what percentage of the stars actually show periodic fluctuations. In a survey reported by Bouvier (1987) about 1/3 of the 26 stars observed displayed periods while Fischerström (1987) found one in a sample of 9 stars. Obviously, if small amplitude variations of a few tenths of a magnitude in V are present they are difficult to detect if disturbed by much larger irregular variations of a different nature.

Periodic variations of relatively large amplitude have been found in a few stars like RY Lupi (Hoffmeister, 1965) and SY Cha (Kappelmann and Mauder, 1981, Schaeffer, 1983). In these cases it may be very difficult, even impossible, to model the variations by the introduction of spots/plages as was realised for RY Lup by Liseau et al. (1987a). We will return to this star in Chapter 3. In fact, Mauder (1977) hinted at a possibility proposed by R. Kippenhahn that infalling circumstellar material may give rise to pulsations of a star like SY Cha. In the following we will investigate the possibility of pulsations in the context of T Tauri variability.

As described by Bouvier (1987) and Gahm and Liseau (1988) the pulsational periods that can be expected among T Tauri stars are generally smaller than the observed photometric periods described above. Given a position on a model track for a contracting star of mass M in the HR-diagram ($\log L/L_\odot$ versus $\log T_{eff}$) the corresponding expected pulsational period π (in days) can be calculated as:

$$\log (\pi/0.0703) = 0.75 [\log L/L_\odot - 4 \log T_{eff} - (2/3)\log (M/M_\odot) + 15.02]$$

This formula holds for fully convective stars (polytropes with index 1.5) but more detailed calculations by Toma (1972) differ very little from this result. In Figure 1a it is seen that *provided a mechanism exists* to drive the pulsations, the expected periods range from 1 hour to 1 day for the low mass stars. Up to now very few series of observations have been designed as to detect periodicities of a few hours.

For one case, RY Lup mentioned above, Liseau et al. (1988) found indications of a period of 19.5 hours in addition to the well established period of 3.75 days. The location of RY Lup in the HR diagram (see Fig. 1a) does not support the view that pulsations are responsible for the fluctuations of short duration.

As discussed by Toma (1972), the on-set of deuterium-burning may cause pulsational instabilities in the stars. Fig. 1b reproduced from Gahm and Liseau (1988) shows the expected location of the beginning of intensive deuterium burning. Shu and Adams (1986) has reported on similar calculations and drawn the attention to the fact that this locus falls along the so called birth-line (Stahler, 1983). It would be of interest to search stars in this region of the HR diagram for short-term fluctuations.

In closing this section we note that the evidence for rotational modulation of light from spots and plages on T Tauri stars is very strong for many stars and that the implied total angular momenta in general are smaller than the total angular momentum of our present solar system. The evidence of pulsational modulation of the stellar light is practically null but it could be rewarding to search for such modes, particularly at evolutionary phases with expected deuterium burning. It is also clear that quite different mechanisms than these are responsible for the variability of many young stars.

The ultraviolet fluxes can behave quite erratic on short time-scales, even on stars that show relatively quiescent regular variations at longer wavelengths. We will continue by considering the possibility that very energetic flarelike events take place on, or at, the stars.

2.3. Flarelike Activity

At X-ray energies several flarelike events have been recorded. The details of these observations will be presented elsewhere at this NATO course but we note that total X-ray energy contents of 10^{34} ergs, at least 100 times more than released during a large solar flare of type 3, have been reported. In an extreme case, the source ROX 20 observed by Montmerle et al. (1983), close to 10^{36} ergs must have been released. The time-scales for these events are a few to several hours. It would not be surprising if such events were noticeable also at optical wavelengths.

In a compilation by Gahm (1986) all published photometrically determined ultraviolet fluxes or magnitudes of young stellar objects were taken under consideration. A selection was made so that all events were listed where the ultraviolet flux had been found to vary by more than 20% (0.20 magnitudes) during one night with at least 3 or more observations of the same star.

In this way 22 stars (of which 21 are T Tauri stars) with a total number of 886.5 patrol hours turn out to have shown significant nightly variations in the U band during an average of 11% of the total patrol time. As can be seen in Table 1, this figure is larger for stars with prominent emission-line spectra (emission classes 4 and 5) than for those with weak emission. Rapid flarelike events are rare, however, and on the average it happens only 3% of the time that a star changes its ultraviolet flux by move than 20% within 3 hours. Since it is sometimes thought that variability on T Tauri stars is extremely dramatic it may be surprising to find that in this compilation only two flarelike events were found where both the increasing and the decreasing branches of the light curve were observed. In the same study an effort was made to evaluate the frequency of flarelike X-ray events with the result that such events occur only 5% of the time on the average. The frequency of X-ray flares and strong fluctuations in the U band are similar and it may be permitted to speculate whether they may be manifestations of the same phenomenon.

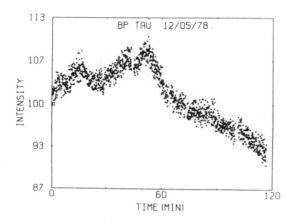

Figure 2: Relative U band intensity for BP Tauri on an arbitrary scale plotted versus time for nearly two hours of 5 sec integrations (from Schneeberger et al., 1979a).

U band variability below the 20% threshold shows a completely different pattern, however. As is seen in Figure 2 from Schneeberger et al., (1979a) fluctuations on a few percent over time-scales of 10 minutes occur repeatedly for the star BP Tau. This happens for several, but far from all, T Tauri stars according to the surveys by Kuan (1976), Schneeberger et al. (1979b), Worden et al. (1981), Schevschenko and Shutemova (1981) and Zaitseva et al. (1985). It should be noted that an increase of 5% in the U band over 10 minutes typically corresponds to a total extra energy of 10^{34} ergs which is of the same order as derived from X-ray flares. The two types of events must be totally unrelated, however, both because of the frequency by which they occur and the time-scales. It is not excluded that the rapid U fluctuations have a counterpart in X-ray variations at a level below the detection limit of the X-ray satellite Einstein.

Rapid changes in the U band are quite often accompanied by similar, but smaller changes in B and V. It is quite puzzling though that there are stars that show the signature of a flarelike event but reversed. On these occasions a star can undergo a sudden drop and recover in periods of hours. It is likely that stellar flares occur on T Tauri stars. Even so, one must be very careful in using the term flares for the rapid variations discussed here. Some events may have a completely different origin like instabilities in a hot boundary layer between a circumstellar disk and the star or shock heated circumstellar cloudlets that enter the inner region of a strong stellar wind.

For many stars the variability is not at all accounted for by spots, plages and flarelike events. As an example we will select RY Lupi for a case study.

TABLE 1. Statistics on U band variability

Total number of patrol hours	886.5
Fraction of time when the stars change in ultraviolet flux by move than 20% (whole material)	0.11
Corresponding fraction for stars with strong emission line spectra (e=4 and 5)	0.20
Fraction of time when the stars change in ultraviolet flux by more than 20% within 3 hours (whole material)	0.30
Fraction of time with significant short-term X-ray variability	0.05

3. RY LUPI - A CASE STUDY

This star has been extensively followed by several observers starting with Hoffmeister (1943, 1958, 1965). Over the last decade photometric broad- and narrow-band data have been collected and discussed by Evans et al. (1982), Kilkenney et al. (1985), Hutchinson et al. (1987) and Fischerström (1987) in addition to extensive visual observations collected by the Variable Star Section of the Royal Astronomical Society of New Zealand. Photometry in combination with polarimetry has been made by Bastien (1987).

Figure 3a shows a compilation made by Fischerström (1987) of more than 6000 observations in the visual. Each point may represent several observations. The star is seen to drop in brightness from a maximum at around V= 10 by sometimes several magnitudes. As shown in Figure 3b the star has slowly become fainter over the years. According to Fischerström (1987) this effect is real and not due to selection effects dependent on instrumentation or comparison stars.

Observations of colour changes that accompany the visual variations exist from the U band to the L band at 3.4 μ. Examples can be found in Figure 4a and 4b where V is plotted against (B-V) and (V-I) respectively. With dereasing brightness the star is seen first to become redder in the optical, then at deep minima it flips, sometimes on time-scales of hours, from being red to blue and the reverse.

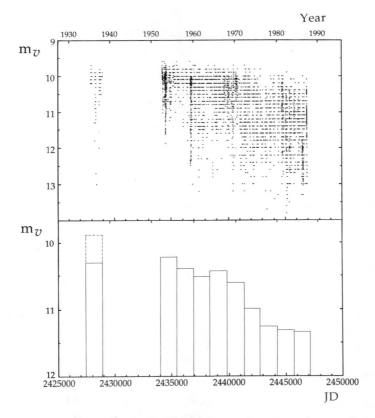

Figure 3 a (above): Compilation of more than 6000 optical observations of RY Lup. Data before 1940 are given as photographic (blue) magnitudes, while data from later than 1950 are transferred to m_{vis}.
Figure 3 b (below): Mean values of m_{vis} over 4 year periods. The dashed level represents a correction of photographic (full-drawn) data to m_{vis}. The star has decreased slowly in brightness with time.

The observational facts about RY Lup are many and we refer to the references quoted above for a full account. In the following we will summarize some of the remarkable findings about this star based on the recent discussion by Fischerström et al. (1987).

A/ The photometric period of 3.75 days has been present and constant over at least 52 years.
B/ A tendency for an additional period of about 20 hours appears when the star becomes faint.
C/ The star shows a normal stellar photospheric absorption line spectrum of spectral type G8 (metallic lines). At maximum light Hβ and Hγ are in enhanced absorption relative to spectral type G 8.
D/ Only Hα is in emission in the optical spectral region at maximum brightness. With decreasing brightness Hβ turns into emission as well.

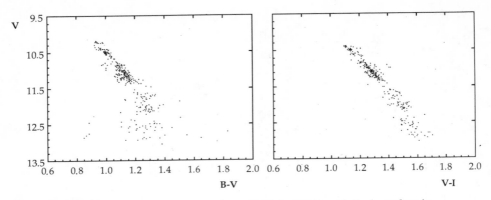

Figure 4 a (left): V magnitude against colour (B-V) for RY Lupi (all photoelectric observations) and
Figure 4 b (right): V magnitude against (V-I).

E/   The *flux* in Hα increases with decreasing brightness of the star.
F/   The spectral type of G 8 is *constant* over the entire range of variations, at least as derived from the wavelength region 4700-9050 Å.
G/   The L band flux is constant when optical variations occur.
H/   The far ultraviolet spectrum is rich in emission lines and has not changed with V during the few IUE-observations that have been made. The spectrum resembles those of Herbig-Haro objects rather than typical T Tauri stars.
I/   The degree of linear polarisation varies periodically with the 3.75 day period and is larger when the star is faint.

Presently, one can conclude that the observed pattern of variability is inconsistent with any spot or plage model and also with the hypothesis by Appenzeller and Deaburn (1984) of a variable global magnetic field that would surpress the photospheric fluxes.

On the other hand, the observed general slope of the colour changes are consistent with variable obscuration by circumstellar dust in front of the star. In this case it is, however, necessary to find a mechanism so that when dust appears in large quantities in front of the star, then the Hα-emitting region produces enhanced fluxes in optical emission lines but not in the far-UV lines. Also, the flarelike "flips" from red to blue and back in Fig. 4a must be explained. There seems to be two directions in which to find a possible solution to this problem.

On one hand, one can attribute the rapid colour flips to energetic flarelike events. Then they cannot be flares on the stellar surface since in this picture the whole star is occulted by dust. If they indeed are energetic eruptive phenomena, then these "flares" are circumstellar and lie outside the region of major obscuration. Unfortunately, we lack blue spectroscopy on these occasions, but at around 5000 Å there are no signs of any additional hot component in the G 8 spectrum during a "flare". Moreover, the change in colour from blue to red goes equally fast as changes in the opposite direction contrary to what is found for flare phenomena in general.

Another direction of thought is to assume that at deep minima we do not see the star any more but only the scattered light from the star coming from the backside of a flat disk which is inclined so that the scattered light sometimes passes unocculted over the obscuring frontside, sometimes suffers from obscuration as well. Combining measured υ sin i with the observed period of 3.75 days, we arrive at an inclination of 60 degrees which is very reasonable in this context.

Because of the scattering properties of grains, the colour (B-V) may very well turn bluer than the star, still seen as a G 8 in the reflected spectrum (see Fig. 4a). In this case the 20 hour period must be connected to a characteristic dimension of structures in the disk which by rotation move in front of the star. If the increase in Hα flux is connected to enhanced infall of foreground material, this infall cannot be radial to the star in which case the infrared emission from dust would be expected to rise - which it does not (L is constant). The 20 hour period could of course be connected to the rotation of the inner parts of the disk in which dust spirals inwards.

Finally, the far-UV spectrum does not change much which means that in the picture just outlined it originates (in a shocked collimated wind?) far out from the star and the disk.

Obviously, more observations of this fascinating object is needed. For instance blue spectroscopy during deep minima can distinguish between the hypothesis of circumstellar flarelike events and variable transmission of scattered stellar light. Also, it is important to investigate carefully whether any pecularities may be connected to the presence of a binary component. The star is not unique. RY Tauri shows a similar and yet very different behaviour. That is what makes T Tauri stars so fascinating.

References

Appenzeller, I., Deaborn, D.S.P.: 1984, *Astrophys. J.* **276**, 689.

Badaljan, H.S.: 1958, *Contr. Burakan Observatory*, **25**, 49.

Badaljan, H.S.: 1962, *Contr. Burakan Observatory*, **31**, 57.

Badaljan, H.S.: 1964, *Contr. Burakan Observatory*, **36**, 71.

Bastien, P.: 1986, private communication.

Bouvier, J.: 1987, in *Molecular Clouds and Protostars* (eds. T. Montmerle
    and C. Bertout), in press.

Bouvier, J., Bertout, C., Benz W., Mayor, M.: *Astron. Astrophys.* **165**, 110.

Cohen, M.: 1976, *Mon. Not. Roy. Astron. Soc.* **174**, 137.

Evans, A., Bode, M.F., Whittet, D.C.B., Davies, J.K., Kilkenny, D.,
    Baines, D.W.T.: 1982, *Mon. Not. Roy. Astron. Soc.* **199**, 37P.

Fischerström, C.: 1987, preprint.

Fischerström, C., Gahm, G.F., Lindroos, K.P., Liseau, R.: 1987, *Astron.
    Astrophys.* submitted.

Fridlund, M.V., Nordh, H.L., Aalders, J.W.G.: 1984, *Stockholm Obs. Report* No. **26**.

Gahm, G.F.: 1970, *Astron. Astrophys.* **8**, 73.

Gahm, G.F.: 1986, in *Flares: Solar and Stellar* (ed. P.M. Gondhalekar),
    Rutherford Appleton Laboratory RAL-86-085, p. 124.

Gahm, G.F., Liseau, R.: 1988, in *Activity in cool star envelopes*, Reidel
    Publ. Co., in press.

Hartmann, L., Hewett, R., Stahler, S., Mathieu, R.D.: 1986, *Astrophys. J.* **309**, 275.

Herbig, G.H.: 1962, *Advances Astron. Astrophys.* **1**, 47.

Herbig, G.H., Rao, N.K.: 1972, *Astrophys. J.* **174**, 401.

Hoffmeister, C.: 1943, *Kleinere Veröff. Univ. Berlin-Babelsberg* No. **27**, 3.

Hoffmeister, C.: 1958, *Veröff. Sternwarte Sonneberg*, **3**, 339.

Hoffmeister, C.: 1965, *Veröff. Sternwarte Sonneberg*, **6**, 98.

Hutchinson, M.G., Evans, A., Davies, J., Bode, M., Whittet, D., Kilkenny, D.: 1987, in *Circumstellar Matter*, (eds. I. Appenzeller, C. Jordan) IAU Symp. No. **122**, p. 109.

Kappelmann, N., Mauder, H.: 1981, *The Messenger, ESO*, **23**, 18.

Kilkenny, D., Whittet, D.C.B., Davies, J.K., Evans, A., Bode, M.F., Robson, E.I., Banfield, R.M.: 1985, *South African Astron. Obs.Circ.* No. **9**, 55.

Kuan, P.: 1976, *Astrophys. J.* **210**, 129.

Liseau, R., Fischerström, C., Gahm, G.F.: 1988, in *Activity in cool star envelopes*, Reidel Publ. Co., in press.

Liseau, R., Lindroos, K.P., Fischerström, C.: 1987, *Astron. Astrophys.* **183**, 274.

Mauder, H.: 1977, in *The interaction of variable stars with their environment* (eds. R. Kippenhahn, J. Rahe, W. Strohmeier) IAU Colloq. No. **42**, p. 66.

Montmerle, T., Koch-Miramond, L., Falgarone, E., Grindlay, J.E.: 1983, *Astrophys. J.* **269**, 182.

Nandy, K., Pratt, N.: 1976, *Astrophys. Space Sci.* **19**, 219.

Plagemann, S.: 1967, *ESRO Scientific Note*, SN-13.

Rydgren, A.E., Vrba, F.J.: 1983, *Astrophys. J.* **267**, 191.

Rydgren, A.E., Zak, D.S., Vrba, F.J., Chugainov, P.F., Zaitseva, G.V.:1984a, *Astron. J.* **89**, 1015.

Rydgren, A.E., Schmelz, J.T., Zak, D.S.: 1984b, *Publ. Naval Obs.* **25**.

Rydgren, A.E., Vrba, F.J., Chugainov, P.F., Shakhovskaya, N.I.: 1985, *Bull. Am. Astron. Soc.* **17**, 556.

Schaeffer, B.E.: 1983, *Astrophys. J. Lett.* **266**, L45.

Schevchenko, V.S., Shutemora, N.A.: 1981, *Astrofizika*, **17**, 286.

Schneeberger, T.J., Worden, S.P., Africano, J.L.: 1979a, *IAU Inf. Bull. Variable Stars*, No. **1582**.

Schneeberger, T.J., Worden, S.P., Africano, J.L.: 1979b, *Bull. Am. Astron. Soc.* **11**, 439.

Shu, F.H., Adams, F.C.: 1987, in *Circumstellar Matter* (eds. I. Appenzeller, C. Jordan) IAU Symp. No. **122**, p. 7.

Stahler, S.W.: 1983, *Astrophys. J.* **274**, 822.

Toma, E.: 1972, *Astron. Astrophys.* **19**, 76.

Weaver, Wm.B., Frank, J.L.: 1980, *Mon. Not. Roy. Astron. Soc.* **191**, 321.

Worden, S.P., Schneeberger, T.J., Kuhn, J.R., Africano, J.L.: 1981, *Astrophys. J.* **244**, 520.

Zaitseva, G.V.: 1985, *Peremennye Zvezdy*, **22**, 181.

PRE-MAIN SEQUENCE BINARIES

BO REIPURTH
European Southern Observatory
Casilla 19001, Santiago 19
Chile

ABSTRACT. The observational work which has been done to date on young low-mass binaries is reviewed. Much effort is currently put into surveys of such binaries, and results from - as well as some limitations of - various techniques are discussed. Statistical analysis of a sufficiently large sample of pre-main sequence binaries will eventually provide essential information on how stars form.

1.INTRODUCTION

We owe important parts of today's astrophysical knowledge to studies of binary stars. Fundamental stellar parameters are derived by careful studies of binaries on or slightly evolved from the main sequence. Binaries in late stages of stellar evolution have opened up new areas in high-energy astrophysics. And yet, in marked contrast to this, even simple questions regarding the formation and early evolution of binary stars are still left unanswered. However, the field of binary star formation processes is finally starting to develop, and a number of observational and theoretical studies are under way. This review gives a presentation of the most important observational work done up to now. A very useful review, with more emphasis on theoretical work, is given by Zinnecker (1984), who also lists a large number of the early references to the subject.

Observational work on pre-main sequence binaries is still largely in the discovery-phase, where various surveys attempt to identify companions to known young stars. This is being done with direct imaging, optical and infrared speckle techniques, lunar occultation observations and radial velocity measurements. With the present sensitivity limits, these various techniques mostly complement each other, and, with sufficient effort, hold promise to provide a sample of young binaries which can be studied statistically. Outstanding questions which need to be answered for young stars are, among others, a determination of the global frequency of binarity, as well as the separation distribution, the mass-ratio distribution and the angular momentum distribution functions. Such information has direct impact on our understanding of how stars form.

305

A. K. Dupree and M. T. V. T. Lago (eds.), Formation and Evolution of Low Mass Stars, 305–318.
© 1988 by Kluwer Academic Publishers.

2.FORMATION MECHANISMS OF BINARY STARS

There are several different mechanisms by which binary stars, in theory, can form, as well as various types of combination of these. First among these is fragmentation. Fragmentation is expected to occur during the isothermal collapse phase, and indeed shows up in 3-D collapse calculations (for a more detailed discussion, see Boss 1987 and references therein). However, when the collapse proceeds into the non-isothermal regime, density perturbations tend to be damped, and fragmentation is not any longer expected to occur. Depending on the initial conditions, this limit may set in at scales as small as 100 AU (Bodenheimer, this conference). In other words, fragmentation could possibly account for binaries with a separation wider than roughly 100 AU. A second mechanism, fission, is commonly appealed to in order to explain very close binaries. Fission may occur in a rapidly rotating protostar, which proceeds through a series of peanut shaped configurations to the bifurcation into two very close stars. Reviews with detailed references are given by Ostriker (1970), Lucy (1981) and Durisen and Tohline (1984). However, in a recent major study, Durisen et al.(1986) used three different 3-D hydrodynamics codes to study fission instabilities. For the n=3/2 polytropes considered, the calculations agree in that they do not produce a detached binary. Instead, mass and angular momentum is ejected through spiral arms, which eventually develop into a ring or an extended disk. It is currently not clear if fission can be considered a viable mechanism of close binary formation.

A third mechanism is capture. A binary could form in this way if two molecular clumps or two protostars experienced a grazing collision and became bound by dissipating energy through viscous interaction (Silk and Takahashi 1979). This formation method is generally expected to form wide binaries. Capture at a later evolutionary stage, when the stars no longer have envelopes to provide drag, is only possible if a third body can carry away energy, and such three-body encounters are very unlikely, except under special conditions like in globular clusters. Finally, binaries can be the end product of evaporation from a small cluster or group of stars. Within a time interval of typically less than 100 times the characteristic time scale of the system, such groups can disintegrate through close encounters which eject members, leaving behind a binary or a stable multiple system (van Albada 1968, Szebehely 1977, Harrington 1977, Contopoulos 1988).

From these models it thus appears that from a theoretical point of view there are difficulties in explaining the very close binaries with separations less than roughly 100 AU. But recent studies of the role of disks in the formation of stars have opened up scenarios in which close companions may form in a sufficiently massive disk, while less massive disks possibly could form planetary systems (see the review of Shu et al.1987 and Frank Shu's contribution to this volume).

The binaries which form from these formation mechanisms may change with time because of various effects like tidal forces or mass transfer, and a determination of the distribution functions mentioned in sect.1 for more evolved stars may therefore not fully reflect the situation at their time of formation.

3. DUPLICITY ON THE MAIN SEQUENCE

Despite the possible dynamical effects which may take place during the evolution of binaries towards the main sequence, studies of main sequence binary stars are still our best means to gain insights into binary formation mechanisms.

A large number of studies of duplicity among various types of stars along the main sequence have been carried out, and these are summarized in reviews by Abt (1983) and Herczeg (1984). The main problem in all such studies is to make a proper allowance for 'missed objects'. Taking such corrections into account, the overall fraction of duplicity among main sequence stars appears to be about 60-65%, that is, if two stars are studied more closely then, on average, one of them should turn out to be a binary. This fraction does not appear to show substantial variation with spectral type.

Of particular relevance for the formation of solar-type stars is the study by Abt and Levy (1976) and Abt (1979). They obtained 20 coude spectra of 123 bright F3-G2 IV/V stars, and combining spectroscopic and visual binaries and common proper motion pairs, they found observed ratios of single, double, triple and quadruple systems to be 45:46:8:1 (%). After an incompleteness study, the number of companions (88 observed) grew to an estimated 172. Abt and Levy then found that for periods smaller and larger than 100 years, the secondary-mass distributions are very different. In the shorter-period sample, the systems are more likely to have almost equal mass, while for the wider pairs the frequency distributon as function of secondary mass follows a van Rhijn distribution, so small mass ratios (m2/m1) are more common. This is interpreted by Abt and Levy as a result of two types of binary formation: the shorter period binaries are formed by fission of a protostar into two most commonly equal components, while the longer period systems are due to capture, which would imply that the secondaries should follow field star distributions. For solar-type components, a period of 100 years corresponds to a sum of the semi-major axes of 27 AU. This dividing line should, however, just be considered an order-of-magnitude estimate.

Recently, Morbey and Griffin (1987) have presented an improved analysis of the 25 spectroscopic binaries which Abt and Levy discovered. Their re-analysis does not support the binary nature of 21 of the stars. Abt (1987) argues that if the limiting detectable orbital amplitude must be increased, then this also affects the incompleteness estimates, and concludes that the basic division into two groups of binaries above and below a period of 100 years is unchanged. Obviously, the next step must be to obtain new data with still higher velocity resolution.

In a statistical survey of 205 main sequence spectroscopic binaries, Halbwachs (1987) found that the frequency of close binaries is about twice as large for mass ratios around 0.4 than around 1, in disagreement with the results of Abt and Levy. He found a similar distribution among visual binaries in the Yale Catalogue of Bright Stars (Halbwachs 1986), and concludes that from an observational point of view there may after all not be a need for several binary formation processes.

4. BINARY SEPARATION DISTRIBUTION FUNCTION

The nearest clouds which are actively producing low-mass stars are
the Taurus, Chamaeleon, Lupus, Ophiuchus and Corona Australis clouds,
which are all at distances between 100 and 200 pc. It is of interest
to consider the expected incidence of pre-main sequence binaries as
detected by various observational techniques. In the following, we
shall consider 150 pc as a typical distance to a star forming cloud.
    Heintz (1969) studied the frequency of binaries as a function of
semi-major axes, incorporating binaries of all spectral types with m(v)
< 9, with a separation of less than 16", and making corrections for
incompleteness. Fig.1 shows Heintz' binary separation distribution
function, as presented by Retterer and King (1982). It is clearly a

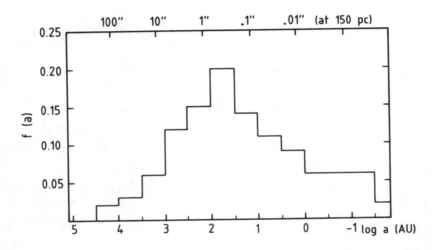

Fig.1. The relative frequencies of binaries as function of their
semi-major axes. The ordinate is number of companions per star. After
Heintz (1969) and Retterer and King (1982).

rather smooth, single-peaked function from spectroscopic binaries to
common proper motion pairs. Let us assume the function is more or less
the same for PMS binaries. The upper abscissa gives the angular scale
assuming a distance of 150 pc. We see that for the nearby clouds
mentioned above the peak occurs at sub-arcsecond scales. In the visual
range, say from 1" to 10", roughly one in five pre-main sequence stars
should be a visual binary. For the closer binaries, various techniques
can be employed. The resolution of speckle interferometry is given by
the diffraction limit of the telescope. For optical speckle
observations at a 4m size telescope this limit is a few hundreths of an
arcsecond, while in the infrared at 2.2 micron it is about 0.15".
Given the distribution in fig.1 and that pre-main sequence stars often
are bright infrared objects, then infrared speckle interferometry seems

the ideal way of detecting young binaries. The Space Telescope is expected to have a resolution of 0.05" to 0.1" with direct imaging, and should be equally successful for such detections. For further resolution, one must turn to lunar occultations, although this applies only to the direction of the lunar motion. For the very closest binaries, radial velocity measurements can reveal their orbital motion, unless the stars move in the plane of the sky.

In the following sections, an overview is given of the present status (October 1987) of observational work on different categories of pre-main sequence binaries, starting with the very closest.

5. SPECTROSCOPIC AND ECLIPSING PMS BINARIES

Spectroscopic binaries are important, because of the information contained in a velocity curve, among others the period P, the excentricity e and the half-range of the radial-velocity variation of one or both of the components, K1 and K2. If the spectrum is double-lined, the mass-ratio is directly obtained as m1/m2 = K2/K1. If single-lined, then only the so-called mass function can be derived (see e.g. Batten 1973 for further details).

Only very few spectroscopic PMS binaries are known. The first one to be confirmed, found by Mundt et al.(1983) using high-dispersion spectroscopy, is V826 Tau (also known as FK1 or P2), a so-called naked T Tauri star (Walter 1986). It is an X-ray source (Feigelson and DeCampli 1981), has weak H-alpha emission and a spectral type of K7 (Feigelson and Kriss 1981), and shows quasi-sinusoidal light-variations from star-spots with a full V-amplitude of 0.06 magnitude and period 4.05+/-0.2 days (Rydgren and Vrba 1983). The high-resolution spectroscopic data shows double lines and gives a period of 3.9063+/-0.0006 days, excentricity e=0.01+/-0.02, K1=18.0+/-0.3 km/s and K2=17.3+/-0.9 km/s. The stars are thus moving in circular orbits, have equal masses and, from the consistency between spectroscopic and optical periods, at least one component must rotate synchronously. Additional information is needed to get the individual masses, but if placed on convective tracks in the HR-diagram, one gets 0.6 solar masses for each component. Assuming these masses, the inclination becomes 7.2 degrees to the plane of the sky.

A second spectroscopic binary was recently announced by Marschall and Mathieu (1987). This is Parenago 1540, near the Orion Trapezium. The star shows strong Li absorption at 6707A, intense Ca II H and K emission, and it is also a weak X-ray source. It is double-lined with a period of 33 days.

Finally, a third candidate was found independently by de la Reza et al.(1986) and Byrne (1986). It is V4046 Sgr (HDE 319139). The star shows very strong and variable H-alpha emission, it has flares, is a dK5 star, and has considerable far-infrared excess. Its pre-main sequence nature has, however, been uncertain, but the recent detection of strong Lithium at 6707A in both components gives support to the youth of the object (Torres,Quast,de la Reza, this conference), although it may be in a post-T Tauri phase, and is perhaps related to the stars FK Ser (Herbig 1973) or HD36705 (e.g. Vilhu et al.1986). It is located about 1 degree from the nearest visible cloud, but other isolated young stars are known. The object has a period of 2.4213 days, almost equal masses, a circular orbit, and total velocity amplitude over 100 km/s.

These objects are only the first PMS spectroscopic binaries known, in addition a number of possible candidates have been discussed, like RY Tau and SU Aur (Herbig 1977, Bouvier et al.1986, Hartmann et al.1986). Several surveys are currently under way. It is, however, to be expected that even with a large sample of spectroscopic binaries, a statistical discussion will be hampered by severe observational selection effects, f.ex. one towards detection of binaries with mass ratio close to one (for a discussion of selection effects in the detection of spectroscopic binaries see Halbwachs 1987).

If a double-lined spectroscopic binary has an orbit virtually perpendicular to the plane of the sky and eclipses of the components take place, then a wealth of astrophysical information can be derived, in particular individual masses (e.g.Batten 1973). In the solar neighborhood about 0.1-0.2% of stars are eclipsing (e.g.Guinan, this conference). Since T Tauri stars have larger radii than their main sequence counterparts, this should increase somewhat the probability that they would eclipse. But so far none are known. Because of large star spots, erratic variability and circumstellar material, it is, however, unlikely that very accurate analysis of a lightcurve could be made.

Although this meeting concerns low-mass stars, it should be mentioned here that two massive eclipsing PMS binaries are known, both members of the Orion Trapezium. One is Theta Ori B, consisting of a rotationally flattened B2 star of 6 solar masses and a late A star of 1.8 solar masses, orbiting each other with a period of 6.5 days (Popper and Plavec 1976). The second is Theta Ori A, discovered by Lohsen (1975); observational progress on this object is slow, mainly because of its very long period of 65.4 days and its extended eclipses.

6. SPECKLE OBSERVATIONS, LUNAR OCCULTATIONS AND RADIO TECHNIQUES

Speckle observations are a powerful technique for resolving sub-arcsecond binaries, both at optical (McAlister 1977,1985) and infrared wavelengths (Chelli 1984).

The best-known PMS binary detected by speckle techniques is T Tauri itself. Dyck et al.(1982) carried out one-dimensional speckle interferometry of T Tauri at 2.2, 3.8 and 4.8 microns, and found an "infrared" companion at a separation of 0.61"+/-0.04", with much redder colors than T Tauri itself. Because of an ambiguity of 180 degrees in the position angle of the binary components, they proposed two different models that both fitted their data, but for their discussion happened to adopt the wrong one, as demonstrated by further ir speckle observations of Schwartz et al.(1984). The component of T Tauri lies to the south, has a color temperature of 800 K, and is fainter than T Tauri itself at all infrared wavelengths out to and including the M filter at 4.8 micron. The term "infrared" companion is therefore correct only in the sense that the object is much colder than T Tauri, not in the sense of dominating the near-infrared flux of the system, as is the case for some other PMS binaries.

In addition to the southern companion, Nisenson et al.(1985) discovered a third component in the system by optical speckle observations. This companion lies north of T Tauri at a distance of 0.27"+/-0.04", and is 4.33+/-0.09 magnitudes fainter than T Tauri at V. Nisenson et al.(1985) suggest a bolometric luminosity of this optical companion of 0.3-0.5 solar luminosities, and an effective temperature of 3000+/-200 K, corresponding to a M4-M8 star.

In order to gain in limiting magnitude, but at the cost of resolution, one can employ slow infrared slit scans instead of infrared speckle observations (e.g. Perrier 1986). Such a technique has been employed to study regions of high-mass star formation, and several sources have been resolved as doubles (e.g. Dyck and Staude 1982, Eiroa et al.1987 and references in these papers), but such massive objects fall outside the scope of this meeting.

Chelli et al.(1988a) made near-infrared slit scans of low-luminosity infrared sources in the Chamaeleon and Ophiuchus clouds, and found companions to Glass I and Elias 22 at separations of 2.9 and 2.8 arcsec, respectively. Both companions are very red, and in fact completely dominate the flux of the primary at 4.8 microns. Chelli et al.(1988b) made slit scans at J,H,K and L of 16 T Tauri binaries from a survey by Reipurth and Zinnecker(1988), obtaining accurate infrared photometry of the individual components in these systems. An example of such a slit scan at H (1.65 micron) of the newly discovered T Tauri triple system Sz 30 in the Chamaeleon clouds is shown in fig.2. All three components are clearly visible, and relative fluxes can be obtained from Gaussian fitting.

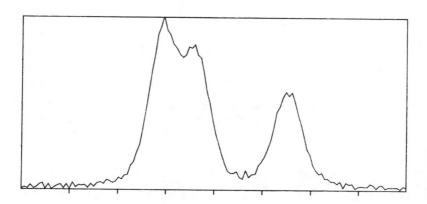

Fig.2. An infrared slit scan at 1.65 micron of the triple T Tauri system Sz 30, made at the ESO 3.6 m telescope. The spacing between tick-marks is 1.5 arcseconds. After Chelli et al.(1988).

Lunar occultation observations are currently the method which yields highest resolutions, and one can detect extended structure or binarity on a scale of milliarcseconds (Longo and Rigutti 1981). Simon et al.(1987) have successfully applied the technique to the detection at 2.2 micron of binaries among PMS stellar populations. Out of 18 stars in the Taurus and Ophiuchus clouds, they found 4 binaries within their detection range of about 5 to 300 milliarcseconds and limiting K-magnitude of 9. An example of one of their detections, S-R 12 in Ophiuchus, is shown in fig.3, which illustrates the reappearance of the source at the lunar limb. Each component gives rise to a Fresnel diffraction pattern, which can be modelled to show possible deviations from a point source pattern. The timespan between egress of the components basically gives the angular separation in the direction of the lunar motion, and the heights of the signal give the flux-ratio. Combined with infrared speckle observations at J,H and K, Simon et al.(1987) deduced that S-R 12 has a separation of 0.30"+/-0.03" at position angle 85+/-10 degrees, corresponding to a projected separation of about 48 AU, and they can deduce color temperatures for each component. Other binaries they found are FF Tau (36.7+/-1.8 marcsec), HQ Tau (4.9+/-0.9 marcsec) and ROX 31 (129.7+/-2.2 marcsec).

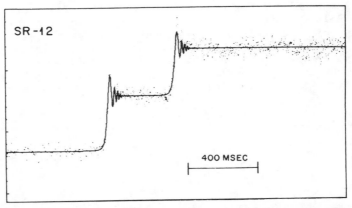

Fig.3. Lunar occultation observation at 2.2 micron of the pre-main sequence binary S-R 12 in Ophiuchus. Dots are data points and the solid line a binary model fit. After Simon et al.(1987).

Radio continuum observations at cm-wavelengths from ionized winds around T Tauri stars are a powerful method of detecting binarity in these stars, because of the high spatial resolution offered by arrays like the VLA. Unfortunately, however, T Tauri stars are generally weak radio continuum emitters, and at present sensitivities only few are detected (Bieging et al.1984,Bieging and Cohen 1985,Brown 1987). Among the positive detections is the T Tauri system. Cohen et al.(1982) noted a discrepancy between the optical and radio positions of T Tauri, and following the discovery of the infrared companion, Schwartz et al.(1984) resolved the system into two components at 6 cm wavelength.

The southern companion completely dominates the radio emission at this wavelength, as well as at 1.3 cm and 2 cm. Bertout (1983) suggested that the radio region is photoionized by radiation from an accretion shock surrounding a protostellar core.

Wootten (1987) observed the very young embedded source IRAS 16293-2422 at 6 cm and 2 cm, and found two sources separated by about 5", which he argues originate in an extremely young binary system.

7. VISUAL PRE-MAIN SEQUENCE BINARIES

At a time of increased interest in PMS binarity, it is worth to remember that the subject is almost as old as the definition of T Tauri stars. The first T Tauri binary to be discovered was S CrA (1" separation) in 1942 and four more of the eleven then known T Tauri stars were subsequently found to be doubles (Joy and van Biesbroeck 1944). In his classical paper on the T Tauri phenomenon, Herbig (1962) listed 29 doubles of which at least one component is a T Tauri star, Rydgren et al.(1976) studied 13 pairs, a couple of which were new, and Cohen and Kuhi (1979) listed 34 pairs, most of which were new.

While doing multi-color CCD imaging in the Bok globule B62, Reipurth and Gee (1986) found a faint companion to the T Tauri star RNO 92 at a separation of 4.4" and noticed how much stronger the companion showed up relative to the primary through a filter around 1 micron. Similarly, Sandell et al.(1987) found two PMS binaries in L1642 through the same very red filter. Given that the companions are likely to be of later spectral type, and thus redder, than their primaries, and the ease of detection with CCD-imaging, Reipurth and Zinnecker (1988) embarked on a red CCD survey of southern low-mass PMS stars for binarity. Examples of objects found are shown in fig.4, all obtained through a red filter around 1 micron. In order to avoid contamination by accidental alignments along the line-of-sight, stars were chosen that are in front of opaque clouds, thus essentially limiting the line-of-sight to a few hundred parsec. In addition to the already known binaries, 28 new binaries have been discovered to date. They are a homogeneous sample, having all been found with the same equipment. Their projected separation distribution is shown in fig.5, and it shows a clear increase towards smaller separations (the opposite of what would be seen for purely optical pairs). Evidently, there is an effect on detection near the resolution limit, which ranges between 1" and 2", depending on the seeing. Photometry in four filters have been obtained of these new as well as of about 25 previously known visual PMS binaries, and infrared photometry and optical spectroscopy of the individual components is presently being carried out.

In the course of this survey, a number of multiple systems have been found. Such systems can be divided into hierarchical systems, in which successive separations increase by large factors, or Trapezium systems, with more or less equal separations (e.g. Mirzoyan and Salukvadze 1984, Abt 1986). Systems of the latter type are dynamically unstable and will develop into hierarchical systems or eject members on time-scales of several million years. This is comparable to the duration of the pre-main sequence phase of low-mass stars.

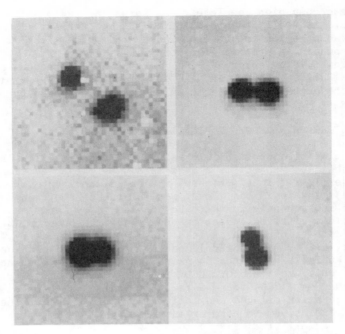

Fig.4. CCD images of four visual T Tauri binaries, obtained through a red filter near 1 micron. From Reipurth and Zinnecker (1988).

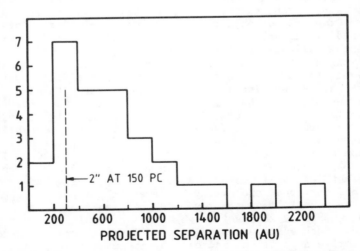

Fig.5. A separation distribution function for a preliminary sample of newly detected PMS visual binaries. From Reipurth and Zinnecker (1988).

8. VERY WIDE PRE-MAIN SEQUENCE BINARIES

The widest binaries are fragile, and susceptible to encounters with other stars and molecular clouds. Ambartsumian (1937) found a dissolution time inversely proportional to the binary separation. It has been suggested that there is a real cutoff in binary frequency for separations exceeding 20000 AU or 0.1 pc. (Bahcall and Soneira 1981, Retterer and King 1982), but this has been contested by Wasserman and Weinberg (1987). Interest in these wide binaries has increased, because they are indirect probes into galactic structure and the missing mass problem (Weinberg et al.1987 and references therein).

It would seem that the long tail of the separation distribution of PMS binaries might be relevant for comparison with such studies. However, most T Tauri stars are members of associations, which become unbound when their ambient molecular material disperses (von Hoerner 1968, Wilking and Lada 1984), and wide binaries with very low binding energies may become unbound when the stars begin to evaporate from their loose clustering. Another problem is simply to identify the widest binaries among the T Tauri stars. A separation of 20000 AU at the distance of the nearest molecular clouds corresponds to angular separations of more than two arcminutes. On these scales, one often finds sub-clustering of T Tauri stars, with little evidence of pairing. For the widest binaries, the final separation distribution may therefore not be developed in their pre-main sequence phase.

The youngest very wide PMS binaries so far identified may have been found in a study by Lindroos (1986 and references therein). He was concerned about identifying a sample of post-T Tauri stars, and therefore studied low-mass companions of OB stars. For 254 systems with OB type primaries he carried out Stromgren photometry and spectroscopy to exclude optical pairs and derive physical parameters for real binaries. Of these, 84 secondaries are likely physical companions and mostly later than spectral type F0. These late-type secondaries are located just above the Zero Age Main Sequence. What is interesting in the present context is that 16 of his physical binaries have projected separations exceeding 18000 AU.

9. SUMMARY AND OUTLOOK

Observations with a variety of techniques have shown the feasibility of detecting pre-main sequence binaries from the very close to the very wide. Only among the visual PMS binaries have a reasonably large number of objects been found and studied so far, particularly using CCD imaging, but ongoing surveys with other techniques are expected to increase the number of known closer binaries. Especially promising for the detection of PMS binaries are speckle observations and lunar occultations at both optical and infrared wavelenghts.

While studies of individual young binaries of course have an interest in themselves, it is the statistical analysis of a large number of binaries which eventually will allow the necessary comparison between observation and theory. This situation is still several years into the future. Particularly interesting for such comparisons are studies of the close PMS binaries, because of the uncertainty in their mode of formation. Is it possible for such objects to form by fission? Or could a companion form by instabilities in a massive disk around a protostar in the making? These are obviously questions to be adressed by theoreticians. But it is hoped that observers one day can provide distributions of separations, mass-ratios and excentricities, as well as f.ex. possible differences in binarity between clouds with strong and weak magnetic support for comparisons with the models.

Observers should focus attention on spectroscopic binaries which are resolvable. Optical speckle interferometry employing photon detecting cameras at 4m class telescopes is capable of resolving 15th magnitude binaries with separations down to 0.025" (McAlister 1985). Consider an ideal case of both components lying in the plane of the sky in a circular orbit perpendicular to the plane of the sky. A separation of, say, 0.1" in a cloud at 150 pc distance corresponds to a semi-major axis of 7.5 AU, which for a total mass of the system of 1 solar mass means a period of 20 years. This yields K1+K2 = 11 km/s, which is measurable with current techniques for a mass-ratio not too different from unity, also over such a long time-scale. While the short-period binaries are attractive because it is easier to find them and to determine their elements, it is the systems with periods of several years which are of particular value, because of the possibility to get multi-color differential photometry, and, with a concerted effort, masses of the components from a speckle-determined orbit.

ACKNOWLEDGEMENTS

I have benefited from discussions with Hans Zinnecker and Peter Bodenheimer. This work has been supported by ESO and the Danish Board for Astronomical Research.

REFERENCES

Abt,H.A.:1979, Astron.J. 84,1591
Abt,H.A.:1983, Ann.Rev.Astron.Astrophys. 21,343
Abt,H.A.:1986, Astrophys.J. 304,688
Abt,H.A.:1987, Astrophys.J. 317,353
Abt,H.A.,Levy,S.G.:1976, Astrophys.J.Supp. 30,273
Ambartsumian,V.A.:1937, Astr.Zh. 14,207
Bahcall,J.,Soneira,R.:1981, Astrophys.J. 246,122
Batten,A.H.:1973, "Binary and Multiple Systems of Stars",
    Pergamon Press
Bertout,C.:1983, Astron.Astrophys. 126,L1
Bieging,J.H.,Cohen,M.,Schwartz,P.R.:1984, Astrophys.J. 282,699
Bieging,J.H.,Cohen,M.:1985, Astrophys.J. 289,L5
Boss,A.P.:1987,in "Proceedings from the Summer School on
    Interstellar Processes", Eds.D.Hollenbach,H.Thronson
Bouvier,J.,Bertout,C.,Benz,W.,Mayor,M.:1986,
    Astron.Astrophys. 165,110
Brown,A.:1987, Astrophys.J. 322,L31

Byrne,P.B.:1986, Irish Astron.J. 17,294
Chelli,A.:1984,in "Very Large Telescopes,their Instrumentation
      and Programs",IAU Coll. no. 79, p.309
Chelli,A.,Zinnecker,H.,Carrasco,L.,Cruz-Gonzalez,I.,
      Perrier,C.:1988a, in press
Chelli,A.,Cruz-Gonzalez,I.,Reipurth,B.:1988b, in preparation
Cohen,M.,Bieging,J.H.,Schwartz,P.R.:1982, Astrophys.J. 253,707
Cohen,M.,Kuhi,L.V.:1979, Astrophys.J.Supp. 41,743
Contopoulos,G.:1988, Proceedings of IAU Coll. no.97, in press
de la Reza,R.,Quast,G.,Torres,C.A.O.,Mayor,M.,Meylan,G.,
      Llorente de Andres,F.:1986,in "New Insights in Astrophysics",
      ESA SP-263, p.107
Durisen,R.H.,Tohline,J.E.:1984,in "Protostars and Planets II",
      Eds.D.C.Black,M.S.Matthews,p.534
Durisen,R.H.,Gingold,R.A.,Tohline,J.E.,Boss,A.P.:1986,
      Astrophys.J. 305,281
Dyck,H.M.,Simon,T.,Zuckerman,B.:1982, Astrophys.J. 255,L103
Dyck,H.M.,Staude,H.J.:1982, Astron.Astrophys. 109,320
Eiroa,C.,Lenzen,R.,Leinert,Ch.,Hodapp,K.-W.:1987,
      Astron.Astrophys. 179,171
Feigelson,E.D.,DeCampli,W.M.:1981, Astrophys.J. 243,L89
Feigelson,E.D.,Kriss,G.A.:1981, Astrophys.J. 248,L35
Halbwachs,J.L.:1986, Astron.Astrophys. 168,161
Halbwachs,J.L.:1987, Astron.Astrophys. 183,234
Harrington,R.S.:1977, Rev.Mexicana Astron.Astrofis. 3,139
Hartmann,L.,Hewett,R.,Stahler,S.,Mathieu,R.D.:1986,
      Astrophys.J.309,275
Heintz,W.D.:1969, J.R.A.S.Canada 63,275
Herbig,G.H.:1962, Advances in Astron.Astrophys. 1,47
Herbig,G.H.:1973, Astrophys.J. 182,129
Herbig,G.H.:1977, Astrophys.J. 214,747
Herczeg,T.:1984, Astrophys.Space Sci. 99,29
Joy,A.H.,van Biesbroeck,G.:1944, Publ.Astron.Soc.Pacific 56,123
Lindroos,K.P.:1986, Astron.Astrophys. 156,223
Lohsen,E.:1975, Infor.Bull.Var.Stars no.988
Longo,G.,Rigutti,M.:1981, in "Photometric and Spectroscopic
      Binary Systems", D.Reidel, p.253
Lucy,L.B.:1981, in "Fundamental Problems in the Theory of
      Stellar Evolution",Eds.D.Sugimoto,D.Q.Lamb,D.N.Schramm, p.75
Marschall,L.A.,Mathieu,R.D.:1987, BAAS 19,707
McAlister,H.A.:1977, Astrophys.J. 215,159
McAlister,H.A.:1985, in IAU Symp. no. 111, p.97
Mirzoyan,L.V.,Salukvadze,G.N.:1984, Astrophysics 20/21,567
Morbey,C.L.,Griffin,R.F.:1987, Astrophys.J. 317,343
Mundt,R.,Walter,F.M.,Feigelson,E.D.,Finkenzeller,U.,Herbig,G.H.,
      Odell,A.P.:1983, Astrophys.J. 269,229
Nisenson,P.,Stachnik,R.V.,Karovska,M.,Noyes,R.:1985,
      Astrophys.J. 297,L17
Ostriker,J.P.:1970, in "Stellar Rotation" Ed.A.Slettebak, p.147
Perrier,C.:1986, The ESO Messenger 45,29
Popper,D.M.,Plavec,M.J.:1976, Astrophys,J. 205,462
Reipurth,B.,Gee,G.:1986, Astron.Astrophys. 166,148

Reipurth,B.,Zinnecker,H.:1988, in preparation
Retterer,J.M.,King,I.R.:1982, Astrophys.J. 254,214
Rydgren,A.E.,Strom,S.E.,Strom,K.M.:1976, Astrophys.J.Supp. 30,307
Rydgren,A.E.,Vrba,F.J.:1983, Astrophys.J. 267,191
Sandell,G.,Reipurth,B.,Gahm,G.:1987, Astron.Astrophys. 181,283
Schwartz,P.R.,Simon,T.,Zuckerman,B.,Howell,R.R.:1984,
     Astrophys.J.280,L23
Shu,F.H.,Adams,F.C.,Lizano,S.:1987, Ann.Rev.Astron.Astrophys. 25,23
Silk,J.,Takahashi,T.:1979, Astrophys.J. 229,242
Simon,M.,Howell,R.R.,Longmore,A.J.,Wilking,B.A.,Peterson,D.M.,
     Chen,W.-P.: 1987, Astrophys.J. 320,344
Szebehely,V.:1977, Rev.Mexicana Astron.Astrofis. 3,145
van Albada,T.S.:1968, Bull.Astr.Inst.Netherlands 19,479
Vilhu,O.,Gustafsson,B.,Edvardsson,B.:1986, in "Cool Stars, Stellar
     Systems and the Sun", Eds.M.Zeilik,D.M.Gibson, Springer, p.268
von Hoerner,S.:1968, in "Interstellar Ionized Hydrogen",
     ed.V.Terzian, (New York:Benjamin), p.101
Walter,F.M.:1986, Astrophys.J. 306,573
Wasserman,I.,Weinberg,M.D.,:1987, Astrophys.J. 312,390
Weinberg,M.D.,Shapiro,S.L.,Wasserman,I.:1987, Astrophys.J. 312,367
Wilking,B.A.,Lada,C.J.:1984, in "Protostars and Planets II",
     Eds.D.C.Black,M.S.Matthews, p.297
Wootten,A.:1987, in press
Zinnecker,H.:1984, Astrophys.Space Sci. 99,41

LONG-TERM VARIATIONS OF STELLAR MAGNETIC ACTIVITY
IN LOWER MAIN SEQUENCE STARS

Sallie L. Baliunas
Harvard-Smithsonian Center for Astrophysics
60 Garden Street
Cambridge, MA 02138 USA

ABSTRACT. Observed chromospheric emission fluxes and their time variation are undoubtedly universal phenomena in lower main sequence stars. Knowledge gained from the
Sun indicates that stellar chromospheric activity is caused by magnetic variations. In
lower main sequence stars, magnetic activity variations can be investigated by proxy with
the chromospheric emission present in the Ca II H and K lines. Cross-sectional and time-
serial studies reveal the dependence of average chromospheric activity levels and their
variations on stellar macroscopic parameters such as mass, age and rotation. Twenty years
of chromospheric Ca II H and K fluxes in cool stars monitored from Mt. Wilson Observatory have revealed long-term variations that are similar to the 11-year sunspot cycle. The
behavior of the long-term activity variations is summarized. Cycle periodicities inferred
for the majority of stars in the Mt. Wilson sample show no obvious dependence on rotation, mass or chromospheric activity level. Subgroups of the sample do, however, contain
insight on magnetic cycle behavior: First, some F- and early G-type dwarf stars display
short (≤4 yr), distinctly non-solar, cycle periods. Second, for stars with extremely obvious
periodicities, the cycle period may depend upon the chromospheric radiative loss expressed as a fraction of stellar bolometric luminosity. Contributing to the tentativeness of
the latter result is the inaccurate determination of cycle periods longer than ten years from
the short, twenty-year baseline of the time series.

1. THE SOLAR MAGNETIC CYCLE

Empirical studies of stellar activity will help provide the impetus for the thorough explanation of the solar magnetic cycle. We have no satisfactory understanding of solar
magnetic activity which is exemplified by an approximate 11-yr period of sunspot number
or the 22-yr period of sunspot number and magnetic polarity (Hale and Nicholson 1925).
During the sunspot cycle, the increase in sunspot number is roughly correlated with
enhanced magnetic activity such as flares and surface coverage by plages.

Spanning several millenia, sunspot records are the longest time series of astronomical measurements. Continuous, modern counts of the sunspot number exist from the 17th
century. During that century, the sunspot cycle ceased and sunspots appeared infrequently
in an epoch labeled the Maunder minimum, which is unlike the current era of cyclic sunspot behavior. Indirect records of solar magnetic activity suggest other sunspot cycle
suppressions or even enhancements compared to modern levels of solar magnetic activity.
Weakened activity and perhaps cycle lulls may have occurred as much as one-third of the
sun's main sequence lifetime (Eddy 1976, 1977, 1983; Damon 1977).

A. K. Dupree and M. T. V. T. Lago (eds.), Formation and Evolution of Low Mass Stars, 319–329.

The magnetic variations that are characterized by the 11-year sunspot cycle are usually interpreted with a magnetic dynamo model. Dynamo models describe the periodic modulation of sunspots by magnetic fields that are twisted and transported by the internal solar motions of convection and differential rotation. In such a model, a poloidal magnetic field can produce a toroidal one and then regenerate a fresh poloidal field, on a timescale of about 11 years (Parker 1955, 1979; Leighton 1969; Howard and Yoshimura 1976; Radler 1976; Moffat 1978; Gilman 1983; Weiss 1983; Weiss et al. 1984).

The magnetohydrodynamic theory behind the dynamo model is complex and the model requires knowledge, which is vague, about the interior motions and magnetism of the Sun: of crucial importance is the knowledge of differential internal rotation and convective zone velocities as well as the internal magnetic field. The study of stellar activity cycles may elucidate their dependence, if any, upon lower main sequence mass (which may dictate convective zone properties), surface rotation (which may govern internal differential rotation), age or chromospheric activity levels (which likely reflect magnetic field strengths). In lower main sequence stars, some of those properties are interdependent (see Hartmann and Noyes 1987; Soderblom, this volume), for example, rotation, age and chromospheric activity. The empirical approach provides hope for understanding an abstruse but universal phenomenon on cool stars, the long-term variations of magnetic activity.

2. LONG-TERM STELLAR CHROMOSPHERIC VARIATIONS

Refining dynamo models will require constraints which may result from stellar observations. Long-term variations of stellar chromospheric activity serve as an indicator of magnetic behavior. For over two decades, research at Mt. Wilson Observatory has been providing time serial chromospheric fluxes in nearly 100 lower main sequence stars. Begun in 1966 at the 100" telescope at Mt. Wilson Observatory (Wilson 1978), transferred to the 60" telescope in 1978 (Vaughan et al. 1978) and continued by a consortium in 1980 (Vaughan et al. 1981), the chromospheric activity program details the long-term behavior of Ca II H and K relative emission fluxes in 99 F-M stars on or near the main sequence. In the range of F-K spectral types, examples of strong-emission line (young and rapidly-rotating) stars and weak-emission line (old and slowly-rotating) stars are included in the sample. (Only one M-type star could be included in the survey.) The H and K emission cores isolated in 1-A passbands are measured relative to the nearby stellar continuum and defined as the observational quantity, S. The progress of the program to monitor the long-term variations has been summarized by Wilson (1978), Baliunas and Vaughan (1985) and Baliunas et al. (1988).

With twelve years of data Wilson (1978) discovered the diversity of behavior of long-term chromospheric variations whose quantitative nature can only be detailed with longer time series. Long, uninterrupted time series of chromospheric activity is one of the goals of the Mt. Wilson 60" telescope's HK survey.

Since the time series of the sample stars are just two decades long we can only begin to assess the long-term chromospheric variations and search for periodicities. We must be realistic about the accuracy of any periodicities apparent in the data. The example of the sunspot record is a telling caveat: the 11-yr sunspot cycle period is an average over the past two centuries. The cycle-to-cycle lengths can be as short as 7 or as long as 15 years. Additionally, during the Maunder minimum the 11-yr period was presumably undetectable. The periodicities in stellar cycles must be verified by averages of numerous cycle lengths or by a statistically large sample of stars. With only a twenty-year timespan, we can verify periods on the order of ten years or less. For shorter periods, however, a greater accuracy can be achieved.

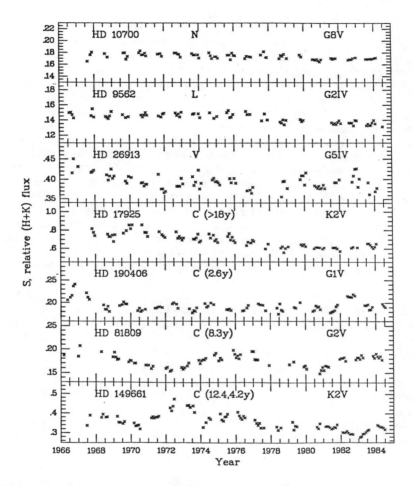

Figure 1--Seven examples of the variety of long-term behavior of chromospheric activity in lower-main sequence stars. The quantity, S, the flux in the Ca II H and K emission cores relative to the photospheric flux, is plotted as a function of time, from 1966-1985. The character of the long-term fluctuations is denoted by a symbol in the middle of each panel which includes the star's HD number and spectral type. From top to bottom panels, symbols represent: no significant variation, "N"; linear trend, "L"; variable with no period, "V"; and possibly or likely cyclic with the period or periods shown in parentheses, "C".

The variety of long-term behavior in the Mt. Wilson sample of lower main sequence stars is shown in Figure 1. The values of the Ca II H and K emission strengths, S, are plotted as a function of time between 1966 and 1985. Through 1977 all the data points, which were collected on a monthly basis by Wilson at the 100" telescope, are plotted. In 1978 the project moved to the 60" telescope, and the monthly observations were continued. In 1980, measurements were scheduled nightly and only monthly averages of those points are shown.

The long-term variations can be grouped according to the results of a subjective period analysis of the time series. In order to search for periodicities, we calculate a periodogram according to Scargle (1982) which is unbiased for unevenly sampled data, that is, the periodogram is unaffected by gaps in the time series caused, for example, by inclement weather. Scargle's technique also includes an estimator of the significance of a peak in the periodogram, called the false alarm probability. The false alarm probability is the likelihood that gaussian noise with the same variance as the data will produce a peak as high as a peak in the periodogram of the data (Horne and Baliunas 1986). Unfortunately, the interpretation of the significance of the periodogram peak is not so straightforward, because noise in the astronomical data is not always gaussian. For example, the growth and decay of Ca II-emitting regions, flares and axial rotation all contribute, along with experimental uncertainty, to the "noise" in the data (Vaughan *et al.* 1981; Baliunas *et al.* 1983; Gilliland and Baliunas 1987). In addition, the cycle period may, as in the case of the Sun, contain other periodicities, although the 11-yr period is the dominant one (cf. Donahue and Baliunas 1988). In addition to the computation of the periodogram and false alarm probabilities, techniques to determine the formal standard deviation of a significant period and to reveal alias frequencies are discussed by Horne and Baliunas (1986).

The plot of seven examples of long-term activity variations (Figure 1) includes the star's HD number, its spectral type, and a symbol characterizing its long-term behavior in each panel. The star HD 10700 (τ Cet) has no significant long-term variation and is labeled, "N". The time series for HD 9562 displays a long-term trend, "L", which, if periodic, would imply a period of at least several decades. Some stars, such as HD 26913, vary significantly but with no clear periodicity, and are labeled, "V". The next four panels display activity variations that are possibly or certainly periodic, "C", and the apparent period is listed in parentheses. The star HD 17925 appears to have a cycle longer than 18 years but possibly near 20 years. The accurate, short cycle period in HD 190406 is 2.6 years. In the solar-like star HD 81809 the average period is 8.3 years. Two significant peaks, neither of which is an alias, correspond to periods of 12.4 and 4.4 years in the star HD 149661. From the examples the problem with short time series for accurate period determinations should be evident.

A summary of the statistics of the long-term behavior of chromospheric activity is presented in Table 1. Most stars in the sample (85%) are variable and most stars in the sample (60% of the *total* sample) are possibly or probably periodic in magnetic activity. The latter number excludes stars with long-term trends, erratic variability or no significant variability, but includes stars with cycles on the order of 20 years. The dominant trait in our sample is that of significant long-term variability, and the majority of stars have likely or possible periodicities. Dynamo models must explain the periodicities that appear in the majority of the time series of our sample of lower main sequence stars.

Table 1. Statistics of Long-Term Stellar Behavior

Symbol	Behavior	Number of Stars
C,C($\geq$ 18)	Cycle or likely cycle	60
L	Linear Trend	13
V	Variation, no trend	12
N	No variation	11
B	Indeterminate	3

Does the long-term behavior of chromospheric variations depend upon one or more stellar properties? Neither obviously nor simply is the answer so far. However, some positive results which may provide insight for dynamo models are becoming apparent. In Figure 2 the average chromospheric emission strength, S, is plotted as a function of (B-V) color index of the stars in the sample. The symbols denote the character of the long-term variations, according to the symbols in Figure 1. This plot shows no clear bifurcation or grouping of certain symbols, with two exceptions. First, the F-stars (small (B-V) color index) with weak activity vary so little if at all that they produce a clump of "N"'s in the lower left corner of the diagram. Because the strength of the emission relative to the nearby continuum is low in those stars, and any variations would border on the instrumental uncertainty, the lack of variation in that region of the diagram may not be physically significant. In contrast, the older, weak emission-line stars of solar spectral type or later all vary significantly, except for one: HD 10700 (τ Cet, see Figure 1). The inactivity in HD 10700 may occur because the star is in a Maunder-minimum state of low activity, similar to the 17th-century Sun.

The average chromospheric emission strength, S, is a function of rotation or, equivalently, age (cf Hartmann and Noyes 1987; Soderblom, this volume). Therefore, Figure 2 also contains information about the behavior of long-term activity as a function of rotation because the young, strong-emission-line stars (high values of S) rotate faster than the weak emission-line stars (low values of S). Long term activity does not clearly depend upon rotation alone with one possible exception: the few stars with two significant periods derived from the periodogram analysis (denoted with a script "C" in Figure 2) all are young or rapidly-rotating, dwarf stars.

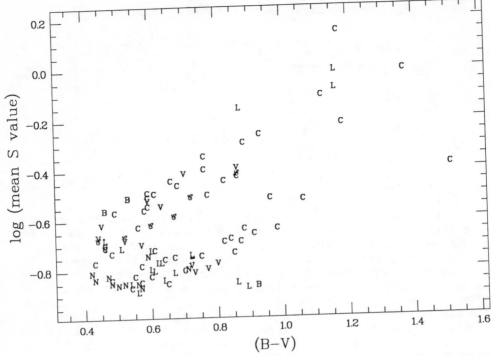

Figure 2--The time-averaged chromospheric emission strength, S, is plotted as a function of (B-V) color index for all the lower main sequence stars in the sample. The symbols characterizing the long-term activity behavior are the same as in Figure 1, except that stars with possibly two independent frequencies (*e.g.* HD 149661, Fig. 1) are denoted with a script "C." The symbol, "B" indicates that the long-term behavior of chromospheric activity is indeterminate.

3. CYCLE PERIODS AND STELLAR PROPERTIES

The majority of stars exhibit long-term variations that are or appear to be cyclic. We next explore the question of the dependence of cycle period (either likely or possible) on stellar macroscopic properties. We will exclude stars that are not characterized as "cyclic" (see Figure 1). Some predictions of the influence of stellar parameters on the cycle period, extrapolated from dynamo models, are listed in Table 2. The assumptions for the extrapolations result in disagreement that is so wide that opposite conclusions about the cycle period dependence can be found. The predictions tabulated are neither exhaustive nor have they been critiqued; Table 2 serves only to emphasize the need for empirical results which could guide dynamo modeling.

Table 2. Estimates of Stellar Dynamo Periods

PREDICTION	REFERENCE
For a fixed rotation period, P_{cyc} increases as mass decreases.	Belvedere *et al.* (1980), Kleeorin *et al.* (1983)
For a fixed rotation period, P_{cyc} decreases as mass decreases.	Robinson and Durney (1982), Durney and Robinson (1982)
For a fixed mass P_{cyc} increases as rotation period increases.	Weiss *et al.* (1984), Noyes *et al.* (1984b)
P_{cyc} increases as $<R'_{HK}>$ decreases.	Noyes *et al.* (1984b)
P_{cyc} increases as Rossby parameter increases.	Noyes *et al.* (1984b)
For a fixed mass and rapid rotation, P_{cyc} is independent of rotation.	Kleeorin *et al.* (1983)
For a fixed mass and intermediate rotation, P_{cyc} decreases as rotation period increases.	Kleeorin *et al.* (1983)
For a fixed mass and slow rotation, P_{cyc} increases as rotation period increases.	Kleeorin *et al.* (1983)
P_{cyc} independent of rotation.	Weiss *et al.* (1984)

The possible or likely cycle periods are plotted against rotation periods in Figure 3. The symbols for the points are represented by the gross spectral type of the star (F, G, K or M). Error bars (with a length of two standard deviations) correspond to the uncertainty of the frequency of the significant period found in the periodogram analysis (Horne and Baliunas 1986). For cycle periods longer than about ten years, those formal error bars may not be good estimates of the accuracy of the period. Uppercase symbols denote stars whose rotation periods are measured directly (Baliunas *et al.* 1983), while lowercase symbols signify stars whose rotation periods are estimated from the chromospheric activity-Rossby number relation of Noyes *et al.* (1984a). The symbols for stars with two significant, independent frequencies are connected by the vertical, dashed lines.

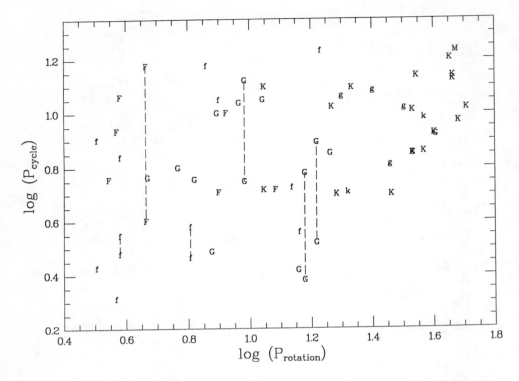

Figure 3--Only the stars with possible or likely cycles are shown above, where the cycle period is plotted as a function of the rotation period. The symbols represent the spectral type of the star. Uppercase symbols signify a directly-measured rotation period and lowercase symbols denote an inferred rotation period (see text). Symbols connected by dashed, vertical lines represent stars with two independent cycle frequencies. Cycle periods longer than about 10 years have an undetermined accuracy. Short (≤4 yr), non-solar cycle periods are found only on stars more massive than the Sun.

No trend of cycle period and rotation period, or mass, is evident. One subgroup of the sample does, however, contain a positive result: some F- and early G-spectral type stars display significant cycle periods that are shorter than 4 years, a distinctly non-solar cycle period. Such short periods are accurately determined and occur in no other spectral class. The stars with short, non-solar cycle periods are both more massive (with presumably shallower convective zones) and more rapidly rotating than the Sun. Not all F- and G-type stars, however, have such short cycle periods. Long cycle periods are also observed in this subgroup.

The cycle periods are not simply dependent upon the Rossby parameter, the ratio of the rotation period to the convective turnover time (Fig. 4), where the symbols are the same as those in Figure 3. Equivalently, for *all* the stars with possible or likely cycles, the periods do not depend upon the chromosperic radiative losses normalized to the stellar bolometric luminosity, R'_{HK}, which is a smooth function of the Rossby parameter (see Noyes *et al.* 1984a for a discussion of that parameterization). The Rossby parameter, which is inversely proportional to the square of the dynamo number, can be used to characterize some dynamo models (Noyes *et al.* 1984a).

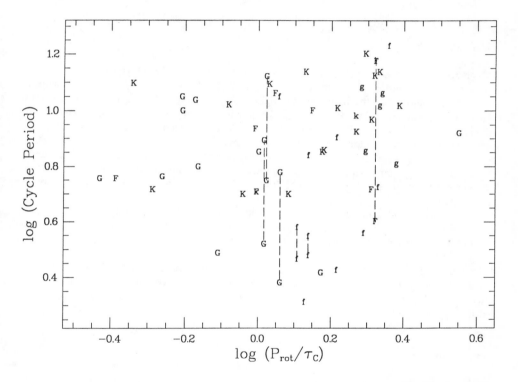

Figure 4--For the stars with possible or likely long-term periods (see Fig. 3), the cycle period is plotted as a function of the Rossby parameter, the ratio of the rotation period to the convective turnover timescale (see text). Cycle periods longer than approximately ten years have an undetermined accuracy.

With shorter time series of a subset of the sample, namely, the slowly-rotating stars with masses less than solar, Noyes *et al.* (1984b) presented a power law relation between cycle period and Rossby parameter. For stars with extremely significant peaks (that is, very low false alarm probabilities) in their periodograms, a power law dependence of cycle period upon normalized chromospheric radiative losses, R'_{HK}, is possible (Baliunas *et al.* 1988). The relation does have significant scatter including a few extremely deviant points, but helps to confirm the Noyes *et al.* (1984b) dependence of cycle period upon Rossby number. While such relations may exist, they cannot be proven unequivocally until the cycle periods, especially the long ones, are determined with greater accuracy from time series longer than just twenty years.

4. SUMMARY

Magnetic activity, specifically, long-term variations similar to the 11-yr sunspot cycle can be seen in time series of chromospheric Ca II H and K fluxes in lower main sequence stars. The majority of the dwarf stars ranging in spectral type from mid-F to late-K that have been observed at Mt. Wilson over the previous twenty years vary with possible or likely cycle periodicities. The apparent periods do not depend simply on mass, rotation, age, average chromospheric flux or related parameters such as the Rossby number. Subsets of the sample do, however, contain positive results: short cycle periods, which have been accurately determined, are observed only on stars more massive than the Sun. The converse statement is *not* true, because long cycle periods are also observed for stars in that group.

A dependency of the cycle period and the normalized Ca II H and K chromospheric losses may exist; the accuracy of long cycle periods must, however, be improved and the exceptions explained before such a result is verified.

I thank my colleagues at Mt. Wilson for their great enthusiasm and endeavors. This material is based on work supported by the National Science Foundation, under grant AST-8616545, the Smithsonian Scholarly Studies Program and the Langley-Abbot Fund of the Smithsonian Institution.

REFERENCES

Baliunas, S.L. *et al.* 1983 *Ap.J.*, **275**,752.

Baliunas, S.L. *et al.* 1988 *Ap.J.*, submitted.

Baliunas, S.L. and Vaughan, A.H. 1985 *Ann.Rev.Astr.Ap.*, **23**,379.

Belvedere, G., Paterno, L., and Stix, M. 1980 *Astr.Ap.*, 91,328.

Damon, P.E. 1977 In *The Solar Output and Its Variation*, ed. O.R. White (Boulder: Colo. Assoc. Univ. Press), p. 429.

Donahue, R.A. and Baliunas, S.L. 1988 *Pub.Astr.Soc.Pac.*, submitted.

Durney, B.R. and Robinson, R.D. 1982 *Ap.J.*, **253**, 290.

Eddy, J.A. 1976 *Science*, **192**, 1189.

Eddy, J.A. 1977 In *The Solar Output and Its Variation, op.cit.*, p. 51.

Eddy, J.A. 1983 *Sol.Phys.*, **89**, 195.

Gilliland, R.L. and Baliunas, S.L. 1987 *Ap.J.*, **314**, 766.

Gilman, P.A. 1983 In *Solar and Stellar Magnetic Fields: Origins and Coronal Effects, IAU Symp. no. 102*, ed. J.O. Stenflo (Boston:Reidel),p. 247.

Hale, G.E. and Nicholson, S.B. 1925 *Ap.J.*, **62**, 270.

Hartmann, L. and Noyes, R.W. 1987 *Ann.Rev.Astr.Ap.*, **25**, 271.

Horne, J.H. and Baliunas, S.L. 1986, *Ap.J.*, **302**, 797.

Howard, R. and Yoshimura, H. 1976 In *Basic Mechanisms of Solar Activity, IAU Symp. No. 71*, ed. V. Bumba and J. Kleczek (Boston:Reidel), p. 19.

Kleeorin, N.I, Ruzmaikin, A.A. and Sokoloff, D.D. 1983 *Astr.Spa.Sci.*, **95**, 131.

Leighton, R.B. 1969 *Ap.J.*, **156**, 1.

Moffat, H.K. 1978 *Magnetic Field Generation in Electrically Conducting Fluids* (Cambridge: Cambridge Univ. Press), 343 pp.

Noyes, R.W., Haartmann, L. Baliunas, S.L., Duncan, D.K. and Vaughan, A.H. 1984a *Ap.J.*, **279**, 763.

Noyes, R.W., Weiss, N.O. and Vaughan, A.H. 1984b, *Ap.J.*, **287**, 769.

Parker, E.N. 1955 *Ap.J.*, **122**, 293.

Parker, E.N. 1976 In *Basic Mechanisms of Solar Activity, op.cit.*, p. 3.

Parker, E.N. 1979 *Cosmical Magnetic Fields: Their Origin and Their Activity* (Oxford: Clarenden), 860 pp.

Radler K.-H. 1976 In *Basic Mechanisms of Solar Activity, op.cit.*, p. 367.

Robinson, R.D. and Durney, B.R. 1982 *Astr.Ap.*, **108**, 322.

Scargle, J.D. 1982 *Ap.J.*, **263**, 835.

Soderblom, D.R. 1988, this volume.

Vaughan, A.H. Preston, G.W. and Wilson, O.C. 1978 *Pub.Astr.Soc.Pac.*, **90**, 276.

Vaughan, A.H. *et al.* 1981 *Ap.J.*, **250**, 276.

Weiss, N.O. 1983 In *Stellar and Planetary Magnetism*, ed. A.M. Soward (New York: Gordon and Breach), p. 115.

Weiss, N.O., Cattaneo, F. and Jones, C.A. 1984 *Geophys.Ap.Fluid Dyn.*, **30**, 305.

Wilson, O.C. 1978 *Ap.J.*, **226**, 379.

ROTATIONAL VELOCITY EVOLUTION OF YOUNG STARS

John R. Stauffer
NASA/Ames Research Center
MS 245-6
Moffett Field, CA 94035

ABSTRACT. Nearly all of the rotational velocity data for stars less massive than the Sun and younger than the Hyades has been obtained in the past six years. The new data show that low mass T Tauri stars typically have rotational velocities of order 15 km-sec^{-1}. These stars, however, apparently arrive on the main sequence as quite rapid rotators, with rotational velocities up to 200 km-sec^{-1}. Once on the main sequence, winds rapidly decrease the rotational velocities to of order 10 km-sec^{-1} in only a few times 10^7 years for G and K dwarfs, with the spindown timescale increasing with decreasing stellar mass. A model which predicts rotational velocity evolution for young stars that roughly matches the observations has been proposed by Endal and Sofia (1981). That model, however, predicts little spindown of the radiative core of solar mass stars, in apparent contradiction to current helioseismology results. New models for the rotational spindown of low mass stars fit the solar interior rotation well, but do not, as yet, provide a satisfactory fit to the surface rotational velocities of young stars (Sofia 1987).

I. INTRODUCTION

Even without reference to observations of other stars, it is possible to infer that there must be considerable angular momentum loss during the star formation process. Observation of the apparent movement of spot groups across the face of the Sun reveal that, at the very least, some stars are quite slow rotators. Because more than 90% of the angular momentum of the solar system is presently found in the planets, the protosolar nebula must have had considerably more angular momentum than the present day Sun. Going one step farther back in time, it is easily shown that quite plausible initial conditions for molecular clouds when contracted to stellar size scales would predict stellar rotational velocities much larger than gravitational stability arguments allow. Hence, just the existence of the Sun and other stars indicates that stars are able to solve their "angular momentum problem".

Struve and his collaborators were the first to definitively measure the rotational velocities of stars other than the Sun via measurement of absorption line profiles in high resolution photographic spectra (see Huang and Struve 1960 for a review of these techniques). Subsequent work by Slettebak (1949, 1954, 1955), Abt (1961, 1965), Abt and Hunter (1962) and others succeeded in determining ro-

A. K. Dupree and M. T. V. T. Lago (eds.), Formation and Evolution of Low Mass Stars, 331–344.

tational velocities for a considerable number of stars more massive than the Sun. Abt and Hunter (1962) showed that the mean rotational velocities vary smoothly with spectral type - B and early A stars have mean rotational velocities of order 200 km-sec^{-1}, while the rotational velocities of F stars decrease sharply towards later spectral subtypes such that by G0 the mean rotational velocity for field stars is of order 10 km-sec^{-1}. The first direct evidence for angular momentum loss on the main sequence was provided by Kraft (1967), who showed that mid to late F dwarfs in the Pleiades rotate systematically faster than stars of the same spectral type in the Hyades, and that old disk, field F dwarfs rotate slower still. Stars hotter than about F0 do not show CaII K emission cores in their spectra (hence they are assumed not to have significant chromospheres) and also show no evidence of rotational braking (Schatzman 1962; Wilson 1966). This suggested that the chromosphere was the ultimate energy source for the rotational braking, with the actual mechanism being either impulsive events such as stellar flares or a stellar analog of the solar wind (Weber and Davis 1967).

A clear summary of the rotational velocity characteristics of stars earlier than G0 was provided by Kraft (1970). In addition to outlining the results described in the preceding paragraph, Kraft also illuminated the potential role that rotational velocity studies could serve for better understanding stellar formation and evolution. Kraft noted that once sufficient observational data were available, star formation theories could not only be required to match the observed luminosity function of open clusters, but could also be required to match the observed angular momentum distribution as a function of mass. Because of the rotational spindown on the main sequence, the observational data for this comparison should come from quite young clusters or T Tauri stars. Kraft also suggested that the rotational velocity distribution of binary and single stars be compared - binary systems either have very different initial conditions or have devised different solutions to the angular momentum problem. A final research topic for the future suggested by Kraft was the determination of the interior rotation of stars, which Kraft circumvented by assuming solid body rotation.

The decade following Kraft's review saw only small advances in understanding the rotational evolution of low mass stars because the observations were difficult and the rewards were few. The small number of late type stars near enough to permit high dispersion photographic spectroscopy appeared to follow the trend for quite slow rotation shown by the F dwarfs. Spectra of sufficiently high quality to derive rotational velocities such slowly rotating stars required high signal-to-noise and very high dispersion given the traditional rotational velocity analyses techniques used. With photographic plates for detectors, exposure times of several hours were necessary even on large telescopes. Projects that simultaneously demand large amounts of telescope time and show little evidence of exciting results tend not to be proposed, and when proposed tend to be turned down by time allocation committees. A kick in the pants was needed.

Significant progress in the observational pursuit of rotation in low mass stars awaited the development of new instrumentation and new analysis techniques and the realization that young, low mass stars are not all slow rotators. The new instrumentation includes photon-counting Reticon and CCD detectors, usually mated with with efficient echelle spectrographs. The new software methods rely on the application of Fourier analysis techniques and/or cross-correlation methods that allow information from hundred of photospheric lines to be used to synthesize an average absorption line profile for a given program star, thus allowing much lower signal-to-noise spectra to be used. Even with these advances, the rotational velocities of low mass stars would have remained challenging to determine if all single, low mass stars later than G0 had rotational velocities of order of less than 10 km-sec^{-1}. Evidence that this is not always the case was first provided by van Leeuwen and Alphenaar (1982), who deserve considerable credit for stimulating much of the research to be described in the next section.

II. OBSERVATIONS OF STELLAR SPINDOWN

Vogel and Kuhi (1981=VK) were the first attempt to determine rotational velocities for a large sample of pre-main sequence stars. These authors showed that most T Tauri stars are not rapid rotators, as had been assumed based on their strong emission lines, but instead generally have rotational velocities less than the VK measurement limit of about 30 km-sec^{-1}. This is particularly true of low mass T Tauri stars that are still on the Hayashi track portion of their pre-main sequence evolution. The relatively small rotational velocities show that stars solve their "angular momentum problem" prior to the T Tauri phase. Several sets of authors (Hartmann et al. 1986; Bouvier et al. 1987; Hartmann, Soderblom and Stauffer 1987; Rydgren et al. 1987; and Walter et al. 1987) have improved upon these results by obtaining rotational velocities for pre-main sequence stars with measurement limits of 10 km-sec^{-1} or less. These studies show that most low-mass pre-main sequence stars, in fact, have rotational velocities of order or less than 15-20 km-sec^{-1}. The Bouvier et al. results are of particular importance because they claim to have measured rotational velocities for their entire sample (i.e. there are no stars with just limits) - the most slowly rotating T Tauri stars have v$sini \simeq$ 8 km-sec^{-1}, while the most rapidly rotating low mass T Tauri star has v$sini \simeq$ 35 km-sec^{-1}. For stars on the Hayashi track with mass less than 1.0 $M_\odot$, there are no obvious correlations between emission line strength, luminosity (and hence age) and rotational velocity. In particular, the "Naked T Tauri" stars (Walter 1986) and weak Ca K emission stars (Herbig, Vrba and Rydgren 1986) do not show a rotational velocity distribution significantly different from the classical T Tauri stars. This suggests that it is not dynamo driven magnetic activity that powers the emission line behavior of T Tauri stars.

The rotational velocities of low mass stars are expected to increase significantly as they contract to the main sequence due to the decrease in radius and increase in central concentration. If the observed pre-main sequence stars are evolved to the main sequence (using evolutionary tracks provided by D. Vandenberg), and no angular momentum loss is assumed, maximum and mean rotational velocities of order 150 and 75 km-sec^{-1}, respectively, are predicted for stars arriving on the ZAMS. With modern spectrographs and rotational analysis techniques, it is now possible to obtain spectra of young, low mass stars approaching the ZAMS in order to test this prediction.

The youngest open cluster for which rotational velocities for low mass (but post T Tauri) stars have been obtained is IC 2391 (age $\simeq$ 3 x 10^7 years). Unfortunately, the cluster is quite sparse, and only a few late-type members have been identified (Stauffer et al. 1988a). Rotational velocities for these stars are given in Figure 1. The K stars in this cluster are still contracting to the main sequence, but some of them have rotational velocities of order 100 km-sec^{-1}. Such large rotational velocities are compatible with the T Tauri observations only if there is very little angular momentum loss between the T Tauri stage and $\tau = 3 \times 10^7$ years. It is peculiar that the other late type IC 2391 members have rotational velocities of order 15 km-sec^{-1}- the same as typical T Tauri stars. This suggests that they have not spun up much, and thus that they must have had considerable angular momentum loss in this time period. That the rapid rotators appear to have lost little angular momentum while the slow rotators have had prodigious winds is in contradiction to the theoretical expectation from wind models that predict dJ/dt $\alpha \, J^n$, with n = 2-3. A possible explanation would be that there is a significant age spread among the IC 2391 stars, and that (for instance) the slow rotators are considerably younger and thus have not yet spun up. This is not the case, since an HR diagram for the cluster shows all of the stars lie along a well defined pre-main sequence isochrone, with a possible age spread less than 10^7 years.

The Alpha Persei cluster (age $\simeq 5 \times 10^7$ years) provides a somewhat older but much richer target for rotational velocity studies (Stauffer et al. 1985). At the nominal age for the cluster, all of the G stars have arrived on the main sequence, and the K dwarf members are only slightly above the main sequence. The rotational velocity distribution for the cluster (Figure 2) is even more startling than for IC 2391, with a few stars having v$sini$ near 200 km-sec^{-1}, and a large fraction having v$sini$ greater than 50 km-sec^{-1}. No strong spectral type dependence is seen in the rotational velocity distribution. As for IC 2391, there are, however, some late-type stars with rotational velocities of order 10 km-sec^{-1}.

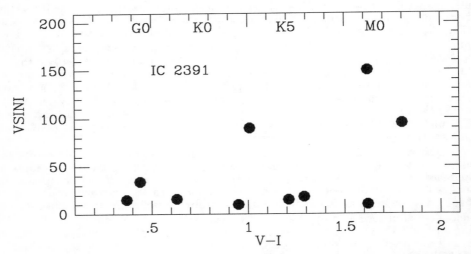

Figure 1: Spectroscopic rotational velocities for late type stars in the 30 million year old open cluster IC 2391.

The next older open cluster for which we have data is the Pleiades, with an age estimated from its upper main sequence turn-off point of about 7 x 10⁷ years. That some of the K dwarfs in the Pleiades have quite large rotational velocities was suggested by the photometric periods derived by van Leeuwen and Alphenaar (1982= VA). Our spectroscopic observations (Stauffer et al. 1984; Stauffer et al. 1987= SH87) were prompted by a desire to confirm that the VA stars were BY Dra variables, in which case the photometric variability is due to asymmetrically distributed spots, and the photometric period should then be equal to the rotation period. The observations amply confirmed this prediction, and thus, for instance, the shortest period VA star (with P = 0.24 days) is also our most rapidly rotating star, with $v\sin i$ = 140 km-sec^{-1}. The rotational velocity distribution for the Pleiades is shown in Figure 2. The primary difference between the Pleiades and Alpha Persei rotational velocity distributions is the lack of rapid rotators among the Pleiades G dwarfs. By contrast, about half of the Pleiades K and M dwarfs have rotational velocities greater than 50 km-sec^{-1}.

The last cluster where good rotational velocity information for low mass stars exists is the Hyades, age $\simeq$ 8 x 10⁸ years. Rotational periods for G stars in the Hyades have been inferred from the Mt. Wilson CaII monitoring program (Duncan et al. 1984). Periods for another set of G and K dwarfs have been derived from a broad-band photometric monitoring program conducted by Radick et al. (1987). We (Stauffer, Hartmann and Latham 1987) have estimated spectroscopic rotational velocities for M dwarfs in the Hyades. The rotational velocity distribution for low mass Hyades stars is shown in Figure 2. For this considerably older cluster, all of the G and K dwarfs are slow rotators. The transition to having a significant number of moderately rotating stars only occurs at spectral type about M0. As noted by Radick et al., for the G and K dwarfs there is very little dispersion among the rotational velocities at a given spectral type, suggesting that there is some characteristic rotational velocity that stars evolve to at a given age.

III. MODELS OF STELLAR SPINDOWN

Endal and Sofia (1981 = ES) derived a model for the rotational velocity evolution of the Sun which we believe provides an explanation for much of the data cited above. ES follow the evolution of a 1.0 mass star from the Hayashi track to the end of the main sequence lifetime. The models incorporate a standard paramaterized wind (Belcher and MacGregor 1976), leading to an angular momentum loss rate proportional to the angular rotation rate to some power. The model stars spin up by angular momentum conservation during PMS contraction, reaching a maximum rotational velocity of 50-100 km-sec^{-1} at 3x10⁷ years (the PMS contraction time for a solar mass star). The angular momentum loss rate is small during the pre-main sequence phase because the large stellar radii imply relatively small angular rotation rates and hence low angular momentum loss rates. The maximum rotational velocity upon arrival on the main sequence also implies a maximum angular momentum loss rate, leading to a rapid decrease in the surface rotational velocity. ES predict that a G2 star spins down from about 70 km-sec^{-1} to 10 km-sec^{-1} in just several x 10⁷ years. After that, the angular rotation rate is again small, and the star's rotational velocity decreases slowly, in approximate accord with a Skumanich (1972) $t^{-1/2}$ decay. It is important to note that ES achieved the quite rapid spin-down after arrival on the main sequence because only the outer convective envelope is spun down. Since the convective envelope of a G2 dwarf contains only about 1% of the mass, or 10% of the moment of inertia of the star, this assumption allows a much more rapid spin down of the surface rotational velocity for a given wind model.

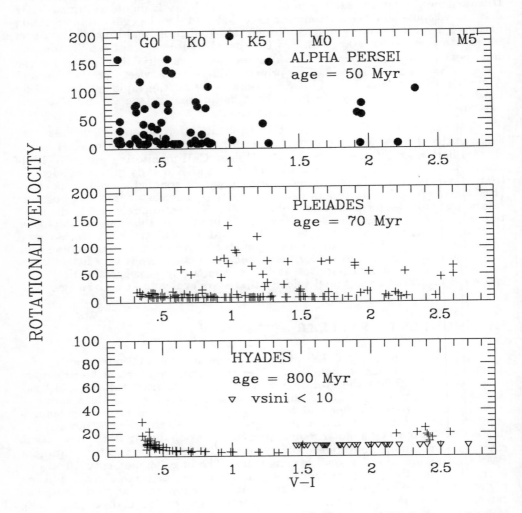

Figure 2a-c: Rotational velocity distribution for low mass stars in three young open clusters. The differences in the three distributions are assumed to illustrate rotational velocity evolution on the main sequence, and not at-formation differences in the initial angular momentum distribution.

ES calculated models only for solar mass stars. We assume that lower mass stars have similar rotational velocity histories, but with longer timescales both because the outer convective envelope contains a larger fraction of the star's total moment of inertia and because the lower mass stars have longer PMS contraction timescales (SH86). The Alpha Persei cluster is sufficiently young so that all of the stars observed have only recently reached the main sequence, or are still contracting to the main sequence. Thus, there are rapid rotators at all spectral types, with a maximum rotational velocity of order 100 km-sec^{-1}. The Pleiades is sufficiently old that the G stars have had time to go through the rapid spindown stage after arrival on the main sequence, leading to the observed lack of rapidly rotating G stars. Because of the longer spindown timescales for lower mass stars, rapidly rotating K and M stars are still present in the Pleiades. By the age of the Hyades, however, the K stars have also reached quite small rotational velocities and rapid rotation is present only for some of the M dwarfs.

Because there is now a considerable body of observational data to constrain models of stellar spindown, it should be possible to improve upon the ES model. Sofia (1987) and MacGregor (1987) are attempting to construct updated ES models for a range of stellar masses in order to attempt to match the T Tauri and open cluster observations.

IV. PROBLEMS WITH CURRENT MODELS

a) How long does it take the core to spin down?

A potentially damaging flaw of the ES model is the nearly complete lack of rotational coupling between the radiative core and the convective envelope. Even after a few billion years of evolution, the core of their model Sun has only lost a few percent of its initial angular momentum, and thus there is a large rotational velocity discontinuity at the boundary between the core and the convective envelope. This type of internal rotation profile is contradicted at least for the Sun by recent helioseismology results (Gough 1987), which show essentially solid-body rotation from the surface down to at least R = 0.3 R$_\odot$. Of course, we do not know for certain that the Sun ever went through a period of rapid rotation on the main sequence. Even a Skumanich law extrapolation predicts that the Sun arrived on the main sequence with a rotational velocity of order 20 km-sec^{-1}, however, so it seems likely that the Sun once had an angular rotation rate at least 10 times larger than is currently observed. This implies that the core has been spun down considerably in a time less than the age of the Sun.

There are no other low mass stars for which we have directly determined the internal rotation. There are several indirect arguments that suggest that the core is spun down on a considerably longer timescale than the outer convective envelope: (1) comparison of the Pleiades and Alpha Per data shows that G dwarfs spin down at least as rapidly as predicted by the ES model (Stauffer et al. 1985; SH87). Therefore, either it is primarily just the envelope that is being spun down at first, or the angular momentum loss rate is a factor of 10 larger than the simple wind model used by ES; (2) the angular momentum loss mechanism has to be such that the wind does not prevent spin-up during pre-main sequence evolution, but allows rapid spin-down on the main sequence. In the ES model, this occurs because the fraction of the moment of inertia in the convective envelope is substantially greater during the PMS phase, and thus a given angular momentum loss rate will have a greater affect on the main sequence star; (3) comparison of F dwarf rotational velocities in Alpha Per and the Pleiades (Stauffer et al. 1988b; Benz, Mayor and Mermilliod 1984) shows that F dwarfs also spin down from large to relatively small surface rotational velocities in a few times 10^7 years after arrival on the main sequence. This is compatible with only the outer envelope being spun down at first, because the considerably weaker winds expected from these stars (due to their much thinner convective envelopes and hence weaker dynamos) are compensated for by the much smaller moment of inertia of the envelope. If the entire star is being spun down, it is difficult to understand how such an efficient angular momentum loss rate is maintained.

The above arguments suggest that the core is spun down on a longer timescale than the convective envelope. It is not possible at present to constrain the core spin down timescale better than to suggest it is more than a few times 10^8 years but less than a few times 10^9 years for G stars. Having no F dwarf for which the internal rotation rate is known, it is not even possible to state with certainty that a solar age F dwarf would or would not still have a rapidly rotating core.

b) Why are a large fraction of the low mass stars in young clusters slow rotators?

While a large fraction of the late type stars in the Pleiades and Alpha Per clusters are rapid rotators ($v\sin i \simeq 100$ km-sec^{-1}), approximately half of the stars in the same spectral type range have rotational velocities of order 10 km-sec^{-1}or less (Stauffer et al. 1985; SH87). This is not expected, since their precursors (the T Tauri and Naked T Tauri stars) do not show a large range in initial angular momenta or a rotational velocity distribution that is so strongly skewed toward low rotational velocities. In particular, Bouvier et al. (1987) claim to have measured rotation rates for all of their T Tauri stars, with no upper limits. The projected rotational velocities for their sample range from 6 to 35 km-sec^{-1}, with a mean of about 15 km-sec^{-1}. Evolved to the main sequence without angular momentum loss, they would have rotational velocities of order 30-170 km-sec^{-1}. This nicely fits the rapid rotator population of the young clusters, but fails entirely to explain the slowly rotating stars. A comparison of the predicted (T Tauri stars evolved to the ZAMS without angular momentum loss) and observed (Pleiades) rotational velocity distributions is provided in Figure 3. Any traditional angular momentum loss model, with dJ/dt proportional to the angular rotation rate to some power, would primarily decrease the rotational velocities of the most rapidly rotating stars - thus compressing the predicted distribution but not leading to many stars with rotational velocities less than 10 km-sec^{-1}.

Three alternative mechanisms have been proposed to explain the excess of slowly rotating late type stars in the Pleiades and Alpha Per clusters. One (SH87) assumes that the angular momentum loss rate for rapidly rotating stars is essentially independent of angular rotation rate. A Skumanich law braking mechanism is assumed to operate for slowly rotating stars. This "model" does allow the T Tauri rotational velocity distribution to be compatible with the young cluster data, provided that the turnover to a Skumanich law spin-down occurs at about 10-20 km-sec^{-1}. A prediction of this model is that the fraction of slow rotators should decrease greatly toward later spectral types. A second alternative is that there is a significant age spread among the late type stars in these clusters. The older stars have had time to spin down to relatively small rotational velocities, while the younger stars are still rapidly rotating. Because spin down timescales are longer for lower mass stars, this model also predicts that the lower mass stars should show fewer slow rotators. One problem with this model is that an age spread of order 10^8 years seems necessary in order to fit the Pleiades data. This age spread for stars in a bound cluster seems uncomfortably long. Also, as already noted, the IC 2391 stars show a large range in rotational velocities and essentially no age spread ($\Delta\tau < 10^7$ years). A third possible explanation for the large fraction of slowly rotating stars in the young open clusters is that their progenitors are slowly rotating, inactive PMS stars that have not yet been discovered because they lack the chromospheric and coronal emission used to locate T Tauri and Naked T Tauri stars. However, the known PMS stars do not show an obvious correlation between chromospheric emission and rotation, and so it is not clear that these hypothetical "skinless T Tauri" stars would be slow rotators.

c) Why Aren't F Dwarfs in Open clusters all Slowly Rotating?

F dwarfs have very thin convective envelopes, and thus store only a relatively small fraction of their angular momentum there. If only the outer convective envelope is spun down at first, it would seem plausible that most F dwarfs should have rotational velocities less than 10 km-sec^{-1}. One common explanation for why this is not the case is that the thin convective envelopes lead to weak winds, and thus considerably lower angular momentum loss rates relative to later type dwarfs. The observations do not support this view. F dwarfs in the Alpha Persei cluster have rotational velocities up to 100 km-sec^{-1} or more. As shown best by Benz, Mayor and Mermilliod (1984), by the age of the Pleiades, mid-F dwarf rotational velocities have dropped considerably, with a mean of only about 30 km-sec^{-1} and a relatively small dispersion at a given spectral type. By the age of the Hyades, the mean rotational velocity has only decreased by a small amount more, while the dispersion about the mean has become very small, and perhaps consistent with a unique value for a given effective temperature (Benz, Mayor and Mermilliod 1984; Radick et al. 1987).

The initial spin-down of the F dwarfs from large to relatively small rotational velocities occurs in only a few times 10^7 years, and is thus comparable to G dwarf timescales. This does not suggest a "weak" wind. Gray (1982) has suggested that a rotational "brake" determines the F dwarf rotations. That is, for a given spectral type, there is a unique rotational velocity, above which angular momentum loss rates are large, and below which they are much lower. Gray suggests that the character of the dynamo changes at this velocity. With this model, presumably stars that drop below the critical velocity should have much weaker chromospheric emission than stars above that rotational velocity. The Pleiades F dwarfs would provide a reasonable sample for such a test.

Another possible explanatiion of the F dwarf rotational velocity distributions observed in open clusters is that there is significant angular momentum input from the core to the envelope. That is, the observed rotational velocity envelope for Hyades F dwarfs (for example) delineates the locus of velocities where dJ/dt carried away by the wind equals dJ/dt transferred from the core to the envelope. As the core spins down via this process, the angular momentum transfer from the core to the envelope would plausibly decrease, leading to the observed slow decline in surface rotational velocities for F dwarfs older than the Pleiades.

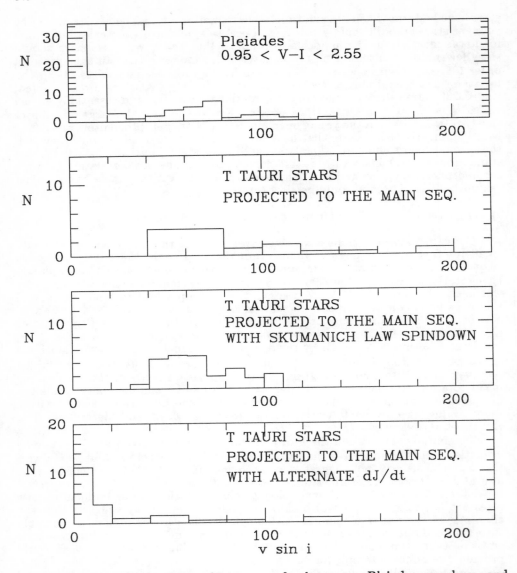

Figure 3: Rotational velocity histograms for low mass Pleiades members, and for T Tauri stars using three ways of projecting the observed rotational velocity distribution to the main sequence. The first T Tauri histogram assumes no angular momentum loss during subsequent pre-main sequence evolution; the second assumes a Skumanich-law type braking; while the third assumes dJ/dt is independent of J for $v_{eq} > 10$ km-sec^{-1}.

d) Other Problems and Puzzles

The rotational velocities derived for low mass stars can potentially be used to test theoretical models of angular momentum loss via winds. For relatively old stars, it is often argued that the Skumanich law implies that for rotation periods greater than about 3 days, $dJ/dt \, \alpha \, \Omega^3$ (c.f. Vilhu and Moss 1987). However, if the cores of G dwarfs spin down on a considerably longer timescale than the convective envelopes, it is not possible to infer that a Skumanich law relation for the surface rotational velocities implies anything definitively about the angular momentum loss rate because the observed change in rotation rate is not directly connected to the primary angular momentum reservoir. For shorter rotational periods, the Skumanich law apparently breaks down, and it is often assumed that the wind changes character in some fashion. As one explanation for the K dwarf rotational velocity distribution in the Pleiades and Alpha Per clusters, we (SH87) suggested that the angular momentum loss rate becomes approximately independent of Ω for large rotational velocities. This effect could be due to the magnetic field saturating above a certain rotational velocity, as suggested by Walter's (1982) X-ray data for late type stars. Gray's (1982) rotational brake idea posits that there is a dramatic increase in the angular momentum loss rate above some spectral type dependent critical velocity. The brake is responsible for the focusing of F dwarf rotational velocities to a small dispersion at fixed effective temperature evident in the data of Benz, Mayor and Mermilliod (1984).

Binary star data may provide some useful information regarding rotational velocity evolution. The Pleiades K dwarf rapid rotators are much less often photometric binaries than the slow rotators (SH87). This may simply reflect the fact that binary star systems have an additional sink to deposit their excess angular momentum which is not available to single stars. Among the approximately 150 stars in the Pleiades and Alpha Per cluster for which we have spectra, none are composite in the sense of having a slow and rapid rotator (of similar luminosity) in a binary system. Once higher signal-to-noise spectra are obtained, such data might eventually provide useful constraints on the formation and evolution of these stars. For instance, the Pleiades G dwarfs are all now relatively slow rotators, while many of the K dwarf are rapid rotators. Some of the G dwarfs presumably have K dwarf secondaries, some of which should still be rapidly rotating if our simple ideas are correct. High signal-to-noise spectra in the red might usefully test this idea.

Finally, the recently discovered lithium abundance gap found in mid-F dwarfs in the Hyades and elsewhere (Boesgaard and Tripicco 1985) should be considered when attempting to model the rotational evolution of F dwarfs. The most commonly referenced explanation for the lithium gap is the diffusion mechanism proposed by Michaud (1986) which relies on the variation in the depth of the convective zone with spectral type to reproduce the gap. The diffusion mechanism will not work if the mass loss rates from the F dwarf winds are too large, hence the presence and shape of the gap places constraints on angular momentum loss rates from these stars if the diffusion mechanism is responsible for the lithium depletion. Alternatively, it should be noted that the lithium gap coincides in position (in an HR diagram) with the steep gradient in rotational velocities between early type and late type stars. Is it possible that rotation and the lithium gap are connected? That is, stars in the gap have sufficiently strong winds that over the age of the Hyades most of the original mass in the envelope is lost. Mass (and angular momentum) would have to be transferred from the core to the envelope, and if the core is sufficiently well mixed, the transferred matter would be lithium poor. Stars hotter than the gap stars may have essentially no wind, and thus retain their primordial lithium abundances. Stars cooler than the gap have massive enough convective envelopes so that the matter brought up from the core only constitutes a fraction of the total envelope mass, and thus only a small lithium depletion is observed. Further constraints on the mechanism producing the lithium gap have recently been provided by lithium observations of Pleiades and Alpha Persei F dwarfs (Boesgaard, Budge and Ramsey 1987; Pilachowski, Booth and Hobbs 1987; Balachandran, Lambert and Stauffer 1987), which show little or no lithium depletion in the gap temperature range. Thus, most of the lithium depletion occurs between the age of the Pleiades and the age of the Hyades.

V. FUTURE OBSERVATIONAL DIRECTIONS

To date, we have determined the rotational velocity distribution for low mass stars in only three open clusters. That is insufficient to sample the evolutionary timescales for different masses, and forces us to assume that all cluster to cluster differences are evolutionary and not at-formation variations. We have already surveyed the richest, nearby clusters, and it will be necessary to go to more distant clusters in order to increase the sample size. The vastly better observing efficiency offered by fiber-fed, multi-object, echelle spectrographs will allow longer exposures, and thus make it possible to survey these more distant clusters. For clusters younger than 5×10^7 years, accurate CCD photometry should provide a direct measurement (or significant upper limit) on the age spread of star formation in the cluster, thus eliminating one of the free parameters in the search to explain the rotational velocity distributions. The combined photometric and spectroscopic data should also allow a better determination of the interrelationship between duplicity and rotation. For the Pleiades, a multi-object echelle will allow us to determine the rotational velocity distribution for cluster M dwarfs. This should help identify the mechanism that produces the slowly rotating K dwarfs, as previously discussed.

Further work on the rotational velocities of very young stars should be pursued. The Bouvier et al. (1987) and Hartmann et al. (1987) studies do not provide rotational velocities for a complete (e.g. magnitude limited) sample of T Tauri stars - the strong emission line and "continuum" T Tauri stars still have unknown rotational velocities because they generally show no photospheric absorption lines. Perhaps these stars are the progenitors of the most rapidly rotating, low mass open cluster stars. Shu (1987) predicts that stars arrive on the "birth-line" rotating at close to break-up velocities; rotational velocities have not yet been estimated for these stars because they are still deeply embedded. According to the model, the centrifugally driven winds from these stars power the bipolar outflows observed in CO as well as some of the optical jets and HH objects observed in star forming regions. High resolution, near-IR spectrographs now being constructed may allow rotational velocities for these stars to be measured, and thus provide a test of Shu's model. At a slightly later (?) stage of evolution, FU Ori stars have recently been shown to have moderate rotational velocities (Hartmann and Kenyon 1987a,b). The variation in derived rotational velocity between the optical and IR strongly suggests, however, that it is just the disk in these stars that is being observed. It would be useful to find evolved FU Ori stars, where the disk has faded substantially, in order to be able to observe the star and determine its rotational velocity.

The determination of the internal rotation rate of F dwarfs (particularly in the Hyades) via astroseismology would provide extremely useful information on the timescale for the radiative core spindown. So far, astroseismology measurements have been limited to only a handful of the brightest stars, and even then the analyses have been controversial. However, bigger telescopes and more stable spectrographs should provide useful data in the not too distant future, and we may soon be able to determine the internal as well as external rotation of at least a few stars. A less direct indication of the spindown time for the core may be provided by rotational velocity measurements of post-main sequence stars. Peterson (1983) has measured relatively large spectroscopic rotational velocities for horizontal branch stars in some globular clusters. The simplest explanation for this result is that at least for some approximately solar mass stars in these clusters the cores did not spindown completely during the star's main sequence lifetime.

It is now nearly two decades since Kraft's review of stellar rotation. We have just now begun to develop the instrumentation and software necessary to answer the questions posed in that review. Perhaps in another decade we will have those answers.

REFERENCES

Abt, H.A. 1961,*Ap.J.Supp.*,**6**,37.
Abt, H.A. 1965,*Ap.J.Supp.*,**11**,429.
Abt, H.A., and Hunter, J.H. Jr. 1962,*Ap.J.*,**136**,381.
Balachandran, S., Lambert, D., and Stauffer, J. 1987, preprint.
Benz, W., Mayor, M., and Mermilliod, J. 1984,*A.A.*,**138**,93.
Belcher, J., and MacGregor, K. 1976,*Ap.J.*,**210**,498.
Boesgaard, A., Budge, K., and Ramsay, M. 1987, preprint.
Boesgaard, A., and Tripicco, M. 1986,*Ap.J.Lett.*,**302**,L49.
Bouvier, J., Bertout, C., Benz, W., and Mayor, M. 1987,*A.A.*,**165**,110.
Duncan, D.K., Baliunas, S.L., Noyes, R.W., Vaughan, A.H., Frazer, J., and Lan-
    ning, H.H. 1984,*Pub.A.S.P.*,**96**,707.
Endal, A., and Sofia, S. 1981,*Ap.J.*,**243**,625.
Gough, D. 1985,*Sol.Phys.*,**100**,65.
Gray, D. 1982,*Ap.J.*,**261**,259.
Hartmann, L., Hewett, R., Stahler, S., and Mathieu, R. 1986,*Ap.J.*,**309**,275.
Hartmann, L. and Kenyon, S. 1987a,*Ap.J.*,**312**,243.
Hartmann, L., and Kenyon, S. 1987b,*Ap.J.*,**322**,393.
Hartmann, L., Soderblom, D., and Stauffer, J. 1987,*A.J.*,**93**,907.
Herbig, G., Vrba, F., and Rydgren, A. 1986,*A.J.*,**91**,575.
Huang, S.-S., and Struve, O. 1960, in Stars and Stellar Systems, Vol. VI, Stellar
    Atmospheres, ed. J.L. Greenstein (Chicago: University of Chicago Press),
    p. 321.
Kraft, R.P. 1970, in Spectroscopic Astrophysics, ed. G. H. Herbig (Berkeley:
    University of California Press), p. 385.
MacGregor, K. 1987, priv. comm.
Michaud, G. 1986,*Ap.J.*,**302**,650.
Pilachowski, C., Booth, J., and Hobbs, L. 1987, preprint.
Radick, R., Thompson, D., Lockwood, G., Duncan, D. and Baggett, W. 1987,*Ap.J.*,**32**

Rydgren, A., Vrba, F., Chugainov, P. and Shakhovskaya, N. 1985,*B.A.A.S.*,**17**,556.
Schatzman, E. 1962,*Ann.d'Ap.*,**25**,18.
Skumanich, A. 1972,*Ap.J.*,**171**,565.
Slettebak, A. 1954,*Ap.J.*,**119**,146.
Slettebak, A. 1955,*Ap.J.*,**121**,653.
Sofia, S. 1987, Fifth Cambridge Cool Star Conference.
Stauffer, J., and Hartmann, L. 1987=SH87,*Ap.J.*,**318**,337.
Stauffer, J., Hartmann, L., Burnham, J., and Jones, B. 1985,*Ap.J.*,**289**,247.
Stauffer, J., Hartmann, L., Jones, B., and McNamara, B. 1988a, preprint.
Stauffer, J., Hartmann, L., Jones, B., and Balachandran, S. 1988b in preparation.
Stauffer, J., Hartmann, L., and Latham, D. 1987,*Ap.J.Lett.*,**320**,L51.
Stauffer, J., Hartmann, L., Soderblom, D., and Burnham, J. 1984,*Ap.J.*,**280**,202.
van Leeuwen, F. and Alphenaar, P. 1982,*E.S.O.Messenger*,**28**,15.
Vilhu, O., and Moss, D. 1987 preprint.
Vogel, S. and Kuhi, L.V. 1981,*Ap.J.*,**245**,960.
Walter, F.M. 1982,*Ap.J.*,**253**,745.
Walter, F.M. 1986,*Ap.J.*,**306**,573.
Walter, F., et al. 1987 in preparation.
Weber, E.J., and Davis, L. Jr. 1967,*Ap.J.*,**148**,217.
Wilson, O.C. 1966,*Ap.J.*,**144**,695.

KINEMATIC CLUES TO THE ORIGIN AND EVOLUTION OF LOW MASS CONTACT BINARIES

Edward F. Guinan
Department of Astronomy & Astrophysics
Villanova University
Villanova, PA 19085

David H. Bradstreet
Department of Mathematics and Physical Science
Eastern College
St. Davids, PA 19087

ABSTRACT. We present new results concerning the rather old age (8 - 10 Gyr) of the majority of the low mass contact binaries known as W UMa systems inferred from their space motions, velocity dispersions, and memberships in old clusters such as NGC 188. The great age and expected relatively short lifetimes of the W UMa binaries imply that they have evolved into the contact state from detached or semi-detached progenitors with orbital periods of a few days. Angular momentum loss may be provided by magnetic torques in a stellar wind, the same mechanism that brakes the rotation of single, cool dwarf stars during their main sequence lifetimes. According to Webbink, W UMa binaries ultimately coalesce into single stars. We speculate that a significant number of presently single stars evolved from the coalescence of short period binaries.

1. Introduction

Low-mass contact binaries, known as W Ursae Majoris stars, are the most common close binaries. The relative incidence of W UMa systems appears to be high, with recent estimates in the range of 1 - 2 out of 1000 stars of all types (Budding 1982; Van't Veer 1975). However, the frequency of W UMa stars may be even higher when the contact binaries are compared with field stars of similar spectral types. Van't Veer (1975) estimates that ~ 1 out of 100 dwarf F and G stars may be W UMa systems. These dumbbell-shaped binaries have orbital periods typically less than 2/3 day and have both components in contact with their Roche limiting surfaces. The components of W UMa systems generally have spectral types between F and K and lie near or just above the main sequence. The total mass of most W UMa-type binaries is $(M_1 + M_2) < 2.5\ M_\odot$, and while the spectral types of the components are similar, the masses are often quite dissimilar. Because of their strong tidal interactions, the rotational periods of the

345

A. K. Dupree and M. T. V. T. Lago (eds.), Formation and Evolution of Low Mass Stars, 345–375.
© 1988 by Kluwer Academic Publishers.

components are synchronized with their orbital periods. Their rapid rotation and convective atmospheres apparently contribute to the observed strong chromospheric and coronal emissions which are among the strongest in surface flux of cool stars. In addition, the light curves of the shorter period W UMa systems are often markedly asymmetric and have been modelled by assuming the presence of large starspots as shown in Figs. 1 and 2. The origin of the strong chromospheric and coronal emissions as well as the presence of large starspots appear to be manifestations of strong, dynamo-generated magnetic activity. The strong UV chromospheric and transition region emission features are shown in Fig. 3 from Dupree and Preston (1981) for a sample of W UMa stars. W UMa stars also exhibit strong x-ray emissions which are thought to originate in magnetically heated coronae (Cruddace and Dupree 1984). While the x-ray and chromospheric emissions of W UMa stars are very strong, they are not as strong as expected from an extrapolation of the period-activity relation for RS CVn and single stars (e.g. Vilhu and Heise 1986). This leveling off of activity at extremely high rotation rates may be caused by the total filling of the stellar surface by magnetically active regions (Rucinski 1983; Vilhu 1984).

We will not address the problem of early-type contact binaries. These high mass systems are considerably rarer than W UMa binaries. They typically have orbital periods of 1 - 3 days and consist of nearly identical components of spectral type O, B, or early A. Because of their presumed radiative envelopes, these systems are not expected to have dynamo driven magnetic fields, although they might possess remnant magnetic fields inherited from their formation from the ISM. The lifetimes of the early-type contact binaries should be short due to their expected rapid nuclear evolution and large mass loss rates resulting from strong stellar winds.

2. Theories on the Origin and Evolution of W UMa Binaries

At present we do not have a completely satisfactory theory to account for the origin, structure, and evolution of the W UMa binaries. These are controversial subjects and have been discussed recently in reviews by Mochnacki (1985) and Rucinski (1985).

According to one scenario W UMa systems form as contact binaries at the end of the pre-main sequence contraction by a fission process (Roxburgh 1966). Their structure and evolution as zero-age main sequence contact binaries have been described by either the thermal relaxation oscillation (TRO) theory (Lucy 1976, 1977) or by the contact discontinuity (DSC) theory (Shu, Lubow, and Anderson 1976, 1979). These theories assume that the angular momentum of the binary system is conserved.

In another scenario, W UMa binaries evolve into contact systems from initially detached systems by angular momentum loss via magnetic torques from a stellar wind in which the spin angular momentum and the orbital angular momentum are coupled through tides. This hypothesis for the origin of W UMa stars has been discussed by Huang (1966), Mestel (1968), Okamoto and Sato (1970), Van't Veer(1979), and Vilhu and Rahunen (1980). In addition to magnetic braking angular momentum is lost through gravitational radiation when the orbital period of the binary system becomes small (P < 0.5 d). Because of angular momentum loss, these systems ultimately coalesce into single stars (Webbink 1976, 1985; Mochnacki 1985).

The determination of the age of W UMa-type systems is an important criterion for discriminating between the different theories. In the case of no angular momentum loss,

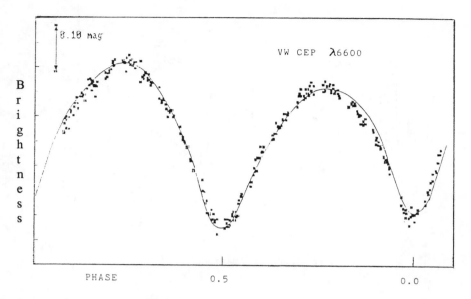

Figure 1. The asymmetrical light curve of the W UMa-type system VW Cep obtained in October 1986 at Villanova. The solid curve represents the theoretical light curve with starspot as depicted in Fig. 2.

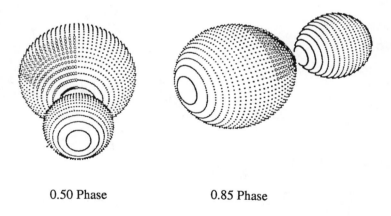

0.50 Phase   0.85 Phase

Figure 2. The three dimensional representation of the VW Cep model with the spotted region viewed at phases 0.50P and 0.85P, respectively.

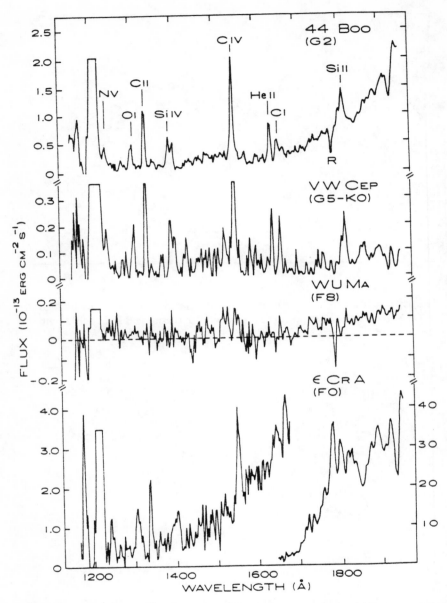

Figure 3. IUE ultraviolet spectra of representative W UMa stars taken from Dupree and Preston (1981). The strongest chromospheric and transition region line emissions are indicated.

the systems form as contact binaries on the ZAMS, and remain contact systems for as little as 0.1Gyr (Van't Veer 1979; Vilhu 1982) and as long as several Gyr (Mochnacki 1985). Therefore, as a group, the age of W UMa-type binaries should have the same age distribution as single stars of similar mass and spectral type - i.e., Pop I up to young disk stars. In the case of angular momentum loss, the binary starts out as a detached close binary and evolves into the contact configuration via magnetic braking. This occurs from the tidal interaction of the binary components when the spin angular momentum of the stars is coupled to the orbital angular momentum. According to Mestel (1984) the orbital angular momentum is fed into the spin angular momenta of the components by the process of tidal friction so that a strong braking torque on the individual stars causes their orbit to shrink and their rotations to increase. The angular momentum from the binary is lost as the plasma in the magnetic winds of the stars transfer angular momentum to the interstellar medium. The time scale for reaching contact depends on the initial orbital separation, the magnetic activity of the stars, the nature of the braking mechanism, and the extent of the stellar wind. As discussed in Section 7, binary systems with initial orbital periods in the range of 2 to 4 days would lose enough angular momentum to reach contact in 0.5 - 7.0 Gyr. If the initial orbital period is 1 day or less then the time to reach contact is extremely short (t < 10 - 100 Myr).. Thus the age of the group as a whole would tend to be mixed, but with a bias toward older systems because the progenitors of W UMa systems can be old before reaching contact.

Furthermore, according to current ideas about stellar dynamo theory (eg., Parker 1986), it is necessary for the stars to have convective envelopes and significant rotation to generate and maintain strong magnetic fields. For main sequence stars convective envelopes develop for stars having masses < 1.5 $M_\odot$ which correspond to spectral types of ~ F2-5 V and later. Thus, for magnetic braking to be effective, the components of the initially detached binary systems would have masses of < 1.5 $M_\odot$. In addition, tidal effects on the components would have to be large enough so that the star's rotational angular momentum would be coupled to the orbital angular momentum. Binary systems having components with initial masses > 1.5 $M_\odot$ should lose little angular momentum during their main-sequence lifetimes because dynamo generated activity should be negligible due to the absence of convective envelopes. These systems remain as detached binaries until the more massive star evolves off the main sequence and ultimately fills its Roche lobe. These stars could become "RS CVn-type" binaries before becoming semi-detached systems as the more massive component evolves to the point where its atmosphere becomes convective and when tidal coupling occurs. These higher mass systems could evolve into Algol-type binaries as the more massive component fills its Roche lobe and undergoes mass loss and mass exchange with its companion.

On the other hand, the low-mass progenitors of the W UMa-type stars would start out as detached binaries with initial periods of < 4 days and as angular momentum loss occurs their orbital periods would shorten as the orbit decreases in size. Before becoming W UMa-type stars they could pass through a stage of being short period detached binaries such as ER Vul and then semi-detached or contact, mass-exchanging systems.

3. The Space Motions of W UMa Stars

The U, V, and W velocity vectors of a star are measured with respect to the Sun, where, following the notation of Eggen (1965), the U component is directed away from the Galactic Center (in the direction of $l = +180°$; $b = 0°$); V is measured in the direction of galactic rotation (i.e., toward $l = 90°$, $b = 0°$); and the W component is directed toward the North Galactic Pole (i.e., toward $b = +90°$). The resultant space motion S is defined by S $= (U^2 + V^2 + W^2)^{1/2}$. The expressions for computing the U, V, and W space motions are given by Mihalas and Binney (1981) and by Atanasijevic (1971). The quantities needed to compute the space motions are the star's right ascension and declination, the proper-motion components (μ_α, μ_δ), the radial velocity (V_0), and the distance from the Sun. This work

updates and expands the earlier studies of the space motions of W UMa stars by Eggen (1967) and Rucinski and Kaluzny (1981). In Table I are listed the W UMa systems for which these quantities are presently available. The celestial coordinates and proper motion components were obtained from the SAO catalogue and from Artiukhina (1964) except where noted otherwise. We list the proper motion components from the SAO catalog and those given by Artiukhina. In calculating the space motions we used SAO values when both were available. In most cases, the agreement between the proper motion determinations were very good. Most of the distances were taken from Rucinski (1983) and the radial velocities correspond to the motion of the center of mass of the binary. These were obtained chiefly from Batten *et al,* (1978) unless otherwise indicated. The greatest uncertainties in the calculation of the space velocities arise chiefly from uncertainties in the values of the radial velocity of the barycenter and the distance estimates. Additional information on the W UMa systems in our sample is given in Table II. This information was chiefly obtained from Mochnacki (1985), Rucinski (1983), and the *General Catalog of Variable Stars.* The U, V, W, and S values for these stars are given in Table III. The mean value of S for the 34 W UMa stars in our sample is S ~ 65 km s^{-1}. Also given in Table III are the velocity components *u, v, w* and resultant space motion *s* transformed to the local standard of rest (LSR), where the velocity components of the Sun relative to the LSR are assumed to be $U_\odot = -9.2$ km s^{-1}, $V_\odot = +12.0$ km s^{-1}, and $W_\odot = +6.9$ km s^{-1} (Mihalas And Binney 1981). *u, v,* and *w* are also defined as Π, Θ, and Z in other kinematic studies. The *u* and *v* velocity components (relative to the LSR) of the W UMa binaries are plotted in Fig. 4. A velocity circle of radius ~ 65 km s^{-1} is shown which

is the nominal boundary between high and low velocity stars (Oort 1926). As shown in the figure, a significant fraction of these stars have velocities in excess of 65 km s^{-1}. As shown, VZ Psc, TY Pup, and RW Com have resultant space motions which exceed 100 km s^{-1}. Also plotted in the figure is the rapidly rotating star FK Com. FK Com has been shown by Guinan and Robinson (1986) to be a high velocity star with a resultant space velocity *s* ~ 66 km s^{-1}. Bopp and Rucinski (1981) have suggested that FK Com has formed from the coalescence of a W UMa-type binary, as predicted by evolutionary models of Webbink (1976).

In Fig. 5, adapted from Delhaye (1965), we compare the velocity distributions in the U, V plane of the W UMa stars in our sample with those of dwarf F, G, and K stars within 20 pc of the Sun. These solar-like main sequence stars are similar in spectral type and luminousity to most of the components of W UMa binaries. Being in the solar neighborhood, the F, G, and K stars are chiefly evolved main-sequence stars which range

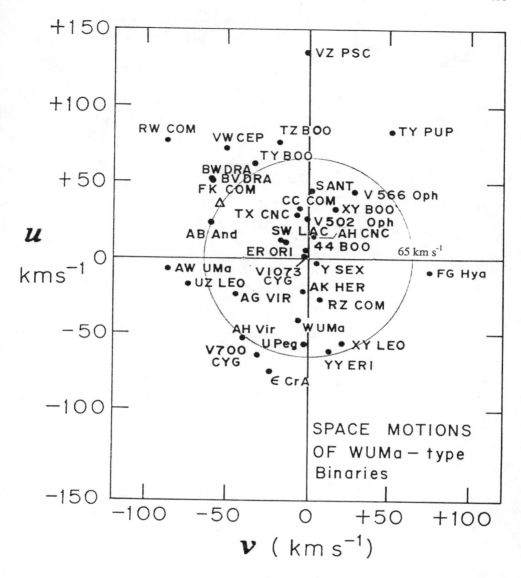

Figure 4. Plot of the *u* and *v* velocity components of W UMa stars relative to the Local Standard of Rest. The circle is the 65 km s^{-1} loci relative to the LSR. Also shown for comparison is the position of the peculiar, rapidly rotating single giant FK Com.

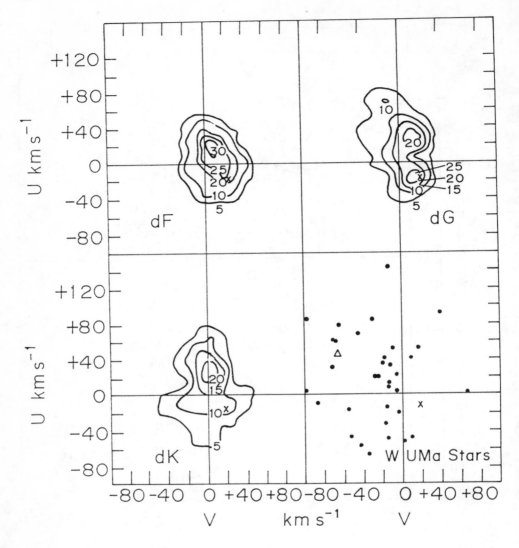

Figure 5. The UV space motions of the W UMa systems are plotted together with samples of nearby F, G, and K dwarfs. Note the large dispersion in velocities of the W UMa-type systems compared to single non-evolved stars of similar masses. This figure is adapted from Delhaye (1965).

in age from young to old disk. As shown in the figure, the distributions of the single stars are more compact than the sample of W UMa stars. The velocity dispersions are discussed in the next section.

4. The Kinematic Age of W UMa Binaries

The velocity dispersions $<u^2>^{1/2}$, $<v^2>^{1/2}$, and $<w^2>^{1/2}$ were computed for our sample of W UMa stars and the following values were found: $< u^2 >^{1/2} = 51$ km s^{-1} , $<v^2>^{1/2} = 36$ km s^{-1} and $<w^2>^{1/2} = 18$ km s^{-1}. The dispersion of the resultant space velocity, defined as $<s^2>^{1/2} = [<u^2> + <v^2> + <w^2>]^{1/2}$ is 65 km s^{-1}. These velocity dispersions are listed in Table IV along with the velocity dispersions of other representative groups of stars obtained from Delhaye (1965). The velocity dispersions of nearby F, G, K, and M dwarfs are given as representative of intermediate age disk-population stars. These stars are on or near the main sequence and are a mixture of a few young stars and mostly old stars. Stars with spectral types earlier than F have significantly smaller velocity dispersions and, from considerations of stellar evolution, are mostly young stars. Also listed in the table are the velocity dispersions for planetary nebulae, white dwarfs and short period (P < 0.45 d) RR Lyrae variables which are intermediate-to-old disk population stars. For comparison, the velocity dispersions for subdwarfs and long period (P > 0.45 d) RR Lyrae variables are included in the table as representatives of (Pop II) spheroidal-component stars. As seen from the table, the velocity dispersions of the W UMa stars are about twice those of of F, G, K, and M dwarfs and about 1/2 to 1/3 less than the Pop II representatives. On the other hand, the velocity dispersions of the W UMa stars closely match those of old disk objects which are evolved objects with mean evolutionary ages in the range 8 - 10 Gyr. (Wielen 1974; Mihalas and Binney 1981) A good correlation between velocity dispersion and stellar age has been found by Byl (1974), Mayor (1974), and Wielen (1974). As discussed in Mihalas and Binney (1981) the mechanism most likely responsible for the increase in the peculiar velocity dispersions of disk stars with their ages is encounters with other stars or interstellar clouds. Over time these encounters tend to randomize and increase the velocity distribution. Spitzer and Schwarzschild (1951, 1953) were first to conclude that encounters with interstellar cloud complexes, now often identified with giant molecular clouds, are more important in increasing the velocity dispersions because of the large masses of the clouds. The interstellar clouds have vastly greater kinetic energy than an individual star, so in star-cloud interactions the stars tend to gain energy and move away at higher speeds. Spitzer and Schwarzschild showed that the time dependence of the root-mean-square (rms) speed of a star, v_{rms}, is described approximately by the formula

$$v_{rms} \sim v_{rms}(0) \left[1 + (t / \tau_E) \right]^{1/3} \qquad (1)$$

where $v_{rms}(0)$ is the initial rms speed set by the internal velocity dispersion (~ 10 km s^{-1}) in the ISM in which the star originally formed, and where τ_E is the characteristic energy-exchange time for the collision process. Spitzer and Schwarzschild estimate that $\tau_E \sim 0.2$ Gyr from the observed increase in the velocity dispersions of disk population stars of known age. For our sample of W UMa stars we find $v_{rms} = 65$ km s^{-1} where v_{rms}

$= <s^2>^{1/2}$ in our notation. This equation yields an unrealistically old age for disk stars.

The relationship between velocity dispersions and age has been empirically determined by Wielen (1974). Wielen finds that the velocity dispersions of stars with known ages appear to follow a relationship nearly identical to eq. (1) but with an exponent of 1/2 rather than 1/3, i.e.

$$v_{rms} \sim v_{rms}(0) \left[1 + (t / \tau_E) \right]^{1/2} \tag{2}$$

where the constants are the same as for eq. (1). As discussed by Mihalas and Binney (1981), the better fit to the observations given by eq. (2) compared to eq. (1) from Spitzer and Schwarzschild may possibly imply that the assumption of Spitzer and Schwarzschild that a star will be perturbed by only one cloud at a time may not be correct, i.e., multiple collisions may be important especially when the star encounters a spiral arm. Using eq. (2) we calculate a mean age for the W UMa stars of ~ 9 Gyr. The results of Wielen are shown in Fig. 6 adapted from Mihalas and Binney (1981) in which an increase of the $< u^2 >^{1/2}$, $< v^2 >^{1/2}$, $< w^2 >^{1/2}$, and $< s^2 >^{1/2}$ with stars of increasing average stellar age. The line in the upper part of the figure shows the increase in velocity dispersion expected from the theory of stellar encounters with interstellar complexes ($v_{rms} \sim t^{1/3}$) by Spitzer and Schwarzschild. When the velocity dispersions of the W UMa stars are plotted in this figure between the ages of ~ 8 - 10 Gyr, they are found to be consistent with the observed velocity dispersions of the stars of known age in Wielen's study.

We have also computed the mean distance from the galactic plane for this sample of W UMa stars and find $|\bar{Z}| = 110$ pc. Despite the fact that our sample is composed chiefly of relatively nearby stars, this result is consistent with membership of the W UMa stars to old-disk population stars.

5. Direct Age Determinations of W UMa Binaries in Clusters and Moving Groups

Another important age indicator for W UMa stars is the membership of at least 7 of these systems in the old open cluster NGC 188 (Baliunas and Guinan 1986; Kaluzny and Shara 1987a). The six W UMa stars in NGC 188 for which color indices have been measured are plotted in Fig. 7 taken from Kaluzny and Shara (1987b). The evolutionary age of NGC 188 has been determined to lie between 5 and 11 Gyr (Sandage 1962; Janes and Demarque 1983; Van den Berg 1983). However, the most recent determination of its age by Adler and Rood (1985), using up-to-date opacities and mixing length parameters, yields an age of $t = 10 \pm 1$ Gyr for the cluster.

AH Cnc is a member of M67 ($t \sim 3-4$ Gyr) and AH Vir is a probable member of the Wolf 630 moving group ($t \sim 3$ Gyr). The U, V, W space motions of VW Cep (+80, -64, -10) are similar to those determined by Eggen (1965) for the 61 Cyg moving group (+91, -53, -8). According to Eggen the evolutionary age of the 61 Cyg moving group is $t \sim 3 - 4$ Gyr, and we assume that age for VW Cep. RZ Com has been suggested as a candidate for membership in the Coma cluster. However, its space motions (U = -19 km s^{-1}; V = -5 km s^{-1}) differ from those of the cluster (U = +5 km s^{-1}; V = -6 km s^{-1}) as given by Eggen (1963). Moreover, if it were a cluster member at a distance of ~ 80 pc, RZ Com would be

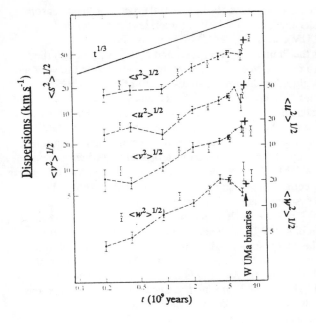

Figure 6. This plot shows the variation of velocity dispersions of disk stars of different age groups indicating that velocity dispersion increases with age. When the velocity dispersions of the W UMa stars (denoted by "+") are plotted with an assumed mean age of ~ 8 Gyr, they are found to be consistent with stars of similar age. The solid line at the top of the figure shows the predicted increase in velocity dispersion with time due to stellar encounters with interstellar complexes from Spitzer and Schwarzschild (1951; 1953). This figure is based on Wielen's (1974) study and adapted from Mihalas and Binney (1981).

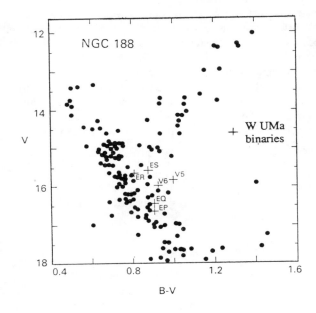

Figure 7. The color-magnitude diagram of NGC 188 showing the W UMa star members. This figure is adapted from Kaluzny and Shara (1987b).

located ~ 2 mag below the cluster's main sequence for its spectral type. For these reasons it is unlikely that RZ Com is a cluster member. Table V summarizes the cluster and moving group memberships for W UMa stars. No W UMa stars have been found to be members of young clusters such as the Pleiades, h and χ Per, and α Per even though they have been thoroughly studied. Also no W UMa stars have yet been detected in the very old spheroidal population stars such as globular clusters. The discovery of a W UMa binary in a globular cluster might be difficult, however, because of the tendency of these binaries to be located in the cores of these clusters due to their higher combined masses compared to single stars of comparable spectral type. Although the data on memberships of W UMa stars in clusters or moving groups with good age determinations is small, the large number of systems in NGC 188 and the paucity of them in young clusters is in accord with the space velocity results in that most W UMa stars are old. The apparent lack of W UMa stars among spheroidal component stars could be consistent with the hypothesis proposed by Webbink (1976) in which W UMa systems coalesce into single stars.

Recently Kaluzny and Shara (1987b) have made CCD surveys of six intermediate to old open clusters searching for W UMa binaries. The clusters surveyed range in age from 1-6.5 Gyr. About 1600 cluster members were examined for light variability and no definite contact binaries were found. This result is not surprising if most W UMa stars are as old as their space motions indicate (t ~ 8 - 10 Gyr). Kaluzny and Shara suggest that the large number of W UMa stars identified in NGC 188 is unique and unexplained. However, the large number may be consistent with the mean age of W UMa stars found from their space motions if NGC 188 is ~ 10 Gyr old as determined by Adler and Rood rather than the 5 Gyr adopted by Kaluzny and Shara.

Although NGC 188 is not presently a rich cluster (Sandage and Eggen 1969), it must have originated with a large number of stars to still exist today. This is because the gravitational binding of open clusters is relatively weak and they are tidally disrupted from outside gravitational forces so that few remain with ages greater than ~ 5 Gyr. In addition, equipartition of kinetic energy occurs among the cluster members in which the lower mass members diffuse away from the center of mass. In time this diffusion process would tend to leave the higher mass stars near its center while the lower mass stars would have a higher probability of being lost from the cluster. In the process of gravitational diffusion and disruption, close binary systems would tend to be located near the center of the cluster because of their higher combined masses. After billions of years the relative number of close binaries might grow with respect to single stars with masses comparable to the masses of the individual binary components.

Thus, the large relative number of W UMa binaries observed in NGC 188 can be explained if the cluster has lost many of its low mass single members. The existence of W UMa systems in such an old cluster and the expected short lifetimes of W UMa stars implies that they have evolved into contact from detached systems through angular momentum loss. This will be discussed in more detail in Section 7.

6. Age Dependent Properties of W UMa Binaries

Having established the old mean age of the W UMa stars, we next investigated whether correlations exist between space velocity (i.e., age) and the physical properties of the stars such as orbital period (P), mass ratio (q), system type (A or W), fillout factor (F), ultraviolet excess (δc_1), and metallicity index (δm_1). Because of the small size of the sample, however, it is only possible at present to draw tentative conclusions from these analyses.

6.1. Period (P) / Space Velocity (s)

When all the stars are included no statistically significant correlation was found between orbital period and space velocity. However, if the two extreme, long period systems with small mass ratios are excluded (ε CrA and AW UMa), then a weak correlation seems present in the sense that space velocity tends to increase with decreasing orbital period.

6.2. Mass Ratio (m_2/m_1) / Space Velocity

According to most theories of evolution of close binaries (Webbink 1985), the systems should evolve toward low mass ratios $q = m_2/m_1$ where m_2 and m_1 are the masses of the

smaller mass and greater mass star, respectively. In comparing space motion to mass ratio of the stars in our sample, we find no correlation between them. There are many W UMa stars with high velocities, such as $s \sim 80$ km s^{-1}, having mass ratios $q \sim 0.1$ to 0.8 (Mochnacki 1985). As an illustration, VZ Psc and SW Lac both have mass ratios of $q \sim 0.85$, yet the space motion of SW Lac is low ($s = 17$ km s^{-1}) and therefore it is relatively young while VZ Psc ($s = 144$ km s^{-1}) is probably one of the oldest stars in the sample.

6.3. Binary Type (A or W) / Space Velocity

W UMa stars are classified into A- or W-type systems from their light curves and radial velocity curves (Binnendijk 1970). The W-type systems are those like W UMa itself in which the hotter component (the star eclipsed at the primary minimum) is the smaller, less massive star. These systems tend to have components of $\sim$ F5 spectral types and later and have shorter periods than A-types. The A-type systems are those in which the hotter star is the larger, more massive component of the system. The A-type systems quite often have primary components with spectral types of A5V to F5V. The prototypes of this class are AW UMa or ε CrA. A-type systems tend to have longer periods than W-types and have mass ratios in which $m_h/m_c > 1.0$ while the W-types have mass ratios $m_h/m_c < 1.0$. Some

interesting ideas have been proposed concerning the nature of the two types of systems. Some theories suggest the possibility that W-type systems evolve into A-type through mass-exchange while others suggest the opposite (see Rucinski 1985). On the other hand, it has been suggested that W-type systems occur because the larger, more massive star is heavily spotted, making it appear cooler than its smaller, less massive mate (Rucinski 1985).

In our sample of W UMa stars we have identified 11 A-type and 22 W-type systems. The mean resultant space motion (relative to the LSR) for the A-type systems is $s(A) = 63$ km s^{-1} while $s(W) = 57$ km s^{-1}. TZ Boo was excluded because it has been classified as both an A- and W-type system (Hoffman 1978). Thus, there is no significant distinction in the resultant space velocities between the W- and A-type W UMa stars. From this it appears that both groups are approximately the same age. The high space velocities of the

A-type W UMa systems are particularly interesting because of the high masses of the primary components. These systems typically have primaries with spectral types of A5V to F5V and with corresponding masses of 1.8 $M_\odot$ to 1.3 $M_\odot$. The evolutionary main sequence lifetimes of 1.8 $M_\odot$ to 1.3 $M_\odot$ single stars is ~ 1 - 4 Gyr respectively (Iben 1965, 1967). The existence of A-type W UMa systems with mean kinematic ages of 8 - 10 Gyr implies that the primary component originally began life as a star of lower mass and subsequently gained mass at the expense of the now low mass mate when both components came into contact. Once the stars are in contact, mass and energy exchange take place in accordance with Webbink's (1976, 1985) evolutionary models of contact binaries. This would explain the existence of main sequence A and F stars with inferred ages of 2 to 5 times their nuclear main sequence lifetimes. As discussed by Webbink (1985), these systems evolve towards smaller mass ratios and ultimate coalescence of the binaries into rapidly rotating single stars. Perhaps some blue-stragglers and some high velocity A and F type stars are formed from the coalescence of the W UMa systems. Also, some of the W-type systems could evolve into A-type systems if mass loss and mass exchange occur.

6.4. Fillout Factor (F) / Space Velocity

The degree of over-filling of the inner critical equipotential surface is an important parameter for the understanding of contact binaries. The fillout factor (F) ranges from F = 1.0 for systems whose components are just in contact with their inner (Roche) critical surface to F = 2.0 for systems whose components fill out their outer critical envelopes (Binnendijk 1977). As an example, Fig. 8 shows a model of VZ Psc with an assumed fillout of ~ 1.9 (Bradstreet 1985). Mass and energy exchange between the components take place through the connecting throat. In addition, it might be plausible to suppose that as a system evolves with loss of angular momentum through magnetic winds, that the system would grow from contact (F = 1.0) toward an overcontact configuration where F approaches 2.0. Afterwards the binary would coalesce into a single, rapidly rotating star, perhaps by the scenario proposed by Webbink (1976). FK Com is believed by some to be the product of binary star coalescence. (Bopp and Rucinski 1981; Bopp and Stencel 1981). Unfortunately, the fillout factors for W UMa stars are not always well determined due to complications in their light curves and uncertainties in the gravity darkening exponent (e.g., see Mochnacki 1985). In Fig. 9 we have plotted F versus the space velocity s (relative to the LSR) for the 33 systems for which F is reasonably well determined. Least squares fits to these data indicate a possible correlation between F and s. The least squares fit to the data is shown in the figure which indicates a tendency for systems with large fillouts to have high space motions. However, because of the large scatter within the data set, the correlation is only marginally significant. Since space motion is correlated with age, this indicates that the older systems may have larger fillout parameters. Based on its high space velocity of $s = 144$ km s^{-1}, VZ Psc could be the oldest star in this sample. A recent analysis of its light and radial velocity curves (Hrivnak and Milone 1987) indicates a large fillout factor of F ~ 1.7 so that the stars nearly fill their outer critical surface. The large degree of overcontact, its advanced age, and short orbital period (P = 0.261 days) could indicate that VZ Psc is well on its way into coalescing into a single star.

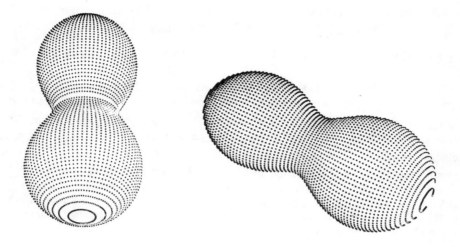

Figure 8. This figure illustrates a large fillout of F = 1.9 for an early model of the peculiar W UMa system VZ Psc taken from Bradstreet (1985).

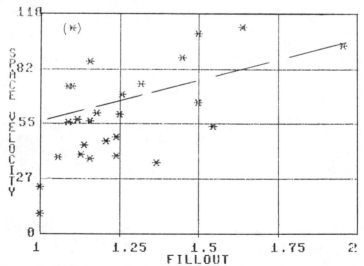

Figure 9. The resultant space velocity is plotted against fillout factor for W UMa-type binaries. A fillout of 1.0 corresponds to the stars just filling their Roche critical surfaces. A fillout of 2.0 corresponds to the extreme case of overcontact with the stars reaching their outermost potential surfaces. There appears to be a weak relation between large fillout (overcontact) with high space velocity (i.e., age).

Additional systems and more reliable determinations of F for these stars are needed before we can draw any definite conclusions at this time.

6.5. Ultraviolet Excess (δc_1) and Metallicity Index (δm_1)/ Space Velocity

Rucinski and Kaluzny (1981) and more recently Rucinski (1983) have studied the ultraviolet excesses (δc_1) and metallicity index (δm_1) for 61 W UMa systems based on Stromgren photometry. Twenty-one of these stars have space velocity determinations. In agreement with these studies, we find no definite correlation of δc_1 and δm_1 with space velocity. However, the mean values of $\delta c_1 = +0.010 \pm 0.036$ and $\delta m_1 = +0.023 \pm 0.029$ indicate no net ultraviolet excess and a weak metal deficiency relative to solar values for the W UMa stars studied. However, from a larger sample of W UMa systems, Rucinski (1983) found a small net ultraviolet excess. Rucinski suggests that these abnormal indices could result from unknown complications in the photospheres of the contact systems and may not be due to actual underabundances in metals.

7. The Formation of Contact Systems from Angular Momentum Loss by Spin-Orbit Coupling

Our results yield a mean age for this sample of W UMa binaries of 8 - 10 Gyr implying that at least some (maybe even all) W UMa systems evolve into contact systems from detached or semi-detached binaries through angular momentum loss due to magnetic torques from stellar winds. This is contrary to the idea of ZAMS contact formation. We now investigate the loss of angular momentum of a detached binary by spin-orbit coupling following the method of Vilhu (1982).

The angular momentum of a rigidly rotating star is given by

$$J_{spin} = k^2 R^2 M \Omega \qquad cm^2 \, gm \, s^{-1} \tag{3}$$

where k^2 is the gyration constant typically ranging from 0.07 to 0.15 for solar-like stars, and where R, M, and Ω are the radius, mass, and angular velocity of the star. The product kR is the radius of gyration. Alternatively eq. (3) can be expressed in terms of stellar rotational velocity v_{rot} by $\Omega = v_{rot}/R$ so that

$$J_{spin} = k^2 R M v_{rot} \tag{4}$$

From studies of the spin-down rates among main sequence solar-like stars determined for members of clusters of different ages, it has been found that the stellar rotational velocity decreases as

$$v_{rot} \sim t^{-1/2} \tag{5}$$

(Skumanich 1972; Soderblom 1985 and references therein). This relation can be roughly calibrated using the mean rotation velocity of solar-type stars in the Pleiades which we assume to be $v_{rot} \sim 17$ km s^{-1} (eg. Kraft 1967; Soderblom 1982) for an age of the cluster of t $\sim$ 70 Myr. The sun, with an age of t = 4.6 Gyr, has a rotation velocity of 2 km s^{-1} ($P_{rot} \sim$ 26 d.). Thus with this calibration the spin down rate is given by

$$v_{rot} \sim 1.4 \times 10^{10} \, t^{-1/2} \tag{6}$$

where time t is expressed in years and the v_{rot} in cm s^{-1}. Numerous studies indicate now that stars lose angular momentum with time, with an accompanying decline in their chromospheric, transition-region, and coronal emissions (Simon, Herbig, and Boesgaard 1985 and references therein). The decrease in stellar activity with decreasing rotation is expected from dynamo-driven magnetic fields.

Differentiating eq. (4) with respect to time yields

$$(dJ/dt)_{spin} = k^2 RM (dv_{rot}/dt) \tag{7}$$

in which we assume that there will be no significant mass loss, or change in the internal structure, or radius of the star. From eq. (6) we obtain

$$(dv_{rot}/dt) \sim -7.1 \times 10^9 \, t^{-3/2} \tag{8}$$

Substituting eq. (8) into eq. (7) and replacing rotational velocity with rotation period (P_{rot}) yields a semi-empirical expression for angular momemtum loss:

$$(dJ/dt)_{spin} \sim -4.5 \times 10^{43} \, k^2 \, MR^4 P^{-3} \tag{9}$$

where the rotation period is expressed in days and M and R are in solar units. For solar-like stars it is assumed that the loss of angular momentum is from magnetic torques in a stellar wind. The torque produced by the magnetic field in the stellar wind depends on the strength of the magnetic field, the Alfven radius, and the extent of the stellar wind. In addition, the angular momentum loss is a function of the stiffness of the magnetic field at large distances from the star (e.g. Mestel 1984). Except for the Sun, this information is not known well. Eq. (6) is approximate since it is calibrated for relatively slowly rotating stars ($v_{rot} < 17$ km s^{-1}) which are mildly active. The $v_{rot} \sim t^{-1/2}$ law on which it is based may not apply for more active, rapidly rotating stars, and thus caution should be used in using this relation to determine angular momentum loss.

With this cautionary note, we next assume that the total angular momentum of the orbit is decreasing through tidal coupling of the rotational angular momentum - i.e., spin-orbit coupling. The orbital angular momentum is given by

$$J_{orb} = 1.242 \times 10^{52} \, q(1 + q)^{-2} (M_1 + M_2)^{5/3} P_{orb}^{1/3} \tag{10}$$

where q is the mass ratio (q = M_2/M_1) and M_1, M_2 are the masses of the stars in solar units and where the orbital period P_{orb} is expressed in days. Taking the derivative of J_{orb} with respect to time yields

$$(dJ/dt)_{orb} \sim 4.1 \times 10^{51}q(1 + q)^{-2}(M_1 + M_2)^{5/3}P_{orb}^{-2/3}(dP_{orb}/dt) \qquad (11)$$

Next we take eq. (11) and set it equal to the angular momentum loss from magnetic braking. This results in the following expressing:

$$4.1 \times 10^{51}q(1 + q)^{-2}(M_1 + M_2)^{5/3}P^{-2/3}(dP/dt) =$$
$$-4.5 \times 10^{43}k^2P^{-3}(M_1R_1^4 + M_2R_2^4) \qquad (12)$$

In this expression the loss of angular momentum of both components is taken into account but where the gyration constant k^2 is assumed to be the same for both stars. Also, $P = P_{rot} = P_{orb}$ because the stars rotational and orbital periods are assumed to be synchronized. As before the period is expressed in days and time t in years. This expression can be simplified to the following form where the change in orbital period with time due to spin-orbit coupled angular momentum loss is:

$$(dP/dt) \sim 1.1 \times 10^{-8}q^{-1}(1 + q)^2(M_1 + M_2)^{-5/3}k^2(M_1R_1^4 + M_2R_2^4)P^{-7/3} \qquad (13)$$

For a given binary the values of q, M, R, and k^2 are fixed, assuming no significant evolution, mass loss, or mass exchange takes place so that we can rewrite this expression as

$$(dP/dt) \sim b\, P^{-7/3} \qquad (14)$$

where b includes all the constant terms on the right hand side of eq. (13). The time needed for the period to change from an initial long period to a shorter value through angular momentum loss is obtained by integrating eq. (14) to obtain:

$$t\ (yrs) \sim 0.30(P_o - P_f)^{10/3}\, b^{-1} \qquad (15)$$

In this expression P_o and P_f are the initial and final orbital periods of the binary system expressed in days and t is the time needed to go from the longer period (P_o) to the shorter period (P_f). To estimate the time needed for a binary system to reach its contact state, the final period is set equal to the critical period at which the two components reach contact. The critical period depends on the mass of the binary.

Using the above expressions we can calculate the time needed for an initially detached binary of a certain mass and initial period to reach a contact state. For purposes of comparison, we assume that the components have masses of $M_1 = 1.11\ M_\odot$ and $M_2 = 0.74\ M_\odot$ which are the same as adopted by Webbink (1976). With these masses the critical period for contact is 0.315 d. We also assume that the gyration constant for each star is $k^2 = 0.10$, the same as used by Webbink. Fig. 10 shows the decrease in the orbital periods over time for five different assumed initial periods P_o from 1 d to 5 d. Systems having initial periods of $P_o > 6$ d may not experience spin-orbit coupling as main sequence stars since the tidal effects are small due to their relatively large separations. As shown in the figure, the time to reach contact depends strongly on the initial period of the system.

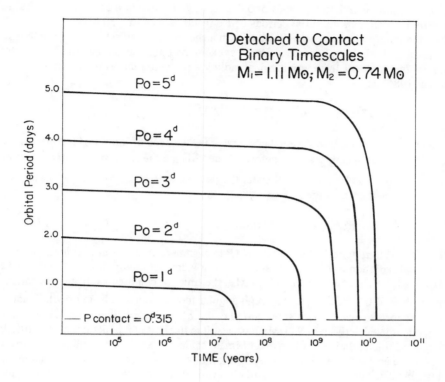

Figure 10. The decrease in orbital period of originally detached binaries with initial orbital periods of 1 to 5 days is shown. The decrease in periods is due to angular momentum loss from magnetic braking, assuming a velocity spindown rate $v_{rot} \sim t^{-1/2}$ and spin-orbit coupling as discussed in the text. For the examples given the period of the system upon reaching contact is $P_c = 0.315$ d.

For example, stars with initial periods of 5 d take ~ 17 Gyr to attain contact whereas at the other extreme binaries with initial periods of 1 d reach contact in ~ 30 Myr.

Once the binary system reaches contact with a short orbital period, general relativistic gravitational radiation can become a significant source of angular momentum loss. Orbital angular momentum loss from gravitational radiation for a circular orbit is given by Webbink (1985) as

$$(\delta ln J/\delta t)_{GR} = -(32/5)G^{-5/3}c^{-5}M_1M_2(M_1 + M_2)^{-1/3}(2\pi/P)^{8/3} \quad (16)$$

where G is the universal gravitational constant, c is the speed of light, P is the orbital period expressed in seconds, and as before M_1 and M_2 are the stars' masses expressed in grams. Without mass loss or mass transfer the two stars (assumed to be point masses) would spiral together in a time

$$(t_{coales})_{GR} = 1/8 \; |\delta ln J/\delta t|^{-1}_{GR} \quad (17)$$

The time for a short period (P < 0.40 d) W UMa system to coalesce into a single star from gravitational radiation is a few to several Gyr. For a contact binary discussed above with P = 0.315 d, $M_1 = 1.11 \, M_\odot$, and $M_2 = 0.74 \, M_\odot$, the time for coalescence from gravitational radiation alone is t_{GR} ~ 3.3 Gyr. Although angular momentum loss from gravitational radiation becomes important for short period contact binaries, judging from the high levels of magnetic related activity in W UMa binaries, it is likely that magnetic braking still plays the dominant role in their evolution and ultimate coalescence. It is difficult at this time to determine the amount of angular momentum lost during the contact stage from magnetic braking. Insufficient information is currently available on the strength and configuration of the magnetic field and the wind from these complex systems. Following Vilhu's (1982) arguments, we speculate that the lifetime of the contact stage is relatively brief,

0.1 Gyr < $t_{contact}$ < 1.0 Gyr.

A possible scenario for the origin and evolution of W UMa stars is illustrated in Fig. 11. Shown to approximate scale are three different detached binary systems consisting of ZAMS components of masses 1.1 and 0.74 $M_\odot$. With angular momentum loss from magnetic braking following a v_{rot} ~ $t^{-1/2}$ law, the times necessary for these stars to reach contact are approximately 4 Gyr, 1 Gyr, and 0.07 Gyr for detached binaries having initial periods of 3.4 d, 2.3d, and 1.2 d, respectively. These detached binary progenitors of W UMa stars would result in a contact binary with a period of 0.315 d, assuming no substantial mass loss. The evolution of the contact system into a rapidly rotating star of 1.85 $M_\odot$ with a possible outflow of gas from its equator was taken from Webbink (1976) in which mass and energy exchange between components occurs, but in which no angular momentum loss due to magnetic braking was considered. As shown in the figure, the end result is a rapidly rotating A-type star which could evolve into a rapidly rotating giant and become an FK Com star. The actual evolution after reaching contact depends upon the amount of nuclear burning the stars have undergone prior to contact. If the primary should exhaust its hydrogen fuel at the core prior to reaching contact, it will fill its Roche lobe and form a semi-detached system. In this case mass exchange will occur on a thermal timescale

365

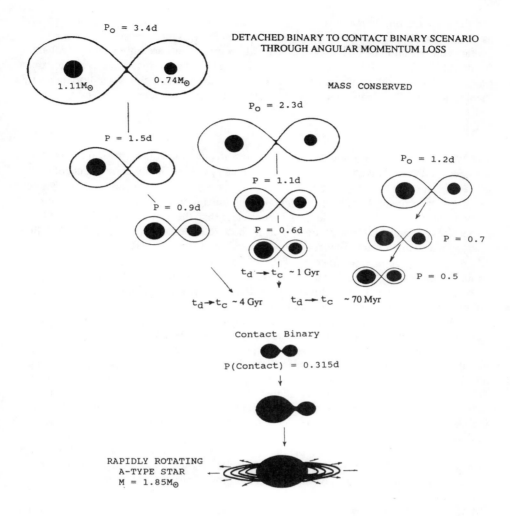

Figure 11. Shown to approximate scale is a possible scenario for the origin and evolution of W UMa stars for components of mass 1.11 M⊙ and 0.74 M⊙. Note the funneling effect whereby detached binaries of diverse initial periods will all become contact systems due to angular momentum loss, although at considerably different rates. These detached binary progenitors of W UMa stars would result in a contact binary with a period of 0.315 d, assuming no substantial mass loss. The evolution of the contact system into a rapidly rotating star of 1.85 M⊙ with a possible outflow of gas from its equator was taken from Webbink (1976) in which mass and energy exchange between components occurs, but in which no angular momentum loss due to magnetic braking was considered. As shown in the figure, the end result is a rapidly rotating A-type star which could evolve into a rapidly rotating giant and become an FK Com star.

and the resulting binary might resemble the short-period "Beta Lyrae" type systems known as V1010 Oph stars. For the shortest period system in the figure, the stars reach the contact configuration in such a short time that the components are only slightly evolved from ZAMS. ER Vul (G0V + G5V; P = 0.69 d) is a system that may have evolved from a progenitor binary with an initial period of 2 - 3 days since it now appears still to be near the main sequence. This is an active system probably in a pre-contact configuration.

8. Conclusions

The average space motion of the 34 W UMa stars in our sample is $s = 61$ km s^{-1} (relative to the LSR) and the relatively large dispersions in the u, v, w velocity components of $< u^2 >^{1/2} = 51$ km s^{-1}, $< v^2 >^{1/2} = 36$ km s^{-1}, and $< w^2 >^{1/2} = 18$ km s^{-1} are consistent with a mean dynamic age of 8 - 10 Gyr. Although there are a few low velocity, young disk stars in the sample, the majority of the stars appear older than the Sun. In fact the velocity dispersions of the W UMa stars are similar to those of the evolved, oldest disk population stars such as white dwarfs, planetary nebulae, and shorter period RR Lyrae variables. Although the space motions of some of the W UMa stars are large with $s > 65$ km s^{-1} and similar to those of spherical population (Pop II) stars, their relatively small w velocity components are not as large as found among the Pop II stars. W UMa stars appear to be old disk population stars, but not Pop II stars. The A-type and W-type W UMa systems have similar mean space velocities indicating no significant difference in age among the stars of the two types. Analysis of the data also indicated no apparent correlation among the properties of the systems such as orbital period, metallicity index, and mass ratio. A possible correlation was found between fillout F and space velocity in the sense that the oldest stars (those with large values of s) tend to have the largest values of F. As noted before the fillout is a measure of the degree of overcontact of the binary. This could indicate that W UMa systems evolve into greater degrees of overcontact with time. Ultimately this could result in the coalescence of the close binary into a single, rapidly rotating star.

From this study it appears possible that W UMa stars are formed from detached binary progenitors. Initially detached short period binaries of low mass appear to *funnel* into the contact stage at different rates depending on their initial periods and angular momentum loss. This funneling effect explains why there are large relative numbers of W UMa stars (Budding 1982; Van't Veer 1975) since all low mass binaries having periods < 5 d would become contact systems. The apparent old age of W UMa stars and expected short time as contact systems is explained by the relatively long lifetimes of the components as detached binaries prior to reaching contact as shown in Fig. 10. This is also supported by the existence of at least seven W UMa systems in the old open cluster NGC 188. The apparent scarcity of W UMa stars in young clusters could be explained if the initial binary period distribution function of their progenitors is similar to that proposed by Kraicheva *et al.* (1979) in which there is a period cut-off near ~ 2 days. With this condition, detached binaries in young clusters may not have had enough time to reach contact. The nature of W UMa systems is complicated by the possibility that contact binaries evolve from detached systems on timescales ranging from 0.02 Gyr for 1 d initial periods to 5 Gyr for progenitors with initial periods ~ 4 days. Thus it is possible to have both very young and very old W UMa binaries, but the old age of most of the systems implied from their galactic kinematics infers that most of the binaries in our sample began with periods of 2-5 days. Detached systems with initial periods < 2 d rapidly evolve into contact, pass through the contact stage, and through coalescence now appear as single stars.

The average old age of W UMa stars of ~ 8 Gyr cannot be explained by constant rate of star formation occurring over the lifetime of the Galaxy. The apparent higher frequency of older systems relative to younger ones is probably a consequence of a more intense era of star formation 5 to 10 Gyr ago. This agrees with the results of Perek (1959) and Blaauw (1965) which indicate that intermediate to old disk population stars comprise ~ 70% of the total stellar mass of the Galaxy. This indicates a broad maximum of star formation producing intermediate-to-old disk stars between 5 to 10 Gyr ago.

This study could have an impact on the formation of single stars since a significant number of single stars could have originated as short period binaries that coalesced into single objects after passing through a contact binary stage. The presence of blue straggler stars in clusters, high velocity A and F stars, and FK Comae stars could be explained by this scenario. Perhaps even the rapidly rotating K dwarfs discovered in the Pleiades clusters (Van Leeuwen, et al. 1986) could be recently coalesced stars that evolved from very short period low mass binaries as suggested by Mestel (1984). These stars would in time evolve through further angular momentum loss into normal slowly rotating main sequence stars.

We wish to thank Craig R. Robinson for help in the data analysis. EFG wishes to acknowledge his appreciation for support of this work from the Langley-Abbot Visiting Scientist program at Harvard - Smithsonian Center for Astrophysics. DHB wishes to thank Eastern College for research support.

368

REFERENCES

Adler, D. S., and Rood, R. T. 1985, *Bull. Am. Astron. Soc.*, **17**, 593.
Artiukhina, N. M. 1964, *Variable Stars*, **15**, 127.
Atanasyevic, I. 1971, *Selected Exercises in Galactic Astronomy*, (New York : Springer-Verlag), p. 49.
Baliunas, S. L., and Guinan, E. F. 1986, *Ap. J.*, **294**, 207.
Batten, A. H., Fletcher, J. M., and Mann, P. J. 1978, *Publ. D. A. O.*, **15**, 121.
Batten, A. H., and Lu, W. 1986, *Publ. Astron. Soc. Pac.*, **98**, 92.
Binnendijk, L. 1967, *Publ. D. A. O.*, **13**, 27.
_____. 1970, *Vistas in Astronomy*, **12**, 217.
_____. 1977, *Vistas in Astronomy*, **21**, 359.
Blauuw, A. 1965, in *Galactic Structure*, eds. A. Blauuw and M. Schmidt
    (Chicago : Univ.of Chicago press), p. 435.
Bopp, B. W., and Rucinski, S. M. 1981, in *Fundamental Problems in the Theory of*
    Stellar Evolution, IAU Symp. No. 93, eds. D. Sugimoto *et al.* (Dordrecht : Reidel), p. 177.
Bopp, B. W., and Stencel, R. E. 1981, *Ap. J. Lett.*, **247**, L131.
Bradstreet, D. H. 1985, *Ap. J. Suppl.*, **58**, 413.
Budding, E. 1982, in *Binary and Multiple Stars as Tracers of Stellar Evolution*, eds.
    Z. Kopal and J. Rahe (Dordrecht : Reidel), p. 351.
Byl, J. 1974, *M. N. R. A. S.*, **169**, 157.
Cruddace, R. G., and Dupree, A. K. 1984, *Ap. J.*, **277**, 263.
Delhaye, J. 1965, in *Galactic Astronomy*, eds. A. Blaauw and M. Schmidt
    (Chicago : Univ. of Chicago), p. 71.
Dupree, A. K., and Preston, S. in *The Universe at Ultraviolet Wavelengths*,
    ed. R. P. Chapman, NASA Conf. Publ. 2171, p. 333.
Eggen, O. J. 1963, *Astron. J.*, **68**, 697.
_____. 1965, in *Galactic Structure*, eds. A. Blaauw and M. Schmidt
    (Chicago : Univ. of Chicago), p. 111.
_____. 1967, *Mem. R. A. S.*, **70**, 111.
_____. 1973, *Publ. Astron. Soc. Pac.*, **85**, 379.
Giclas, H. L., Burnham, R., and Thomas, N. G. 1971, *Lowell Obs. Proper Motion*
    Survey (The G numbered stars), (Flagstaff, Arizona), p. 121.
Guinan, E. F., and Robinson, C. R. 1986, *Astron. J.*, **91**, 935.
Hoffmann, M. 1978, *Astr. Astrophy. Suppl.*, **33**, 63.
Hrivnak, B. J., and Milone, E. F. 1987, *preprint*.
Hrivnak, B. J.,Milone, E. F., Hill, G., and Fisher, W. A. 1984, *Ap. J.*, **285**, 683.
Huang, S.- S. 1967, *Ap. J.*, **150**, 229.
Iben, I. 1965, *Ap. J.*, **141**, 993.
_____. 1967, *Ann. Rev. Astron. Astrophys.*, **5**, 571.
Janes, K. and Demarque, P. 1983, *Ap. J.*, **264**, 206.
Kaluzny, J., and Shara, M. M. 1987a, *Ap J.*, **314**, 585.
_____. 1987b, *Space Tele. Sci. Inst. Preprint Series*, No. 228.
Klemola, A. R. 1977, *Publ. Astron. Soc. Pac.*, **89**, 402.
Kraft, R. P. 1967, *Ap. J.*, **150**, 551.
Kraicheva, Z. T., Popova, E. I., Tutokov, A. V., and Yungelson, L. R. 1979, *Astron. Zh.*, **58**, 520.
Lucy, L. B. 1976, *Ap. J.*, **205**, 208.
_____. 1977, *Astron. J.*, **82**, 1013.
McLean, B. J. 1981, *M. N. R. A. S.*, **195**, 931.
McLean, B. J., and Hilditch, R. W. 1983, *M. N. R. A. S.*, **203**, 1.
Mayor, M. 1974, *Astron. Astrophys.*, **32**, 321.
Mestel, L. 1968, *M. N. R. A. S.*, **138**, 359.
Mestel, L. 1984, in *Cool Stars, Stellar Systems, and the Sun*, eds. S. L. Baliunas and
    L. Hartmann (New York : Springer-Verlag), p. 49.
Mihalas, D., and Binney, J. 1981, *Galactic Astronomy, 2nd ed.* (San Francisco :

W.H. Freeman & Co). pp. 380-463.

Milone, E. F. 1987, *private communication.*

Milone, E. F., Hrivnak, B. J., Hill, G., and Fisher, W. A. 1985, *Astron. J.*, **90**, 109.

Mochnacki, S. W. 1981, *Ap. J.*, **245**, 650.

_____. 1985, in *Interacting Binariy Stars*, ed. P. P. Eggleton and J. E. Pringle (Dordrecht : Reidel), p. 51.

Okamoto, I. and Sato, K. 1970, *Publ. Astron. Soc. Japan*, **22**, 317.

Oort, J. H. 1926, *Kapteyn Astron. Lab. Groningen Publ. No 40.*

Parker, E. N. 1986, in *Cool Stars, Stellar Systems, and the Sun*, eds. M. Zeilik and
       D. M. Gibson (New York : Springer-Verlag), p. 341.

Perek, L. 1959, *Bull. Astr. Inst. Czech.*, **10**, 15.

Popper, D. M. 1948, *Ap. J.*, **108**, 490.

Roxburgh, I. W. 1966, *Ap. J.*, **143**, 111.

Rucinski, S. M. 1983, *Astron. Astrophys.* **127**, 84.

_____. 1985, in *Interacting Binary Stars*, ed. P. P. Eggleton and J. E. Pringle (Dordrecht : Reidel), p. 13.

Rucinski, S. M., and Kaluzny, J. 1981, *Acta. Astron.*, **31**, 409.

Rucinski, S. M., Whelan, J. A. J., and Worden, S. P. 1977, *Publ. Astron. Soc. Pac.*, **89**, 684.

Sandage, A. 1962, *Ap. J.*, **135**, 333.

Sandage, A. and Eggen, O. J. 1969, *Ap. J.*, **158**, 685.

Shu, F. H., Lubow, S. H., and Anderson, L. 1976, *Ap J.*, **209**, 536.

_____. 1979, *Ap. J.*, **229**, 223.

Simon, T., Herbig, G. H., and Boesgaard, A. M. 1985, *Ap. J.*, **293**, 551.

Skumanich, A. 1972, *Ap. J.*, **171**, 565.

Soderblom, D. R. 1982, *Ap. J.*, **263**, 239.

_____. 1985, *Astron. J.*, **90**, 2103.

Spitzer, L., and Schwarzschild, M. 1951, *Ap. J.*, **114**, 385.

_____. 1953, *Ap. J.*, **118**, 106.

Struve, O. 1949, *Ap. J.*, **109**, 436.

Vandenberg, D. A. 1983, *Ap. J. Suppl.*, **51**, 29.

_____. 1985, *Ap. J. Suppl.*, **58**, 711.

Van Leeuwen, F., Alphenaar, P., and Meys, J. J. M. 1987, *Astron. Astrophys. Suppl.*, **67**, 483.

Van Maanen, A. 1942, *Ap. J.*, **96**, 382.

Van't Veer, F. 1975, *Astron Astrophy.*, **40**, 167.

_____. 1979, *Astron Astrophy.*, **80**, 287.

Vilhu, O. 1982, *Astron. Astrophy.*, **109**, 17.

_____. 1984, *Astron. Astrophys.*, **133**, 117.

Vilhu, O., and Heise, J. 1986, *Ap. J.*, **311**, 937.

Vilhu, O. and Rahunen, T. 1980, *IAU Symp. No. 88*, eds. M. J. Plavec*et al.* (Dordrecht : Reidel), p. 491.

Webbink, R. F. 1976, *Ap. J.*, **209**, 829.

_____. 1985, in *Interacting Binary Stars*, eds. J. E. Pringle and R. A. Wade
       (Cambridge : Cambridge Univ. Press), p. 39.

Whelan, J. A. J., Worden, S. P., and Mochnacki, S. W. 1973, *Ap. J.*, **183**, 133.

Whelan, J. A. J., Worden, S. P., and Rucinski, S. M. 1979, *M. N. R. A. S.*, **186**, 729.

Wielen, R. 1974, in *Highlights of Astronomy*, Vol. 3, XVth General Assembly of the
       IAU, ed. G. Contopoulos (Dordrecht:: Reidel), p. 395.

_____. 1977, *Astron. Astrophy.*, **60**, 263.

TABLE I

Star	α (h m s)	δ (° ' ")	μα (s yr⁻¹)	μδ (" yr⁻¹)	V₀ (km s⁻¹)	Dist (pc)
AB And	23 09 09	+36 37 19	+0.0091(S)	-0.059(S)	-45.	127.
			+0.0086(A)	-0.072(A)		
S Ant	09 30 07	-28 24 24	-0.0057	+0.035	+15.	134.
			-0.0068	+0.038		
44 i Boo	15 02 08	+47 50 54	-0.0409	+0.032	+3.4	12.
			-0.0400	+0.026		
TY Boo	14 58 47	+35 19 42			-53.[a]	210.[b]
			-0.0050	+0.028		
TZ Boo	15 06 17	+40 09 30			-36.7[c]	190.
			-0.0062	+0.058		
XY Boo	13 46 48	+20 26 00			+21.1[c]	308.
			-0.0017	+0.023		
AH Cnc (M67-33)	08 48 54	+12 02 12			+22.[e]	766.[e]
			-0.00031[d]	+0.0003[d]		
TX Cnc (Praesepe)	08 37 11	+19 10 30			+26.6[g]	222.
			-0.0023[f]	-0.013[f]		166.[g]
VW Cep (61 Cyg moving group)	20 38 03	+75 24 59	+0.0899	+0.560	-35.[h]	31.
			+0.0906	+0.553	+10.[i]	
CC Com	12 09 34	+22 43 39			-10.2[k]	76.
			-0.0093[j]	+0.010[j]		
RW Com	12 30 32	+26 59 30			-53.4[l]	216.[m]
			-0.0094	-0.0400		
RZ Com	12 32 36	+23 36 54			-12.	258.
			+0.0008	-0.012	-1.8[c]	
Ɛ CrA	18 55 21	-37 10 28	-0.0107	-0.097	+61.9	54.
			-0.0100	-0.096		43.[n]
V700 Cyg	20 29 22	+38 37 30			-30.[o]	145.[o]
			-0.0022	-0.094		
			-0.0023[o]	-0.090[o]		
V1073 Cyg	21 22 55	+33 28 18	+0.0016	-0.013	-8.0	336.
BV Dra	15 10 51	+62 02 33	-0.0245	+0.094	-61.[p]	79.
BW Dra	15 10 50	+62 02 49	-0.0247	+0.088	-61.[p]	79.
YY Eri	04 09 47	-10 35 44	-0.0077	-0.122	-20.	71.
			-0.0077	-0.115		

TABLE I (continued)

Star	α (h m s)	δ (^{o} , ")	μ_α (s yr^{-1})	μ_δ (" yr^{-1})	V_O (km s^{-1})	Dist (pc)
AK Her	17 11 43	+16 24 28	+0.0019	-0.046	-13.0	138.
			+0.0014	-0.045		
FG Hya	08 24 28	+03 39 36	-0.0011	+0.055	-42.[o]	182.
SW Lac	22 51 23	+37 40 19	+0.0058	+0.005	-22.5[q]	82.
			+0.0069	+0.014		
UZ Leo	10 37 54	+13 49 41	-0.0009	-0.037	+3.0[r]	502.
			-0.0015	+0.002		
XY Leo	09 58 56	+17 39 06			-50.[s]	75.
			+0.0042	-0.042		
V502 Oph	16 38 48	+00 36 06			-37.	69.[t]
			-0.0018	-0.013		
V566 Oph	17 54 24	+04 59 31	+0.0046	+0.080	-40.	87.
			+0.0042	+0.075		
ER Ori	05 08 51	-08 36 59	+0.0002	-0.014	+35.	206.
U Peg	23 55 25	+15 40 31	-0.0025	-0.057	+0.0	175.
			-0.0030	-0.050		
VZ Psc	23 25 14	+04 34 42			-11.[v]	71.
			+0.026[u]	+0.182[u]		
TY Pup	07 30 36	-20 41 02	-0.0011	+0.054	+20.0	380.
			-0.0016	+0.033		
Y Sex	10 00 13	+01 20 12			+9.8	311.
			0.000	0.000		
AW UMa	11 27 26	+30 14 35	-0.0077	-0.203	-17.[w]	93.
W UMa	09 40 15	+56 10 56	+0.0030	-0.029	-37.8[w]	64.
			+0.0028	-0.036		
AG Vir	11 58 30	+13 17 13	-0.0006	-0.033	-5.6	385.
AH Vir	12 11 48	+12 05 55	+0.0015	-0.111	+10.0	127.
(Wolf 630)			+0.0027	-0.101		

[a] Milone (1987)

[b] We computed its distance based on its period and spectral type.

[c] McLean and Hilditch (1983)

[d] Mean proper motion of M67 is adapted from Van Maanen (1942)

[e] Whelan *et al.* (1973)

^fMean values of cluster given by Eggen (1973).
^gWhelan *et al.* (1973)
^h Popper (1948)
ⁱ Binnendijk (1967); We used the average value of $V_o = -12$ km s^{-1}.
^j Klemola (1977)
^k Rucinski, Whelan, and Worden (1977); not a member of the Coma cluster.
^l Milone *et al.* (1985)
^mWe estimate the distance from *B-V* and period.
ⁿ from trigonometric parallax
^o Eggen (1967)
^p Batten and Lu (1986)
^q Struve (1949)
^ras given by Rucinski and Kaluzny (1981)
^s Hrivnak *et al.* (1984)
^t as given by Cruddace and Dupree (1984)
^u Giclas, Burnham, and Thomas (1971)
^v Hrivnak and Milone (1987)
^w McLean (1981)

TABLE II

Star	m_v (max/min)	Sp	$(b-y)_o$	P (days)	type	q	F
AB And	+ 9.50 / 10.32	G5V / G5V	+0.481	0.3319	W	0.526	1.454
S Ant	6.4 / 6.92	A9Vn	0.180	0.6483	A		
44 i Boo	5.8 / 6.40	G2V / G2V	0.404	0.2678	W	0.50	1.034
TY Boo	10.81 / 11.47	G3		0.3171	W	0.293	1.16
TZ Boo	10.41 / 11.00	G2Vn		0.2972	W :	0.22	1.0 :
XY Boo	10.3 / 10.61	F5V		0.3705	A	0.185	1.550
AH Cnc (M67)	13.31 / 13.69	F5-7V		0.3604	W :	0.50 :	1.15 :
TX Cnc (Praesepe)	10.00 / 10.35	F8V / F7V	0.360	0.3829	W	0.62	1.21
VW Cep (61 Cyg M. G.)	7.23 / 7.68	G5V / G8V	0.505	0.2783	W	0.41	1.00 :
CC Com	11.30 / 12.21	K3V / K5V	0.746	0.2207	W	0.521	1.235
RW Com	11.00 / 11.70	G5V / G8V		0.2373	W :		
RZ Com	10.42 / 11.13	G0Vn / G2V		0.3385	W	0.43	1.0 :
ε CrA	4.0 / 5.0	F2V	0.238	0.5914	A	0.113	1.320

TABLE II (continued)

Star	m_v (max/min)	Sp	$(b-y)_0$	P (days)	type	q	F
V700 Cyg	11.9(ptg) 12.4	F2V		0.3400			
V1073 Cyg	8.23 8.61	F1V F2IV	0.220	0.7859	A	0.24	1.13
BV Dra	7.88 8.48	F7V	0.350	0.3504	W	0.82	1.16
BW Dra	8.61 9.08		0.405	0.2923	W		1.18
YY Eri	8.1 8.80	G5 G5	0.400	0.3215	W	0.59	1.2
AK Her	8.29 8.77	F2 F6	0.315	0.4215	A	0.233	1.101
FG Hya	9.90 10.28	G0V	0.352	0.3278	A :	0.143	1.65 :
SW Lac	8.51 9.39	G2Vne G3V	0.448	0.3207	W	0.87	1.373
UZ Leo	9.58 10.15	A7Vn	0.223	0.6180	A :	0.249	1.957
XY Leo	9.45 9.93	K0V K0V		0.2841	W	0.79	1.0
V502 Oph	8.34 8.84	G2V F9V		0.4534	W	0.34	1.1
V566 Oph	7.46 7.96	F2Vn F4	0.278	0.4096	A	0.238	1.25
ER Ori	9.28 10.01	F5V F8	0.322	0.4234	W	0.757	1.055
U Peg	9.23 10.07	G2Vn F3 + F3	0.368	0.3748	W	0.81	1.1
VZ Psc	10.20 10.45	K5V	0.657	0.2612	W[a] :	0.80[b]	1.8[b] :
TY Pup	8.40 8.89	A9n	0.224	0.8912	A :		
Y Sex	9.83 10.21	F8		0.4198			
AW UMa	6.83 7.13	F0V F2V	0.225	0.4387	A	0.075	1.62
W UMa	7.75 8.48	F8V G2V	0.391	0.3336	W	0.54	1.13
AG Vir	8.35 (9.58) :	A7V A9V	0.140	0.6426	A :		
AH Vir (Wolf 630)	9.90 10.40	K0V K0V	0.467	0.4075	W	0.42	1.127

[a] We tentatively assign VZ Psc to be a W type based on its late spectral type and short period.
[b] Hrivnak and Milone (1987)

TABLE III[a]

Star	U	V	W	S	u	v	w	s
		(relative to Sun)				(relative to LSR)		
AB And	+32.2	-71.7	-38.1	87.4	+23.	-60.	-31.	71.
S Ant	+52.9	-10.2	-11.5	55.1	+44.	+2.	-5.	44.
44 i Boo	+14.4	-12.9	+13.8	23.7	+5.	-1.	+21.	21.
TY Boo	+70.6	-44.6	-19.0	85.6	+61.	-33.	-12.	70.
TZ Boo	+85.1	-29.9	-5.8	90.4	+76.	-18.	+1.	78.
XY Boo	+41.3	+5.2	+32.5	52.8	+32.	+17.	+39.	53.
AH Cnc	+25.7	-8.9	-0.2	27.2	+15.	+3.	+7.	17.
TX Cnc	+37.5	-19.4	-17.0	45.5	+28.	-7.	-10.	31.
VW Cep	+79.7	-63.6	-9.5	102.4	+71.	-52.	-3.	88.
CC Com	+41.1	-17.1	-16.8	47.6	+32.	-5.	-10.	34.
RW Com	+86.0	-98.6	-62.4	145.0	+77.	-87.	-55.	128.
RZ Com	-19.3	-4.5	-12.1	23.2	-29.	+8.	-5.	30.
Ɛ CrA	-65.4	-35.4	+2.1	74.3	-75.	-23.	+9.	79.
V700 Cyg	-54.8	-42.5	-23.9	73.4	-64.	-31.	-17.	73.
V1073 Cyg	+10.0	-14.2	-34.7	38.8	+1.	-2.	-28.	28.
BV Dra	+57.8	-71.0	-27.1	95.5	+49.	-59.	-20.	79.
BW Dra	+57.8	-71.0	-27.1	95.5	+49.	-59.	-20.	79.
YY Eri	-52.5	+1.5	-28.3	59.6	-62.	+14.	-21	65.
AK Her	-14.4	-15.3	-30.9	37.3	-24.	-3.	-24.	34.
FG Hya	-0.2	+64.5	-7.1	64.9	-9.	+77.	-0.	78.
SW Lac	+20.6	-28.2	-2.6	35.0	+11.	-16.	+4.	17.
UZ Leo	-8.9	-85.7	-36.1	93.4	-18.	-74.	-29.	81.
XY Leo	-47.5	+9.3	-28.9	56.4	-57.	+21.	-22.	65.
V502 Oph	+35.2	-11.9	-9.3	38.3	+26.	+0.	-2.	26.
V566 Oph	+52.8	+17.4	-19.9	59.1	+44.	+29.	-13.	54.
ER Ori	+20.2	-26.0	-18.3	37.7	+11.	-14.	-11.	26.
U Peg	-47.5	-14.3	-25.8	55.9	-57.	-2.	-19.	60.
VZ Psc	+144.3	-12.5	-0.7	144.8	+135.	-1.	+6.	135.
TY Pup	+92.1	+40.4	+22.9	103.1	+83.	+52.	+30.	102.
Y Sex	+3.8	-6.2	+6.6	9.8	-5.	+6.	+13.	15.
AW UMa	+1.6	-98.2	-18.3	99.9	-8.	-86.	-11.	87.
W UMa	-32.3	-18.4	-23.3	43.9	-42.	-6.	-16.	45.
AG Vir	-15.5	-55.9	-23.4	62.5	-25.	-44.	-17.	53.
AH Vir	-44.7	-51.9	-6.4	68.8	-54.	-40.	+1.	67.

[a] All velocities are given in km s^{-1}.

TABLE IV

Stellar Type	Space Motion (km s⁻¹) S	Dispersions (km s⁻¹)			
		$<u^2>^{1/2}$	$<v^2>^{1/2}$	$<w^2>^{1/2}$	$<s^2>^{1/2}$
F dwarfs	17	26	15	14	33
G dwarfs	25	29	18	18	39
K dwarfs	23	32	18	14	39
M dwarfs	20	32	22	18	43
Intermediate-old disk populations					
Planetary nebulae	31	45	35	20	60
White dwarfs	38	50	30	25	63
RR Lyrae variables (P < 0.45 d)	56	45	40	25	65
Spheroidal-component stars (Pop II)					
Subdwarfs	150	100	75	50	135
RR Lyrae variable (P > 0.45 d)	225	160	100	120	224
This study of W UMa binaries					
W UMa binaries	66	51	36	18	65

TABLE V

Age Determinations of W UMa Systems and Membership in Clusters and Moving Groups

Cluster/Group	Age	Number of W UMa's
h and χ Per	0.02 Gyr	0
Pleiades	0.06-0.08 Gyrs	0
Hyades	0.7 Gyr	0
Coma	0.5-0.7 Gyr	0
M44 (Praesepe)	0.5-0.7 Gyr	1 TX Cnc
NGC 2360	~1 Gyr	0 *
NGC 6802	~1.5 Gyr	0 *
Wolf 630	3 Gyr	1 AH Vir
M67	3-4 Gyr	1 (or 2) AH Cnc
NGC 2506	~3.4 Gyr	0 *
NGC 2420	~3.8 Gyr	0 *
NGC 6819	~4.0 Gyr	0 *
61 Cyg moving group	3-5 Gyr	1 VW Cep
Mel 66	6.5 Gyr	0 *
NGC 188	10 Gyr	7 EP, EQ, ER, ES Cep, V5, V6, V7
M92	13 Gyr	0

* Kaluzny and Shara (1987b)

ROTATIONAL EVOLUTION OF LOW-MASS STARS

S. Catalano[1], E. Marilli[2], C. Trigilio[1]
1) Institute of Astronomy University of Catania
2) Astrophysical Observatory of Catania
   Viale A. Doria 6
   Città Universitaria
   95125 Catania Italy

ABSTRACT. From observational data on rotation periods and ages we deduce the evolution of surface rotational velocity of low-mass main sequence stars from the age of the Pleiades to the solar age. The $t^{1/2}$ increase of the rotation period is shown to apply to stars of man between 1.1 and 0.5 solar masses with a rate larger for smaller masses. It is shown that time scale of the solar wind braking action is consistent with an angular momentum transport from the radiative core to envelope but with a rate not sufficient to justify the internal rotation deduced from solar oscillations.

1. INTRODUCTION

It is well known that rotation of low-mass main sequence stars decreases with time (Wilson 1966, Kraft 1967, Skumanich 1972). Schatzman (1962) has already shown that angular momentum losses of rotating stars could be large enough, in the presence of a stellar wind and a moderately strong magnetic field, to brake the star in a time scale comparable to the main sequence life-time. The coincidence of stellar rotation drop and the appearance of chromospheric activity (CaII H and K line emission core) and the onset of predicted deep hydrogen convective zone around spectral type F4-F5 led O. C. Wilson (1966) to the suggestion of strong connection between convection, magnetic activity and rotation braking. In addition a large body of evidence indicating that stellar activity, as measured from chromospheric, transition region and coronal emission is tightly connected to the star rotation has been growing on recent years (Noyes et al. 1984, Pallavicini et al. 1981, Catalano and Marilli 1983, Marilli Catalano and Trigilio 1986a). The present scenario of the relations and possible interplay of the motion of convection

377

A. K. Dupree and M. T. V. T. Lago (eds.), Formation and Evolution of Low Mass Stars, 377–387.
© 1988 by Kluwer Academic Publishers.

and rotation which generate surface magnetic flux denunced by enhanced chromospheric emission (CaII, MgII) through the action of magnetic dynamo, is reported in the lectures of Baliunas and Gough. The evolution of activity along the main sequence is discussed by Soderblom while the rotation of pre-main sequence and young main sequence is thoroughly described in Stauffer's lecture. The global picture shows that:

a) along the lower main sequence, below the sharp decline in angular momentum with respect to the $\langle J/ \rangle \propto M^{2/3}$ dependence at about 1.5 $M_{\odot}$, the behaviour of stellar rotation with age is radically different than for higher-mass stars;

b) for low mass stars rotation decreases with decreasing mass along isochrones corresponding to chromospherically active and low-active stars and at a given mass rotation decreases with increasing age (see also, Baliunas 1984, Catalano and Marilli 1983).

c) The rotation of pre-mainsequence is more complicated because at a given spectral type and age low and fast rotators coexist. However from comparison of clusters of different ages it seems that as stars seat on the main sequence they tend to reach a surface rotation appropriate for their mass and age. The net effect is that rapid rotators desappear first among G stars and later among K and M dwarfs (see Stauffer lecture).

    In this scenario we think it is important to look for quantitative analyses of observational relations that can give constraints in the magnetic braking models and in the magnetic field regeneration. In this paper we will report on some on-going work on the rotational evolution of low-mass main sequence stars. The aim of this work is to explore quantitative analyses and correlations of rotational evolution of low-mass stars along the main sequence for different mass and ages. In particular we will try to get insight on:

- rotation evolution law v/s stellar mass
- time scale for angular momentum loss
- internal rotation

2. OBSERVATIONAL DATA

Quantitative analyses need to be based on un-biased and as accurate as possible data. Our analysis is mainly confined in studying the surface rotation as a function of the mass and age, so that we need accurate values for these three

parameters. Since not always we have accurate values of all
of them for a given star, data of different level of accuracy
have been included in order to bring the sample of the stars
to a reasonable level of statistical significance.
The data have been selected as follows:

- Rotation Periods

i) from CaII H and K line modulation (Vaughan et al. 1981,
Noyes et al. 1984) and broad band photometry. This
method gives directly measures of the rotation periods
with accuracy of few %, without the effect of the
unknown axial projection factor on V sin i. The main
bulk of data comes from Noyes et al. (1984)
ii) from the correlation between chromospheric emission and
rotation. We compute the rotation period from the empirical
correlation of the emission luminosity in the CaII H and K
emission component and rotation period defined by Marilli,
Catalano and Trigilio (1986a)

$$\log L_{HK} = - a(B-V) * P_{rot} + L_{HK}(B-V)$$

This method allows to deduce rotation periods with an

accuracy on the average better than 20%
iii) from projected equatorial velocity V_e sin i. Average

equatorial radii of main sequence stars for the
appropriate B-V have been adopted to deduce the rotation period.

Age

The age of a specific star is a very undefined parameter.
We attribute the age to stars in our sample as:

i) the age of the cluster, if the star belong to an open
   cluster
ii) The age of kinematic moving group, if the star belong
   to a group. This age can be highly uncertain because
   in many case the belonging to the group is question-
   able and also because the age of the group is not
   well defined.
iii) from CaII emission-age correlation and Li abundance-
   -age correlation. Ages estimated from these two
   method have been adopted only when concordant values
   have been obtained. Ages from litium abundance have
   been taken from Duncan (1981) and Soderblom (1983).
   Ages from CaII emission are deduced from an
   unpublished calibration by Trigilio et al. (1987).

3. RESULTS AND DISCUSSION

- Rotation law

From the observed dependence of the CaII chromospheric emission from the rotation period, the age and the mass Catalano and Marilli (1983) predicted that the rotation period of main sequence stars later than F7 should be an unique function of the mass and age. Simon et al. (1986) arrived to a similar conclusion from the analysis of IUE observations of chromospheric and transition region lines, while Duncan et al. (1983) showed that for Hyades main-sequence stars having B-V between 0.4 and 0.85 there is a nearly linear relationship between P_{rot} and B-V (i.e., the mass). The relation within this range of B-V appears to be well defined with a scatter in P_{rot} of roughly 1-2 days in either directions.

Using observed rotation periods we have attempted a more quantitative analysis of the rotation as a function of mass and age for main sequence stars (Marilli, Catalano and Trigilio 1986b, Trigilio, Catalano and Marilli 1986). Rotation periods P_{rot} plotted as a function of the stellar

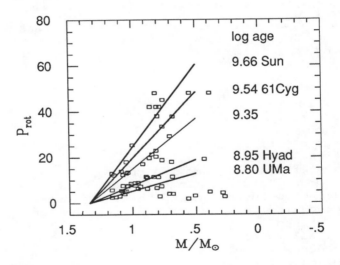

Figure 1. Observed and rotation periods versus mass. The straight lines are loci of equal age. Data points for low mass and short period refer to young flare stars.

mass in Figure 1, clearly show that the data points follow
fairly defined linear sequences. These sequences can be
distinguished by the age of a number of stars identified
along each of them. They include the Sun and stars of solar
age (4.6×10^9 years), 61 Cyg A and B (3.45×10^9 years), the
Hyades cluster ($9.9.10^8$ years), the Ursa Major group (6.3×10^8 years) and a possible group of age about 2.3×10^9 years.
Two main features are apparent in Figure 2: i) the
isochronous lines seem to converge to a mass $M/M_\odot \simeq 1.34$,

which is appropriate for F3 stars; ii) the isochrones widen
towards lower mass stars indicating that the decay of
rotation is faster for lower mass stars.

The convergence of isochrones towards stars of type F3
would indicate that these star should not suffer appreciable
rotation spin-down during the main sequence life time. On the
other hand, as already pointed out, this is the spectral type
of the onset of deep hydrogen convection zones and of the
break down of rotation (Wilson 1966, Kraft 1970).

More informations can be inferred plotting rotation
periods as a function of the age, in particular of the square

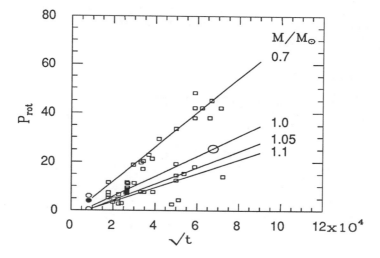

Figure 2. Observed rotation period, in days, versus square
root of age. Lines refer to constant mass stars.

root of the age. In figure 2 the straight lines represent the
linear best fit to stars of well established age for fixed
mass, as indicated to the right. From Figure 3 it is apparent
that the $t^{1/2}$ increase of the rotation period applies to
stars in the range 1.1 to 0.7 solar masses with a rate
increasing as the mass decreases. A plot of data including

rotation period deduced from the chromospheric emission-rotation dependence and age from lithium and chromospheric emission shows that the $t^{1/2}$ rotation period increase can be extended to stars in the range from 1.1 to 0.5 solar mass.

Inclusion of rotation periods of Pleiades stars derived from Vsin i measurements (Benz et al. 1984, Stauffer et al. 1984) indicated in Figure 2 by large open and filled symbols clearly shows that a simple $t^{1/2}$ relation for the rotation period does not hold for solar type and K type as young as the Pleiades. A relation of the type

$$P \propto a(M) \times (t+c)^{1/2} \tag{1}$$

with c also depending on the mass would be more appropriate.
On the other hand this is the kind of relation one should expect for braking due to a magnetic wind with radial structure and field strength proportional to the angular velocity ω (see Gough).
For ages significantly larger than the Pleiades age, say larger than the Coma or Hyades ones the increase in the rate of rotation decay as a function of the mass, i.e. the coefficient a (M) in relation (1), is well represented by

$$a = 1.32 \times 10^{-3} - 0.92 \times 10^{-3} \ M/M_{\odot}$$

shown in Figure 3, when the rotation period is given in days.

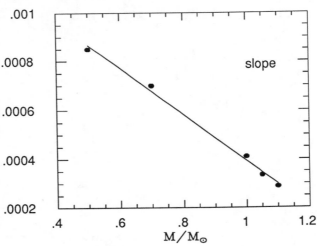

Figure 3. Slopes of linear relations in Figure 2 are plotted as a function of the mass. A value for $M/M_{\odot}=0.55$ not shown in Figure 2 is also included.

We can summarize the results as follows:
i) the decay of rotation with the square root of the age
   holds for low-mass stars at least in the range of 1.1-0.5
   solar masses;
ii) the rate of spin-down (i.e. the increase of rotation
    period) is larger for lower mass stars;
iii) the rotation period for a given low-mass star is tightly
    determined by its mass and age.

For very young stars, equation (1) predicts that the change in the rotation period should be much less than for the older stars, as indicated by the curved line in Figure 4, since the constant term dominates.

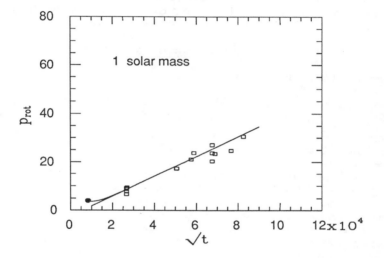

Figure 4. Rotation period in days, versus the square root of the age for one solar mass stars. The filled point represent an average rotation period from V sin i for Pleiades G2 V stars.

While this considerations apply to G stars and slow rotating K stars it is not true for fast rotating K stars. The problem of the fast spin-down just before the main sequence has been discussed by Stauffer, however let us briefly bring the attention to the evolution of slow rotating Pleiades stars.
Belcher and Mac Gregor (1976) have shown that this effect can be explained in terms of changes of the Alfven radius which is smaller for rapid rotators than for slow rotators. This would reduce the efficiency of the angular momentum loss for rotation period shorter than 2.5 days, for

solar type stars. The applicability of their theory is very uncertain and implies accurate modelling of the stellar wind. In addition average rotation periods of slow rotating G Pleiades stars and older ones are longer than the limits setled by Belcher and Mac Gregor for fast magnetic rotators. As a matter of fact evolutionary models for the spin-down of one solar mass stars by Endal and Sofia (1981), essentially based on Belcher and Mac Gregor (1976) analysis lead to an increase of the rotation period faster for young stars, just after the age of the Pleiades than for older star like the Sun. If the magnetic energy of the surface magnetic field generated by the dynamo action is larger in young stars due to their faster rotation a higher rate of angular momentum loss could be expected. In this case a possible explanation of the slow change of rotation period between the age of Pleiades and Hyades can be given assuming a high angular momentum transfer from the core to the envelope, so that a large portion of the stellar interior is slowed down during the early stage of main sequence evolution.

- Time scale and internal rotation

Low-mass stars have outer convective envelopes whose relative depth, $d=R_c/R$, increases as the mass decreases along the main sequence. Models of magnetic wind require that the effect of the wind torque is to decelerate the rotation of the convective envelope which is assumed to rotate rigidly at the same angular velocity as the surface. The flow of angular momentum within the radiative interior and across the envelope-interior interface has been modelled under different assumptions which lead to relative high transfer rate with time scale of 10^7 years (Endal and Sofia 1978) and very low transfer rate with time scale larger than ten times the age of the Sun (Tassoul and Tassoul 1984). A very high angular momentum transfer between core and envelope has been ruled out by Dicke (1970) on the base of the time scale for the spin-down of the whole Sun in rigid rotation under the effect of the measured Solar wind torque (Weber and Davis 1967).

Recent estimates of the solar angular momentum loss rate in the ecliptic plane derived from the "Helios" spacecraft (Pizzo et al. 1983) give a total angular momentum flux near the solar equator of $0.2 - 0.3 \times 10^{30}$ dyn cm sterad^{-1}, about one forth of the previous estimate by Weber and Davis (1967). The total wind torque depends on the latitude and the area of the angular momentum loss region. For isotropic wind the correction factor is $8\pi/3$, so a reasonable estimate leads to a total wind torque of about $1-2\times10^{30}$ dyn cm.

If the outer convective shell, $0.3 R_O$ according to standard solar models, is slowed-down in its rotation the e folding decay time

$$\tau_w = J/(dJ/dt)$$

turns out to be about $4-2 \times 10^9$ years.

We can estimate the present time scale for the slowing down of the solar rotation from $t^{1/2}$ relation. For the age of the Sun we can neglect the constant term c in relation (1) so that we have $P = a(M) x t^{1/2}$. Then the time scale for the angular momentum losses,

$$\tau = \omega/\dot{\omega} = (2 \ P \cdot t^{1/2})/(dP/d(t^{1/2})$$

Substitution of the quantities $P = a(M) \ X t^{1/2}$ and $dP/d(t^{1/2}) = a$ (VI) simply gives

$$\tau(t) = 2t$$

So that we have for the Sun an e folding decay time of the rotation from the $t^{1/2}$ relation

$$\tau_\gamma \simeq 9 \times 10^9 \text{ years}$$

The comparison of the two time scale clearly shows that essentially the convective envelope is slowed down, as already shown by Dicke (1970), without any or marginal transport of angular momentum from the core to the envelope.
If we assume that the difference in the time scales for the slowing down deduced from the solar wind torque and from the $t^{1/2}$ relation is a real one we can estimate an upper limit of the time scale for the angular momentum transport from the core to the convective envelope.

We have $\dfrac{1}{\tau} = \dfrac{1}{\tau_w} - \dfrac{1}{\tau_\gamma}$ which lead to $\tau = 7 \times 10^9$ years.

This value is in better agreement with Endal and Sofia (1978) than Tassoul and Tassoul (1984) estimates. Another consideration does suggest also some angular momentum transport from the core to the envelope. From Figure 4 and relation (1) we see that the rate of increase of the rotation period with age depends linearly on the mass. However among the other terms in the angular momentum loss equation, the most important ones is the momentum of inertia of the braked mass. If we look at the total momentum of inertia along the main sequence we see that between 1 and 0.4 solar masses it varies about as the square of the mass, that is much faster than we have for the slowing-down rate. Conversely the momentum of inertia of the convective envelope increases as the mass decreases along the main sequence by about a factor of two between 1 and 0.5 solar mass. This trend isin the opposite direction with respect to the observed slowing down.

386

A possible explanation can be given in terms of angular momentum transport from the core to the envelope. The amount of the transferred angular momentum is larger for larger mass whose angular momentum in the radiative core is larger and becomes negligible for low mass stars whose core become smaller and smaller as the mass decreases. In conclusion we can say that the observation of the solar wind and the time scale of rotation decay are consistent with a slowing down mainly confined to the convective envelope and an internal rotation law smoothly increasing toward the centre of the Sun. These results rise a very big problem because from solar oscillations it comes out that the internal rotation of the Sun is nearly constant with a value equal to the surface value up to 0.2 of the solar radius where a fast rotating core starts. At the moment there is no way to reconcile the two results unless the estimate of the solar wind torque can be increased by about a factor of ten.

Acknowledgements: This work has been supported by the Ministero della Pubblica Istruzione through the University of Catania, the Osservatorio Astrofisico di Catania and the C.N.R. (Gruppo Nazionale di Astronomia, contract n. 86.00129.02). The extensive use of the computer facilities of the Catania ASTRONET Site is also acknowledged.

References

Baliunas, S.L. 1984, in "Third Cambridge Workshop on Cool Stars, Stellar Systems and the Sun", ed. S.L. Baliunas and L. Hartmann, Springer-Verlag Berlin, p. 114
Belcher, J.W., and Mac Gregor, K.B. 1976, Astrophys. J., 210, 498
Benz, W., Mayor, M., Mermilliod J.C.:1984, Astron. Astrophys 138, 93, 1984
Catalano, S. and Marilli, E. 1983, Astron. Astrophys., 121, 90
Dicke, R.H. 1970, Astrophys. J., 159, 1
Duncan, D.K. 1981, Astrophys. J., 248, 651
Duncan, D.K., Baliunas, S.L., Noyes, R.W. and Vaughan, A.H. 1983, Publ. Astron. Soc. Pacific, 95, 589
Endal, A.S. and Sofia, S. 1978, Astrophys. J., 220, 279
Endal, A.S. and Sofia, S. 1981, Astrophys. J., 243, 625
Kraft, R.P. 1967, Astrophys. J., 150, 551
Marilli E., Catalano, S. and Trigilio, C. 1986a, Astron. Astrophys., 167, 297
Marilli, E., Catalano S. and Trigilio, C. 1986b, in "Fourth Cambridge Workshop on Cool Stars, Stellar Systems and the Sun", ed. M. Zeilik and D.M. Gibson, Springer-Verlag Berlin, p.181

Noyes, R.W., Hartmann, S.L., Duncan, D.K. and Vaughan, A.N. 1984, *Astrophys. J.*, **279**, 763

Pallavicini, R., Golub, C., Rosner, R., Vaiana, G.S., Ayres, T., and Linsky, J.L. 1987, *Astrophys. J.*, **248**, 279

Pizzo, V., Schwean, Q., Marsch, E., Rosenbauer, H., Muhlhausen, K. - H, and Neubauer, F.M. 1983, *Astrophys. J.*, **271**, 335

Schatzman, E., 1962, *Ann. d'Astrophys.*, **25**, 18 Simon, T., Herbig, G. and Boesgaard, A.M. 1985, *Astrophys. J.*, **293**, 551

Skumanic, A., 1972, *Astrophys. J.*, **171**, 265

Soderblom, D.R., 1983, *Astrophys. J. Suppl.*, **53**, 1

Stauffer, J.R., Hartmann, L.W., Soderblom, D.R. and Burnham, T. 1984, *Astrophys J.*, **280**, 202

Tassoul, M., and Tassoul J.L. 1984, *Astrophys. J.*, **286**, 350

Trigilio, C., Catalano, S. and Marilli, E. 1986, *Adv. Space Res.*, **6**, 203

Trigilio, C., Catalano, S. and Marilli, E. 1987, in preparation

Vaughan, A.H., Baliunas, S.L., Middelkoop, F, Hartmann, L.W., Michelas, D., Noyes, R.W. and Preston, C.W. 1981, *Astrophys. J.*, **250**, 276

Weber, E.J. and Davis, L., Jr. 1967, *Astrophys. J.*, **148**, 217

Wilson, O.C. 1966, *Astrophys. J.*, **144**, 695.

THE MAIN SEQUENCE EVOLUTION OF ROTATION AND CHROMOSPHERIC ACTIVITY IN SOLAR-TYPE STARS

David R. Soderblom
Space Telescope Science Institute
3700 San Martin Drive
Baltimore, Maryland 21218 USA

ABSTRACT. The behavior of rotation and chromospheric emission is outlined for solar-type stars. From about 1.2 $M_\odot$ down to 0.8 $M_\odot$, rotation appears to obey a power-law relationship with age: $\Omega \propto t^{-1/2}$. This relation can be derived from simple models of angular momentum loss. For stars more massive than the Sun, there is a sharp decline of Ω with mass at any age, but this levels out for stars less massive than the Sun. The most rapidly rotating solar-type stars rotate at about 10 times the solar rate, and the slowest at about half the solar rate.

Little information is available on stellar differential rotation, a key parameter of magnetic dynamos. There is a clear relationship between chromospheric emission and the overall rotation when expressed relative to the convective turnover time. However, this rotation-activity relation does not extend to stellar activity cycles, where there is no clear dependence of the cycle period on rotation.

Recent observations to determine the relationship between chromospheric emission (CE) and age are also discussed. Although those observations are consistent with a relation of the form CE $\propto t^{-1/2}$ (the Skumanich law), such a power-law leads to a highly non-uniform Star Formation Rate (SFR) when applied to the Vaughan-Preston survey of the solar neighborhood. A curve that corresponds to a constant SFR fits the extant data just as well, and leads to three phases in the evolution of a star's chromosphere.

1. INTRODUCTION

John Stauffer (this volume) has shown us the fascinating phases in the Pre-Main Sequence (PMS) evolution of the angular momentum (AM) of a low-mass star. Others (e.g., Basri, Gahm) have discussed PMS chromospheric activity. By comparison, the main sequence behavior of a low-mass star is staid. We believe that we understand—qualitatively at least—the processes that account for what we see, but our understanding of the details is poor. A major motivation for studying main-sequence stars is to try to understand AM loss and the decline of chromospheric activity under reasonably well-behaved circumstances. We can then hope to extend that understanding to the PMS stars.

In this review, I will briefly discuss AM loss in solar-type stars, and then present some recent results on the relationship between chromospheric activity and age. For these purposes I take a solar-type star to lie between about F6V and K2V in spectral type, which corresponds to about 1.2 to 0.8 solar masses.

A. K. Dupree and M. T. V. T. Lago (eds.), Formation and Evolution of Low Mass Stars, 389–398.

2. ANGULAR MOMENTUM LOSS IN SOLAR-TYPE STARS

There are two classic problems in the study of stellar rotation. The first is why stars form at all, since the angular momentum of protostellar clouds is enormous. The solution may be that stars form in the turbulent eddies between clouds, where the net AM is low (Wolff, Edwards, and Preston 1982). This explanation can also account for the observed random orientations of stellar rotation axes.

The second classic problem is why the Sun rotates so slowly. With a rotation period of about 25 days and $v \sin i \approx 2$ km s^{-1}, its AM is two to three orders of magnitude below that of an early-type star. Only in the last few years have we found any other stars that we are certain rotate more slowly than the Sun.

The Sun itself suggests the answer to this problem, for we observe it to lose AM slowly through the solar wind. But it was Kraft (1967) who placed the Sun in the context of the stars and convincingly showed the relationship between rotation and age for stars like the Sun. I particularly recommend Kraft's Figure 1 for the manner in which it portrays the abrupt decline in rotation at 1.25 $M_\odot$, just where surface convective zones begin to appear.

Kraft also showed that rotation and chromospheric emission (CE) are intimately related in that fast rotators consistently have strong CE. These two relationships—rotation vs. age and rotation vs. CE—lead to a consistent picture: the interaction of convection and rotation (particularly differential rotation) leads to magnetic field generation through the dynamo mechanism. These magnetic fields are manifested as CE, hence the rotation vs. CE relation. The magnetic field enables AM loss by forcing corotation of an ionized stellar wind beyond the star's surface. The presence of a convective envelope is crucial, and accounts for the onset of AM loss at 1.25 $M_\odot$.

Despite the consistency of this scenario, the details are poorly understood. Direct evidence for AM loss has not yet been seen for another solar-type star besides the Sun, with the possible exception of the ultrafast rotator AB Doradus (Collier–Cameron, this volume). We know virtually nothing of differential rotation on other stars, although programs that are now in place may eventually yield that vital information.

Skumanich (1972) attempted to quantify the rotation-age and CE-age relations, using cluster data for young stars and taking the Sun to represent old stars. He found that power-laws of the form rotation (or CE) $\propto t^{-1/2}$ ($t =$ age) fitted those data well. He unfortunately had no data to span the order-of-magnitude difference in age between the oldest cluster (the Hyades) and the Sun. Also, his cluster data were for masses systematically greater than solar.

This $t^{-1/2}$ law is, in fact, consistent with expectation. For a simple model of AM loss (Weber and Davis 1967), we have

$$\frac{dJ}{dt} = -\frac{2}{3}\Omega r_a^2 \frac{dM}{dt}, \tag{1}$$

where r_a is the Alfven radius. But $B_o r_o^2 = B_a r_a^2$ (where the subscripts refer to values at the stellar surface and the Alfven radius, respectively), $B_a^2 = 4\pi\rho u_a^2$ (with u_a the wind speed at r_a), and $M = 4\pi\rho u_a r_a^2$. Combining these,

$$\frac{dJ}{dt} = -\frac{2}{3}\Omega \frac{B_o^2 \, r_o^4}{u_a}. \tag{2}$$

It is reasonable to take the stellar radius and moment of inertia as sensibly constant over most of the main sequence lifetime of a star. If one also arbitrarily takes $du/dt = 0$, then

$$\frac{d\Omega}{dt} \propto \Omega B^2. \tag{3}$$

On the Sun, we observe that the local magnetic field strength and CE are proportional (Skumanich, Smythe, and Frazler 1975), and for stars we see CE $\propto \Omega$. Thus $\dot{\Omega} \propto \Omega^3$, which leads to $\Omega \propto t^{-1/2}$ for older stars, with a flat curve for the youngest stars.

Kraft (1967) pushed to the limit the classical technique for studying stellar rotation: visual examination of line broadening on high dispersion photographic spectrograms. For the Sun, the FWHM of a typical absorption line is about 7 km s^{-1}, much more than the $v \sin i$. Thus the classical technique works only when rotation dominates the line broadening ($v \sin i \gtrsim 10$ km s^{-1}), and does not work for old solar-type stars.

A decade passed before substantial improvements in instruments and techniques were available. Smith (1978) matched computed models to high resolution Reticon observations of individual line profiles and their Fourier transforms (see Smith and Gray (1976) for a review of this technique). Smith found the Sun to rotate more slowly than most of the stars that he observed. He also presented evidence for a decline of a star's macroturbulent velocity (thought to result from convection) with age.

Soderblom (1982) used Smith's technique to extract $v \sin i$ values for a larger sample. He found that $v \sin i$ could be determined to a precision of 0.3 km s^{-1} for high-quality data, although the accuracy was about 0.5 km s^{-1} because of uncertainties in fundamental stellar parameters needed for the models. Soderblom (1983) substantiated Skumanich's $t^{-1/2}$ law for a broad range of ages in solar-type stars, and found the Sun to be typical for a star of its mass and age. He saw no evidence for a relationship between macroturbulent velocity and age.

Even when well-determined, $v \sin i$ values are not entirely satisfactory: There remains ambiguity over the inclination of the rotation axis for individual stars, and the relative uncertainties are high for slow rotators like the Sun. Also, the profile modeling technique is mostly limited to the brightest stars. More recently, $v \sin i$ values have been determined from cross-correlating high resolution spectra. This is particularly effective for observing faint young stars that have relatively broad lines.

Clearly the best technique for observing rotation is the monitoring of CE done by the Mount Wilson group (and now by others as well). From the modulation of Ca II H and K emission they are able to determine a rotation period (P_{rot}) to a few percent, even for stars rotating at half the solar rate. A major result of their work has been the demonstration of a relation between CE and the Rossby number, the ratio of a star's rotation period to its convective turnover time (Noyes et al. 1984).

For a broad look at rotation among solar-type stars of the solar neighborhood, one can invert the procedure of Noyes et al. and transform a survey of Ca II H and K emission into a survey of rotation. The procedure is explained in Soderblom (1985), and the result is shown in Figure 1. Note the steep decline in rotation for stars more massive than the Sun. There is about a one dex vertical spread at any one color, and the slowest stars rotate at about half the solar rate. The lack of stars at intermediate velocities is the so-called "Vaughan–Preston gap," and will be discussed further below. This constant vertical spread suggests

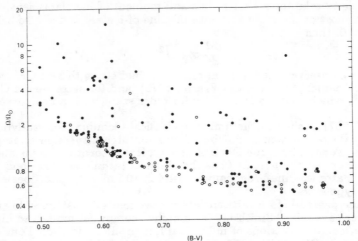

Figure 1. Rotation among solar-type stars of the solar neighborhood (from Soderblom 1985). To produce this figure, the Vaughan-Preston survey of Ca II H and K emission was used. This was transformed to Ω using the relations in Noyes *et al.* (1984). Comparison of calculated to observed rotation periods shows that the former are good to ±20%.

a uniform rotation-age relation for all of these stars. As discussed in Soderblom, the active stars really do rotate rapidly—$v \sin i$ values bear this out—and have other earmarks of youth as well. It is appropriate to warn that this distribution is a derivative of the rotation-activity relation. However, recent refinements to that relation (Radick and Baliunas 1987) do not affect significantly these results. One should also be aware of some evolutionary effects: The oldest F dwarfs have already left this diagram since their main sequence lifetimes are relatively short. Thus it is probably not possible to populate the lower left corner of Figure 1. Similarly, the oldest G and K dwarfs still have many years of main sequence life to go. Thus the base level of rotation at about $1/2 \, \Omega_\odot$ is not a "floor" or minimum value that represents the level of activity in a non-rotating star. In other words, these stars should eventually rotate even more slowly as they gradually lose AM.

3. OTHER ASPECTS OF ROTATION IN SOLAR-TYPE STARS

3.1. Differential Rotation

It's not just rotation, but *differential* rotation, that plays a key role in dynamo models. (I refer here to rotation varying with stellar latitude). To hope to understand magnetic field generation in late-type stars and all the associated magnetic phenomena, we must have some idea of the dependence of differential rotation on the fundamental stellar parameters: mass, composition, and age.

The only star for which differential rotation is known with precision is the Sun, and there it is a subtle effect. It is possible that differential rotation can be extracted from stellar line profiles of exceptional quality (Bruning 1981). However, attempts to date have failed (Gray 1977).

Rotational modulation of chromospheric emission or broad band light presumably arises from spots and other surface inhomogeneities transiting the stellar disk. With the Sun as an example, we expect the latitude of those spots to change systematically over a star's activity cycle. Thus we can hope to measure stellar differential rotation from steady changes in the observed rotation period with time, or if two different period are seen in the same data set. Such double periods have been seen in a few stars (Baliunas *et al.* 1985), but they can be attributed spots moving in longitude (Baliunas, this volume) as well as differential rotation. Decades of observations may be needed to yield useful results. The analysis of solar data by La Bonte (1984) suggests that even then one may not be able to extract the signal of differential rotation.

For a few stars with large $v \sin i$ values, the technique of Doppler imaging can yield the sizes and distributions of star spots (Vogt 1984). Drifts in the positions of these spots are used to infer differential rotation. Similar estimates come from modeling photometric data. In those cases in which evidence for differential rotation is present, it is in the same sense as the Sun (the equator rotates fastest). However, the spots appear to migrate poleward on these stars. This is contrary to the Sun, but is in accord with dynamo models.

3.2. Rotation and Stellar Activity Cycles

The tight relation between P_{rot} and Rossby number (Noyes *et al.* 1984) suggests the importance of rotation in generating magnetic fields on solar-type stars. We then expect to see rotation influencing magnetic activity cycles as well. Recent work on the Elatina sediments in Australia (Sonett and Williams 1987) suggests that the Sun has maintained its 11-year activity cycle for nearly a billion years. It is reasonable to expect that any feature of a star that changes little on evolutionary time scales should be a direct consequence of its mass, composition, and age, as a corollary of the Vogt-Russell theorem.

A number of parameters characterize the solar magnetic cycle, but the most directly observable for other stars is the period of the cycle. The Mount Wilson group has continued a program begun by O. C. Wilson to monitor activity in nearby 100 late-type stars. About 20 years of data are now available (Baliunas *et al.* 1988). Long-term periodicities are seen in most of the stars observed, but there are few or no clear trends in the data (Soderblom and Baliunas 1987). In particular, there is no clear relation between cycle period and rotation period (Fig. 2a), nor between cycle period and the Rossby number (Fig. 2b). This latter result is contrary to Noyes, Weiss, and Vaughan (1984), who found a relation in a smaller sample of data (their relation is shown as a straight line in Fig. 2b).

4. THE DECLINE OF CHROMOSPHERIC ACTIVITY IN SOLAR-TYPE STARS

We would like to better understand the age dependence of chromospheric emission (CE). One motivation is to follow the evolution of stellar dynamos over the main sequence lifetime of a star—this may help us account for the Vaughan-Preston "gap" in some way. We can also hope to use a CE-age relation to derive the ages of individual stars and then apply that information to such Galactic problems as the Star Formation Rate, the Age-Metallicity Relation, and the problem of "disk heating" (the increase of velocity dispersion with age). A CE-age relation could be applied to the low-mass stars, which are intrinsically

394

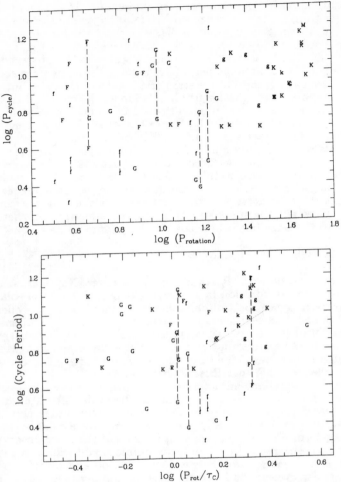

Figure 2. a) Observed stellar activity cycle period versus observed (capital letters) or predicted (lower case) rotation periods (from Baliunas *et al.* 1988).

b) Observed cycle period versus Rossby number (from Baliunas *et al.* 1985). The relation of Noyes, Weiss, and Vaughan (1984) is shown as a short straight line.

numerous, and for which complete samples are available.

Skumanich (1972) first tried to quantify the CE-age relation, using data for young clusters and the Sun. He fitted a power-law relation: CE $\propto$ $t^{-1/2}$. However, Skumanich's data were inadequate for determining the behavior of CE at intermediate ages (about 1 to 3 Gyr). The survey of Vaughan and Preston (1980) suggested that at these ages the CE-age relation might be discontinuous—this is the so-called Vaughan-Preston "gap." There are few clusters in this age range, and those few are too distant to be accessible to the same Mount Wilson instrument that's been used to survey the solar neighborhood. This leaves several questions unanswered:

1) Is there really a one-to-one CE-age relation (is it deterministic), or is it just a tendency (a statistical relation)?
2) If there is a relation (and there is), what is its form?
3) What does it mean?

These questions can be addressed by examining the V-P survey itself, the available data for clusters, and some new data. The new data considered here consists of a sample of widely-spaced binary stars. The primaries in these systems are relatively massive stars whose evolutionary status is determinable from Strömgren photometry. The secondaries are G dwarfs whose CE is measured with the Mount Wilson instrument. A few slightly-evolved F dwarfs have also been observed; these are stars for which ages *and* Ca II H and K emission strengths can be determined.

The details of the observations and reductions will be provided in a forthcoming paper (Soderblom, Duncan, and Johnson 1988). Here I present a summary of the results.

4.1. The CE vs. age relation—deterministic or statistical?

Our scenario of angular momentum loss and magnetic field generation in solar-type stars leads us to believe that the decline of a star's CE with age is inevitable, not just a tendency. But is it? Are there young stars with weak chromospheres or old stars with strong ones?

One can first examine the Vaughan-Preston survey itself (Fig. 3). Note that the high-velocity stars (see Soderblom 1985 for a full description of this diagram) are nearly all along the lower edge. The few exceptions to this rule have either only been observed once, or are only marginally high-velocity. There remains one embarassment: HD 152391. This star is unambiguously high-velocity, and had a very well-measured level of high CE. One can speculate that this star is an undiscovered close binary, since such systems show enhanced CE at all ages (Wilson 1963).

On the other hand, the chromospherically active stars are genuinely young: they have the kinematics of young stars, they rotate rapidly, and have abundant Li. These are nearby, bright stars that are unlikely to be undiscovered close binaries to any significant extent.

Clusters also provide an excellent means of testing the spread and consistency of CE among a group of stars of approximately the same age. Unfortunately, only the Hyades are close enough to be accessible to the Mount Wilson program, but they have been well observed. Among the Hyades G dwarfs, one sees a uniformly high level of CE with minimal scatter (about ±10%) and only a weak color dependence. There are no Hyads with weak chromospheres, which suggests that young stars do not go through a Maunder-minimum-like phenomenon (Soderblom 1985).

Visual binaries where both components are solar-type provide "mini-clusters," and one can apply the same consistency tests. Again, these systems consistently pass the test, whether it be Ca II H and K or Hα emission that is compared.

Thus the weight of the evidence favors the view that there is a deterministic relationship between chromospheric activity and age, with only one star providing an exception.

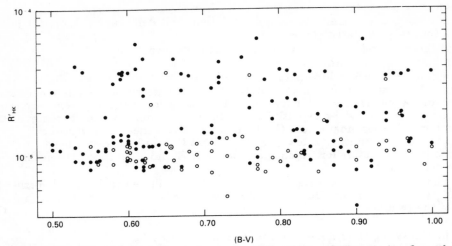

(B-V)

Figure 3. The Vaughan-Preston survey of Ca II H and K emission for solar-type stars of the solar neighborhood (from Soderblom 1985). The R'_{HK} index is the observed H and K emission equivalent width, corrected for photospheric light, and ratioed to the star's bolometric flux (see Noyes *et al.*1984). Open circles are high-velocity stars of the old disk and halo population. They are systematically older than young disk stars (full circles), and have lower HK emission.

4.2. The form of the CE-age relation

One can now take the available data and see what relation best fits it. Figure 4 shows the binary secondaries of Soderblom *et al.* (1988), together with points for the Ursa Major Group ("U," from Soderblom and Clements 1987), the Hyades ("H"), and the Sun ($\odot$). The abscissa, log R'_{HK}, is an index of Ca II H and K emission; its derivation is discussed in Noyes *et al.* (1984). Errors in log R'_{HK} are relatively small and nearly constant. The uncertainties in the ages depend upon how well the available photometry determines the age. The isochrones of Maeder (1976) were used for this purpose.

The simplest fit to these data is a power-law of the form $R'_{HK} \propto t^{-n}$. Depending on which points are fitted and how they are weighted, n varies from 1/2 (the Skumanich law) to 3/4. The rms scatter about a power-law fit is about 0.17 dex. This suggests that one can predict the age of an individual star to ±50% by observing its H and K emission. This is a relatively large uncertainty, but would be adequate for statistical studies of Galactic phenomena.

However, systematic errors may outweight the random errors. Any power-law fit has an important consequence: if it is applied to the Vaughan-Preston survey it lends to a highly non-uniform Star Formation Rate. In particular, one is led to conclude that there has been a recent ($t < 1$ Gyr) burst of star formation at our Galactocentric radius. This would be a fascinating result if it could be substantiated.

At this point, one can invert the problem and construct a relationship between R'_{HK} and age that is consistent with a constant SFR and the distribution of R'_{HK} values in the Vaughan-Preston survey. Such a curve is shown in Figure 5. It fits the binary sample just as well as the power-law does, and fits the young

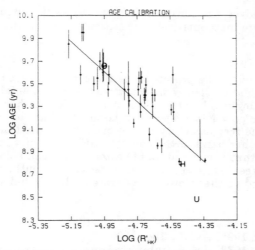

Figure 4. Chromospheric emission versus age for stars of known age. The points for the Sun (⊙), Hyades ("H") and Ursa Major Group ("U") are shown. The remaining points are solar-type secondaries in binary systems or slightly evolved F stars, taken from Soderblom, Duncan, and Johnson 1988. A power-law fit to the data is shown.

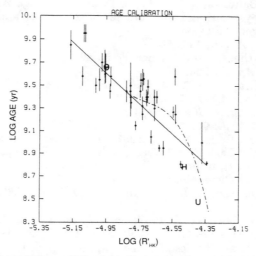

Figure 5. The same data as Figure 4, but now the curve corresponds to a constant Star Formation Rate.

clusters better. Moreover, the constant SFR curve has the virtue of physical simplicity, even if more mathematical parameters end up in the fit.

4.3. Conclusions

We cannot yet tell if the power-law or the constant-SFR curve is the better representation of the data. Clearly we must now observe the faint stars in older clusters that can distinguish between these possibilities. Either way the results are interesting. If the power-law is substantiated, there are important conclusions to be drawn about the recent history of low-mass star formation in the vicinity of the Sun. On the other hand, if the constant SFR rate curve is the correct one, then it appears that the evolution of chromospheres in solar-type stars is a complex one with three distinct phases: slow decline in youth; rapid decline at intermediate ages; and further slow-to-moderate decline in old stars.

REFERENCES

Baliunas, S. L., et al. 1985, Ap. J., 294, 310.
Baliunas, S. L., et al. 1988, Ap. J., submitted.
Bruning, D. H. 1981, Ap. J., 248, 274.
Gray, D. F. 1977, Ap. J., 211, 198.
Kraft, R. P. 1967, Ap. J., 150, 551.
La Bonte, B. J. 1985, Ap. J., 276, 335.
Maeder, A. 1976, Astr. Ap., 47, 389.
Noyes, R. W., Hartmann, L. W., Baliunas, S. L., Duncan, D. K., and Vaughan, A. H. 1984, Ap. J., 279, 763.
Noyes, R. W., Weiss, N. O., and Vaughan, A. H. 1984, Ap. J., 287, 769.
Radick, R. R., and Baliunas, S. L. 1987, in Fifth Cambridge Workshop on Cool Stars, Stellar Systems, and the Sun, eds. J. L. Linsky and R. Stencel (Berlin: Springer), in press.
Skumanich, A. 1972, Ap. J., 171, 565.
Skumanich, A., Smythe, C., and Frazier, E. N. 1975, Ap. J., 200, 747.
Smith, M. A. 1978, Ap. J., 224, 584.
Smith, M. A., and Gray, D. F. 1976, Pub. A.S.P., 88, 809.
Soderblom, D. R. 1982, Ap. J., 263, 239.
_____. 1983, Ap. J. Suppl., 53, 1.
_____. 1985, A. J., 90, 2103.
Soderblom, D. R., and Baliunas, S. L. 1987, in "Secular Solar and Geomagnetic Variations in the Last 10,000 Years," eds. F. R. Stephenson and J. A. Eddy (New York: Reidel), in press.
Soderblom, D. R., and Clements, S. D. 1987, A. J., 93, 920.
Soderblom, D. R., Duncan, D. K., and Johnson, D. R. H. 1988, Ap. J., in preparation.
Sonett, C. P., and Williams, G. E. 1987, Solar Phys., 110, 397.
Vaughan, A. H., and Preston, G. W. 1980, Pub. A.S.P., 92, 385.
Vogt, S. S. 1984, in "The Physics of Sunspots," eds. L. E. Cram and J. H. Thomas (Sunspot, N. M.: Sacramento Peak Obs.), p. 455.
Weber, E. J., and Davis, L. D. 1967, Ap. J., 148, 217.
Wilson, O. C. 1963, Ap. J., 138, 832.
Wolff, S. C., Edwards, S., and Preston, G. W. 1982, Ap. J., 252, 322.

CORONAL MASS EJECTIONS FROM A RAPIDLY-ROTATING K0 DWARF STAR

A. Collier Cameron and R.D. Robinson
Astronomy Centre   Anglo-Australian Observatory
University of Sussex   PO Box 296, Epping
Brighton BN1 9QH   NSW 2121
England   Australia

ABSTRACT. We present a new set of time-resolved Hα spectra of the rapidly-rotating southern K0 dwarf star AB Doradus (= HD 36705). These data show that clouds of neutral hydrogen are forming in the outer corona of the star at the rate of about two per day, and are subsequently being ejected from the system. We propose a cloud formation scenario analogous to solar quiescent prominence formation, and estimate the likely importance of these cloud ejections as a magnetic braking mechanism in young low-mass stars. The magnetic braking timescale resulting from this process alone could be as short as 10^7 years.

1. INTRODUCTION

AB Doradus (= HD 36705) is perhaps the closest of the pre-main-sequence objects similar to the rapidly-rotating highly-active K dwarfs found in the Pleiades cluster. It was identified by Pakull (1981) as the optical counterpart of a flaring soft x-ray source discovered during the EINSTEIN LMC survey. Pakull's follow-up programme of optical photometry revealed rotationally-modulated variability indicating the presence of cool starspots and an unusually short axial rotation period (for an early K star) of 0.51423 day. High-resolution spectroscopic observations by Collier (1982) gave $v\sin i = 85 \pm 5$ km s^{-1}, but showed no evidence of radial velocity variations. Rucinski (1982) found a strong Li I 6707Å feature and proposed that the star was a post-T Tauri object. More recently Innis et al. (1986) have concluded from a kinematic study that the star belongs to the Pleiades moving group. It therefore appears that AB Dor is in the same evolutionary state as the rapidly-rotating K dwarfs in the Pleiades cluster (Stauffer et al. 1984). Being some five magnitudes brighter than its Pleiades counterparts, however, it provides a unique opportunity for detailed studies of the circumstellar environment in such stars.

The discovery of clouds of neutral hydrogen embedded in and corotating with the corona of AB Dor was first reported by Robinson and Collier Cameron (1986). Here we present new optical observations of the cloud system made in 1986 November, which give new insight into the processes by which the clouds are formed in and ejected from the outer parts of the stellar corona.

A. K. Dupree and M. T. V. T. Lago (eds.), Formation and Evolution of Low Mass Stars, 399–408.
© 1988 by Kluwer Academic Publishers.

2. OBSERVATIONS

The observations were made using the Royal Greenwich Observatory spectrograph on the 3.9m Anglo-Australian Telescope (*AAT*), on the nights of 1986 November 14 12:41 to 18:36 UT; November 15 10:28 to 18:23 UT; and November 17 12:08 to 18:37 UT. A 1200-line grating blazed at 1.2 micron was used in its second order in conjunction with the 82 cm camera to give a dispersion of 2.5 Å mm^{-1} at the detector, a GEC CCD. The resolution of the system was ~10 km s^{-1} FWHM. Each exposure was of 150s duration, with an interval of 30s between exposures during which the CCD was read out. The mean signal-to-noise ratio of each individual spectrum was 140.

3. RESULTS

The timeseries of Hα spectra obtained on the nights of 1986 November 14, 15 and 17 are shown in Figs 1 to 3. All three data sets are plotted with stellar rotation phase inceasing up the vertical axis. The rotation phases are calculated from an arbitrarily-chosen epoch at 1986 November 14 06:53 UT. The greyscale is set up so that features brighter than 1.05 times the continuum level appear white, while those darker than 0.75 times continuum appear black; the scale is linear for intermediate values. As in the earlier observations reported by Robinson and Collier Cameron (1986) and Collier Cameron and Robinson (1988), the profile is generally filled-in to continuum level with chromospheric emission. The most striking phenomena present in the spectra are the transient absorption features, which appear as dark diagonal streaks crossing the Hα profiles in Figs. 1 to 3. Individual transients have durations of one to two hours. Initially, they appear blueshifted by some 80 km s^{-1} from line centre. The blue wing of the profile remains stationary as the red wing grows and the feature deepens until it reaches full strength, in some cases depressed 20% to 30% below continuum level at line centre. The entire feature then drifts across the profile, changing its radial velocity linearly with time, until it is displaced 80 km s^{-1} to redward of line centre. The feature then disappears, the blue wing shrinking as the red wing remains stationary.

Collier Cameron and Robinson (1988) argue that this behaviour is consistent with a model in which the absorption features are caused by clouds of neutral material, in enforced co-rotation with the star, transiting the stellar disc and scattering chromospheric Hα photons out of the line of sight. The depth of the absorption profile depends to a certain extent on the cloud geometry, but scales with the fraction of the stellar disc obscured by the cloud. The weakest detectable absorption transients obscure only 1 to 2 percent of the stellar disk. Their profiles, obtained by shifting and co-adding successive spectra, are barely resolved. This places an upper limit of about 10 km s^{-1} on their internal velocity dispersions, and implies optical thicknesses in the range $1 \leq \tau \leq 10$ at line centre in Hα.

The radial distances of the clouds from the star are given directly by the individual rates at which they cross the profile. The velocity shift from line centre at which a feature appears or disappears is always close to 80 km s^{-1}, independent of the drift rate. This indicates that the whole cloud system is in solid-body rotation rather than Keplerian orbit: only in the special case of solid-body rotation does the line-of-sight velocity component of the absorbing material always match that of the obscured stellar limb as the cloud enters and emerges from transit, independently of the rate at which the transit takes place. This model is confirmed by the data shown in Figs. 1 to 3, which show a large and distinctive double cloud complex crossing the disc at rotation phase 0.8, after intervals of two and six stellar rotations.

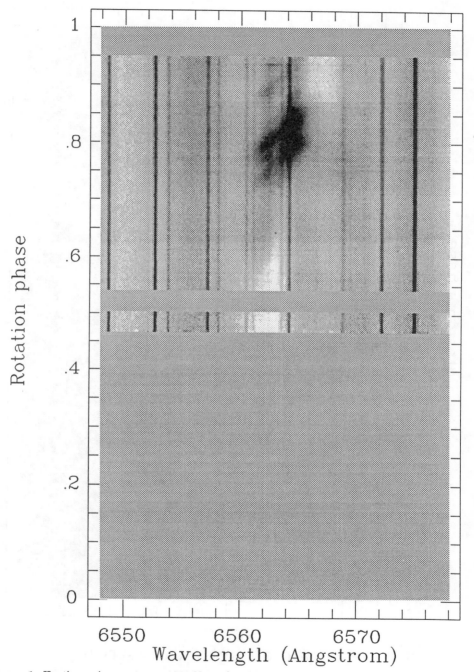

Figure 1. Hα timeseries spectrum, 1986 November 14.

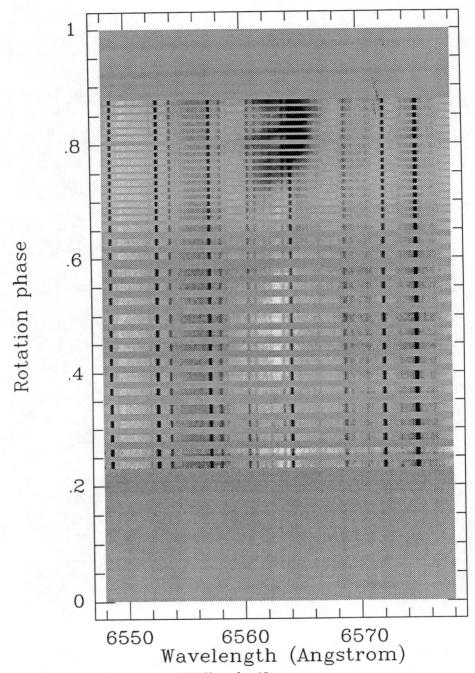

Figure 2. Hα timeseries spectrum, 1986 November 15.

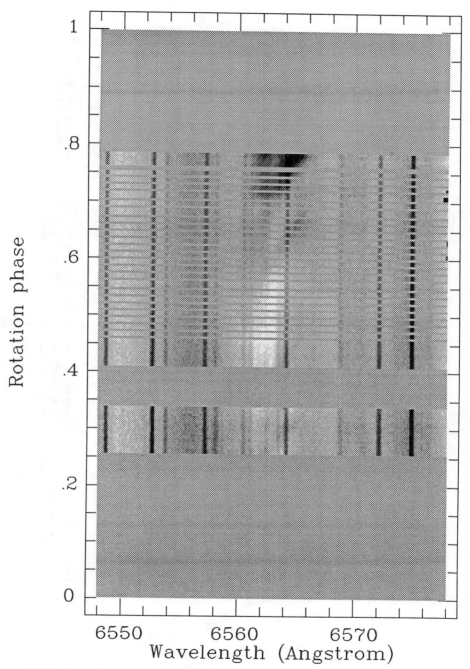

Figure 3. Hα timeseries spectrum, 1986 November 17.

404

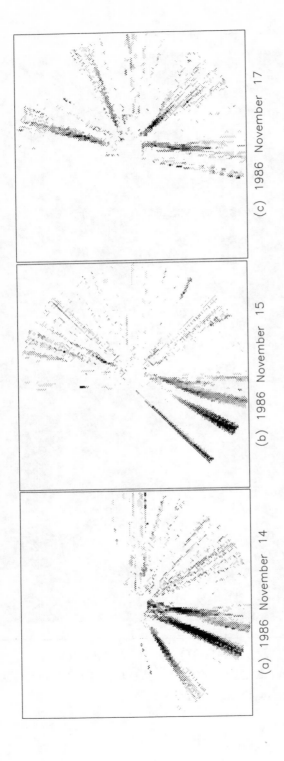

(a) 1986 November 14   (b) 1986 November 15   (c) 1986 November 17

Figure 4. Radial significance maps for absorption transients in the data sets shown in Figs. 1 to 3. The box dimensions are 20 × 20 stellar radii, with the star located at the centre of each box. The system is viewed in projection along the rotation axis, and rotation phase increases clockwise.

In Fig. 4 we use the linear relationship between drift rate and distance from the rotation axis to produce a map, in polar coordinates, of the statistical significance of diagonal features crossing the Hα profile in Figs. 1 to 3 at all observable rotation phases and distances less than 10 R_* from the rotation axis. It can be regarded as a crude map of the cloud distribution in the observable slice of the corona which transits the stellar disc, viewed in projection along the stellar rotation axis. The greyscale is constructed so that white areas represent regions where no transients more significant than 3σ are detected, while black represents a 20σ or stronger detection. A full description of the technique used in producing these maps is given by Collier Cameron & Robinson (1988).

There is a wide spread of drift rates among the clouds, corresponding to perpendicular distances from the rotation axis ranging from 3 to 9 stellar radii, with the mode of the distribution being some 4 stellar radii from the rotation axis. Some clouds also exhibit outward radial motion with respect to the star. This radial motion is greatest in those clouds located furthest from the rotation axis: the leading cloud in the double complex near rotation phase 0.8 moved slowly outward from 3.4 to 4.2 R_* from the rotation axis over the period of observation (a total of six stellar rotations), while the trailing cloud moved out from 5.2 to 9.0 stellar radii over two rotations. Other clouds were seen to form on a timescale of one to two days, between 3 and 4 stellar radii from the axis, but exhibited no measurable radial motion. Note in particular the three significant transients seen at phases 0.3, 0.42 and 0.65 on the night of 17 November (Figs. 3 and 4c) which were not seen (or very much weaker) in the 15 November data.

4. Hα CLOUD DENSITIES AND TEMPERATURES

To date, observations of the absorption transients have only been made at Hα. In order to derive the column densities and hence estimate the masses of the clouds, the degree of excitation in the cloud material must be determined. To this end we carried out approximate statistical equilibrium calculations for a 5-level model hydrogen atom assuming uniform temperature and density throughout a spherical model cloud with a radius of 0.1 stellar radii. The Lyman lines and continuum are considered to be opaque, while the higher continua are transparent and the Balmer lines are treated as being effectively thin.

The observed profiles for the smallest clouds constrain their line-centre optical thicknesses in Hα to be many times smaller than the thermalisation depth, so the source function contains only scattering terms. This is consistent with the observation that the transients appear only in absorption as they transit the disc, with little off-disc emission, and justifies our neglect of detailed radiative transfer effects in the statistical equilibrium calculations. The optical thickness of a cloud in Hα thus depends on the column density of H atoms in the $n = 2$ level, which can be achieved either with a low-density model and a high degree of excitation or a cool, higher-density model, as Fig. 5 shows. The hotter models have very short cooling times (comparable to the disc transit time) but can be maintained in radiative equilibrium through absorption of chromospheric Lα. In this case the upper limit on the cloud density is $n_p + n_H \approx 10^{11}$ cm^{-3} and the temperature must lie in the range $5500K < T < 7500K$. The corresponding cloud masses will lie in the range $10^{-16} < M/M_\odot < 10^{-14}$ for the observed range of linear cloud dimensions. Clouds at higher densities and lower temperatures can also give the observed optical thicknesses in Hα, but absorption of chromospheric Lα is no longer sufficient to maintain radiative equilibrium. However, the cooling times are of order days, so this is not necessarily in conflict with the

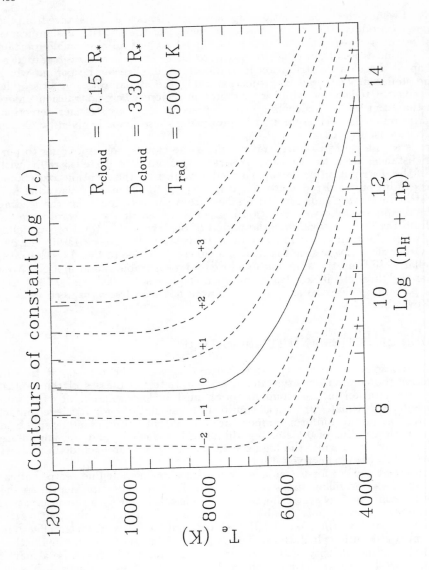

Figure 5. Contours of constant log optical thickness for a model cloud with radius 0.15 R$_*$ located at a distance of 3.3 R$_*$ from the centre of the star. As described in the text, the optical depths of the clouds observed probably lie in the range $1 \le \tau \le 10$.

observed cloud lifetimes of several days.

5. CLOUD FORMATION AND EJECTION MECHANISMS

It is interesting to note that the keplerian corotation radius (KCR) – at which the orbital period of a particle in a circular orbit equals the stellar rotation period – lies some 2.8 stellar radii from the rotation axis in AB Dor. The 1986 November *AAT* data show evidence for clouds forming at distances between 3 and 4 stellar radii from the rotation axis – just outside the KCR – on timescales of 2 days or so. During this time, a weak absorption transient increases its strength, remaining at the same distance from the star. The strongest transients are more likely to be found further from the star, and show outward radial motion which increases with distance from the star while remaining in solid-body rotation.

EXOSAT observations of the 1 to 10 keV x-ray spectrum of AB Dor yield a quiescent coronal temperature of 1.9×10^7 K and a volume emission measure $\int N_e^2 dV = 1.6 \times 10^{53}$ cm^{-3} (Collier Cameron *et al.* 1988). If the emitting plasma is to be confined with loop base pressures comparable with solar active-region pressures (10 to 20 dyn cm^{-2}) it follows that the x-ray corona must occupy a volume extending at least two stellar radii above the photosphere. The clouds therefore appear to form close to or just outside the limits of the x-ray corona.

These data are consistent with a qualitative picture in which the clouds form in the upper parts of closed magnetic loop structures in a manner analogous to the formation of quiescent solar prominences. In a closed coronal loop whose summit extends beyond the KCR, the pressure must *increase* outwards in the upper part of the loop where the outward centrifugal force acting on the loop material dominates over gravity, in order to maintain hydrostatic equilibrium. The increased density enhances the radiative loss rate in the upper part of the loop and can lead to thermal instability near the loop summit (Collier Cameron 1988). When this occurs the coronal plasma cools on a timescale of hours, achieving thermal stability only when the temperature has dropped to temperatures at which hydrogen recombines. The resulting drop in pressure causes material to flow up from the loop footpoints in an attempt to restore hydrostatic equilibrium, provided the magnetic pressure is initially dominant over the gas pressure. The condensed neutral material continues to pool in the loop summit until it reaches a critical density at which the centrifugal force acting on the cloud material dominates over the magnetic tension in the loop. The loop is deformed as the cloud material accelerates outward and eventually escapes from the system.

6. ESTIMATED ANGULAR MOMENTUM LOSSES

The 1986 November observations show at least two new clouds forming per day on average in the observable slice of the corona which transits the stellar disc.

Under the (low-density) assumption that the clouds are in radiative equilibrium and do not escape from the star until they are fully-developed, the masses of the escaping clouds will be on the order of 10^{-14} M$_\odot$. The greatest distance from the rotation axis at which we have observed clouds still held in corotation is close to 9 stellar radii. Under the assumption that the cloud material escapes from the field altogether at 10 stellar radii, and that the moment of inertia of the star is given by $I \approx 0.1 M_* R_*^2$, we obtain for the braking timescale: $J/\dot{J} \approx 10^8$ yr. This is a very conservative estimate, since the Alfven radius at which the magnetic field ceases to impart angular momentum to the escaping cloud material is probably closer to 30 R_* (Mestel and Spruit 1987), in which case the braking timescale would be closer to 10^7 yr.

It appears that this process could be the dominant mechanism by which very rapidly-rotating young MS stars shed angular momentum. If so, it will be effective only as long as the star rotates fast enough to keep the KCR within the outer limits of the closed part of the coronal field, i.e. for rotation periods less than about 2 days. At longer rotation periods the angular momentum losses are more likely to be dominated by centrifugally-driven wind losses along open field lines.

References

Collier, A.C., 1982. *Mon. Not. R. astr. Soc.* , **200,** 489.

Collier Cameron, A., 1988. *Mon. Not. R. astr. Soc.* , submitted.

Collier Cameron A., Bedford D.K., Rucinski S.M., Vilhu O. & White, N.E., 1988. *Mon. Not. R. astr. Soc.* , in press.

Collier Cameron, A. & Robinson, R.D., 1988. *Mon. Not. R. astr. Soc.* , submitted.

Innis, J. *et al.* , 1986. *Mon. Not. R. astr. Soc.* , **223,** 183.

Mestel, L. & Spruit, H.C., 1987. *Mon. Not. R. astr. Soc.* , **226,** 123.

Pakull, M.W., 1981. *Astr. Astrophys.* , **104,** 33.

Robinson, R.D. & Collier Cameron, A., 1986. *Pub. astr. Soc. Australia,* **6,** 308.

Rucinski, S.M., 1982. *Inf. Bull. var. Stars,* **2203**.

Stauffer, J.R., Hartmann, L., Soderblom, D.R. & Burnham, N., 1984. *Astrophys. J.* , **280,** 202.

STELLAR OBSERVATIONS WITHIN THE ESA SCIENCE PROGRAMME

R. M. Bonnet
European Space Agency
8-10, Rue Mario Nikis
75738 Paris Cedex 15
France

ABSTRACT. The main space astronomy missions which ESA is already developing or which are intended to fly in the future are reviewed. Their scientific objectives together with their proposed instrumentation are described in detail. Emphasis is placed on the stellar physics aspects of the various missions. A special attention is given to the cornerstones of ESA's long term plan 'Horizon 2000'.

1. INTRODUCTION

The Space Science Programme of ESA encompasses both the domains of Astronomy and of Solar System exploration, as well as plasma and magnetospheric physics. In this brief article we will concentrate mainly on the astronomy missions. We separate the description into three different chapters: those which are already approved (or even built !), those which are in the process of selection and finally the Cornerstone's of ESA's Long Term Plan, Horizon 2000. Table 1 presents the various missions according to their expected launch date (except in the case of IUE which is this year celebrating its tenth year in orbit) and their main spectral coverage. It can be seen that the overall ESA programme includes at least one mission in each domain and reflects a certain balance from the very long wavelengths to the highest energies, through the infrared, the visible and the ultra-violet. We will review each of them and will pin-point their main characteristics and how they can advance our understanding of stellar physics and in particular of the formation and evolution of low mass stars.

2. PLANNED MISSIONS

Apart from the very successful IUE project which has been operated by ESA, NASA and the British Science Engineering Research Centre, for the last 10 years, ESA is engaged in the the preparation of four missions: Ulysses, the Hubble Space Telescope, Hipparcos and ISO. The first three have been or will be delayed due to the problems encountered with

A. K. Dupree and M. T. V. T. Lago (eds.), Formation and Evolution of Low Mass Stars, 409–431.
© *1988 by Kluwer Academic Publishers.*

the Shuttle and with the Ariane rocket. Of these four only the last
three are of direct relevance to the topic of this lecture and will
therefore be discussed in more detail now.

**Table 1: The various astronomy missions of ESA's Science Programme
presented according to expected launch date and spectral range**

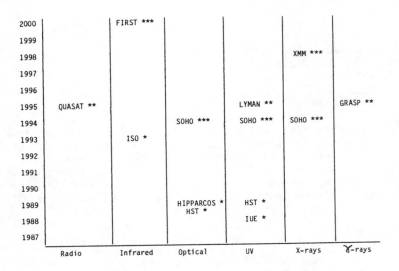

	Radio	Infrared	Optical	UV	X-rays	γ-rays
2000		FIRST ***				
1999						
1998					XMM ***	
1997						
1996						
1995	QUASAT **			LYMAN **		GRASP **
1994			SOHO ***	SOHO ***	SOHO ***	
1993		ISO *				
1992						
1991						
1990						
1989			HIPPARCOS * / HST *	HST *		
1988				IUE *		
1987						

```
*    already approved or in development
**   candidate for the next selection
***  cornerstone of Horizon 2000
```

2.1. The Hubble Space Telescope

The objective of the NASA/ESA Hubble Space Telescope Project is to
place a two-metre class astronomical telescope and its associated
instrumentation in low earth orbit and maintain it over an extended
period of time as an international observatory. It is presently hoped
that ST will be launched by the Space Shuttle in June 1989.
   The focal plane of the Telescope is shared by five scientific
instruments: the Wide Field Planetary Camera, the Faint Object
Spectrograph, the High Resolution Spectrograph, the High Speed
Photometric and the Faint Object Camera.
   The ESA contribution to the NASA ST Project consists of three
elements:
(1) the Faint Object Camera ;
(2) the solar arrays which provide the spacecraft with power ;
(3) scientific and technical personnel for the Space Telescope Science
     Institute in Baltimore, Maryland.
   In return for this contribution, European astronomers from ESA
Member States will be guaranteed a minimum of 15% of Space Telescope
observing time.

Figure 1. Optical Layout of the Faint Object Camera

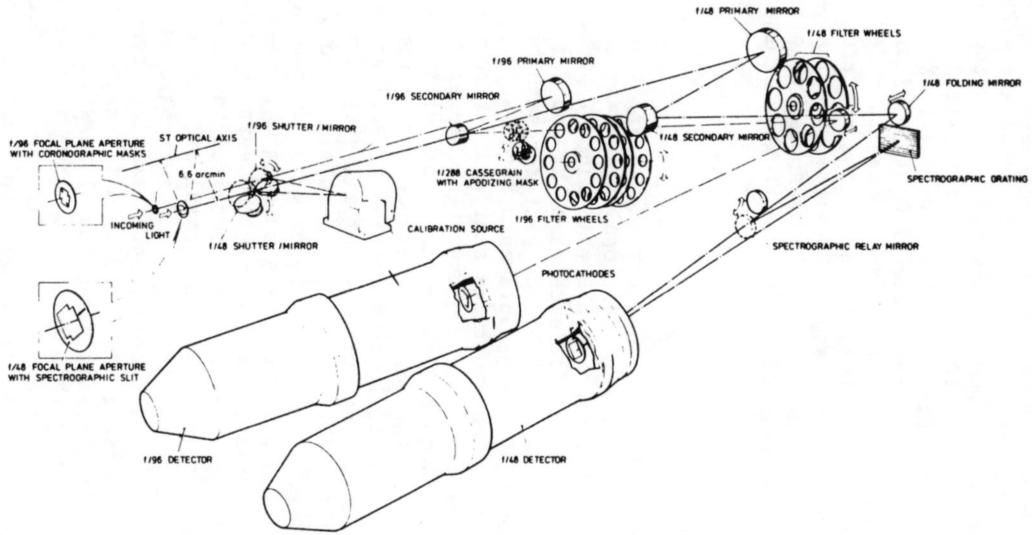

f/48 FOLDING MIRROR

SPECTROGRAPHIC GRATING

f/48 FILTER WHEELS

f/48 PRIMARY MIRROR

f/48 SECONDARY MIRROR

SPECTROGRAPHIC RELAY MIRROR

f/96 PRIMARY MIRROR

f/96 SECONDARY MIRROR

f/288 CASSEGRAIN WITH APODIZING MASK

f/96 FILTER WHEELS

CALIBRATION SOURCE

PHOTOCATHODES

f/48 DETECTOR

f/96 SHUTTER/MIRROR

ST OPTICAL AXIS

6.6 arcmin

f/48 SHUTTER/MIRROR

INCOMING LIGHT

f/96 DETECTOR

f/96 FOCAL PLANE APERTURE WITH CORONOGRAPHIC MASKS

f/48 FOCAL PLANE APERTURE WITH SPECTROGRAPHIC SLIT

The Faint Object Camera (FOC) is capable of operating in four basic modes: direct imaging at $f/48$, $f/96$ and $f/288$ as well as in a 20 x 0.1 arc second (R 1000) long-slit and spectrographic mode. The optical layout of the FOC is shown in Figure 1. The FOC is designed to exploit fully the unique imaging capability of the Space Telescope and provide images with pixel sizes of 0.04, 0.02 or 0.08 arc seconds. The fields of view in the three nominal imaging modes are 22 x 22, 11 x 11, and 4 x 4 arc seconds respectively. The spectral response of the FOC spans the 115 - 600 nm range. Local Group galaxies in FOC images will appear much as the Magellanic clouds appear to ground-based observers today, enabling detailed studies to be made of their stellar populations.

2.2. Hipparcos

The Hipparcos satellite, due for launch by Ariane in July 1988, will be dedicated to the precise positional measurements of some 100 000 selected stars brighter than B = 13 mag. Typical target accuracies will be 0.002 arc sec for each parallax and for each positional component and 0.002 arc sec per year for each proper motion component. As a by-product of the measurements, precise Hipparcos magnitudes will be obtained for each programme star and for each of the 100 or so transits of each star throughout the 2.5-year mission, and a large quantity of data on stellar multiplicity will be obtained. Data from the satellites 'star mapper' attitude-measurement system will also provide positions of lower precision as well as two-colour (424nm and 524nm) photometric measurements of all stars down to a limiting magnitude of B = 10-11 mag. The resulting data will constitute the Tycho Catalogue of some 500 000 or more additional stars.

The mission will provide a uniform whole-sky stellar catalogue suitable for detailed astrometric and astrophysical studies.

The large number of stars contained in the Hipparcos observing programme will fulfil two fundamental and interdependent roles. Irrespective of the star's detailed astrophysical importance, their precise positions will result in a basic reference frame against which all celestial objects, detected in different wavebands, can be identified. At the same time, the changes in position of celestial objects relative to this reference frame will be measured. It is the changes of positions of stars within the reference frame that allows measurement of stellar distances, luminosities etc. (via the parallaxes).

Some of the most spectacular advances to be expected from the mission are likely to come from the fivefold increase in precision of measurements of trignometric parallaxes compared with typical Earth-based observations, and from the very much larger number of stars which will be measurable. A significant outcome of the mission will be a systematic search for, and measurement of, double and multiple stars, and the generation of an enormous body of photometric data both from the main mission and from the star mapper stream which will be of great value in photometric and variability studies.

Figure 2 shows the precision of the Hipparcos reference frame compared with that of some other major catalogues presently available.

The launch of Hipparcos is now scheduled for April 1989, although the spacecraft will be ready as originally foreseen for a launch in July 1988.

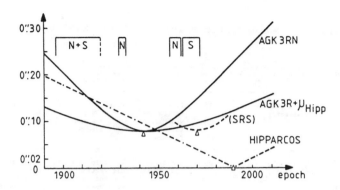

Figure 2. *Accuracy of some main reference star catalogues as a function of time. Epochs of some large photographic catalogues are indicated in the upper part. N = Northern hemisphere coverage ; S = Southern hemisphere coverage. (After de Vegt).*

3. ISO

ISO, the Infrared Space Observatory, is a logical successor to the very successful IRAS mission. It will make detailed observations of selected sources. Compared with IRAS, ISO will have a longer operational lifetime, wider wavelength coverage, better angular resolution, more sophisticated instruments (especially for polarimetry and spectroscopy) and, by a combination of detector improvements and longer integration times, a sensitivity gain of up to several orders of magnitude.

ISO is a facility of unprecedented sensitivity for detailed observations of cold objects, ranging from objects in the solar system right out to the most distant extragalactic sources. Its cryogenically cooled telescope will be equipped with four scientific instruments, which together will permit imaging and photometric, spectroscopic and polarimetric observations at wavelengths from 3 - 200 μm.

Although not specifically devoted to stellar studies, stellar evolution will be studied via observations of circumstellar shells, especially around cool giant stars. How large is the mass loss rate ? What is the nuclear enrichment of the outflowing material as compared with the material from which the star once formed ? Absorption lines of several molecules and their isotopes are predicted to be easily detectable. When studied at high resolution, these lines will permit the conditions in the stellar wind to be determined. Planetary nebulae will be mapped in various forbidden lines. Such lines, together with hydrogen and helium recombination lines can also be studied in Wolf-Rayet stars. IRAS has discovered a large number of stars with thick circumstellar shells in the bulge of our Galaxy. Are these stars truly Asymptotic Giant Branch stars similar to those in the galactic disk ? Are the planetary nebulae in the galactic bulge similar to the planetary nebulae in the disk of the Galaxy ?

Table 2 : Main characteristics of ISO instruments

Instrument and Principal Investigator	Main Function	Wavelength (µm)	Spectral Resolution	Spatial Resolution	Outline Description
ISOCAM (C. Cesarsky, CEN-Saclay, F)	Camera and polarimetry	3 - 17	Broad-band, narrow-band, and circular variable filters	Pixel f.o.v.'s of 1.5, 3.6 and 12 arc secs	Two channels each with a 32 x 32 element detector array
ISOPHOT (D. Lemke, MPI für Astronomie, Heidelberg, D)	Imaging photo-polarimeter	3 - 200	Broad-band and narrow-band filters	Variable from diffraction-limited to wide beam	Four subsystems: i) Multi-band, multi-aperture, photo-polarimeter (3-30 possibly to 110) µm) ii) Far-infrared camera (30-200 µm) iii)Spectrophotometer (2.5 - 12 µm) iv) Mapping arrays (3 bands, 4-22 µm)
SWS (Th. de Graauw, Lab. for Space Research, Groningen, NL)	Short-wave-length spectrometer	3 - 45	1000 across wavelength range and 3×10^4 from 15 - 30 µm	7.5 x 20 and 12 x 30 arc seconds	Two gratings and two Fabry-Pérot inter-ferometers
LWS (P. Clegg, Queen Mary College,London,GB)	Long-wavelength spectromeeter	45 - 180	200 and 10^4 across wavelength range	1.65 arc minutes	Grating and two Fabry-Pérot interferometers

Rich rewards are expected from ISO studies of protostars and regions with star formation. IRAS has discovered point sources and small-scale structures in a large number of dark clouds. The nature of these sources will probably be revealed in the next few years through ground-based and airborne observations. However, especially for the many weak sources, such follow-up observations will be inadequate. ISO offers colour/colour diagrams for large numbers of weak point sources and it may also discover multiple objects or small clusters of stars. At the present time no generally accepted examples of stars in their Helmholtz contraction phase are known, but they may be discovered through detailed spectroscopy in the far infrared.

The satellite is 5.2 m high, 2.3 m wide and will weigh around 2300 kg at launch. The down-link bit rate is 43.7 kbps of which 33 kbps are dedicated to the scientific instruments.

The payload module (Figure 3) is essentially a large cryostat.
Inside the vacuum vessel is a toroidal tank filled with about 2000
litres of superfluid helium, which will provide an in-orbit lifetime of
at least 18 months. Some of the infrared detectors are directly
coupled to the helium tank and are at a temperature of around 2 K.

Figure 3. *Schematic view of ISO payload module.*

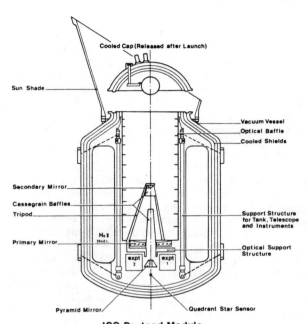

ISO Payload Module

Suspended in the middle of the tank is the telescope, which is a
Ritchey-Chrétien configuration with an effective aperture of 60 cm.

The instrument complement of ISO consists of a camera, an imaging
photopolarimeter and two spectrometers. In keeping with the
observatory nature of ISO, the individual instruments are being
optimised to form a complete, complementary and versatile package. As
shown in Table 2, the total payload provides photometric and imaging
capabilities at various spatial and spectral resolutions from 3 μm to
200 μm and spectroscopic capabilities at medium and high resolution
from 3 μm to 180 μm. The four instruments view adjacent areas on the
sky and only one instrument will be active at a time.

A new, higher orbit was adopted for ISO in May 1987. The
parameters of this orbit are: apogee, 70 000 km; perigee 1 000 km;
inclination 5 deg; and period, 24 hours.

The launch of ISO is presently scheduled for the end of 1992 or
early 1993.

4. NEW SELECTION

In the second half of 1988, ESA will go through a new cycle of selection for a project which would fly in the 1995 time frame. Five projects are competing for this selection, three of them are astronomy missions which we discuss in more detail below. The two others concern the exploration of Asteroids (Vesta, in cooperation with CNES and the USSR Academy of Sciences) and the landing of a probe on the surface of Saturn's moon, Titan (Cassini, in cooperation with NASA).

4.1. Lyman

The Lyman region of the UV spectrum is not observable with IUE or the Hubble Space Telescope and has only previously been explored - in the mid-seventies - by NASA's highly successful Copernicus (OAO-3) satellite. Although the capabilities of this satellite were rather modest by modern standards (limiting sensitivity 6-7th magnitude), this mission nevertheless led to dramatic advances, especially in our understanding of the physical conditions in the interstellar medium and the violent mass loss and superionisation that takes place in the stellar winds of early-type stars. For cost saving reasons and in order not to overlap with the objectives of the longer-wavelength spectrographs onboard the Hubble Space Telescope, it was decided that the Lyman payload should be fully optimised to tackle the science goals that are unique to the prime 90 - 120 nm spectral region covered by the mission. Table 3 contains a list of scientific objectives for the Lyman mission along with the corresponding sensitivities needed to achieve these different goals.

The Lyman mission concept is presently undergoing Phase A study with participation from Australia and Canada and potential participation from the USA. The mission concept envisages a spacecraft ejected into a highly efficient 120 000 km x 1 000 km, 48-hour Exosat-type orbit from a shared Ariane launch into an initial geostationary transfer orbit. This orbit is preferred because it is of low cost, and yet will permit both real-time observing and uninterrupted observations exceeding 38 hours during the low-background portion of the orbit spent above the Earth's radiation belts.

4.2. Quasat

The Quasat mission concept, proposed to ESA in 1982, is to operate an orbiting radio telescope in conjunction with a world-wide very-long-baseline-interferometry (VLBI) network of ground-based telescopes.

The resolution achieved with Quasat will permit studies of several aspects of stellar evolution both through the stellar and interstellar masers and through studies of flare and X-ray stars.

In addition, studies of the proper motion of the point water-maser features will, through statistical parallax considerations, permit measurements of the distance to the sources. Such direct measurements are of particular importance in the case of masers associated with

Table 3 :

A selected set of Lyman science objectives

Science objective	Targets	Magnitudes
Hot stars		
Hot plasma in stellar winds	OB stars	4 - 15
Abundances and T of highly evolved stars	sdO and WR	6 - 15
Mass loss in stars of different galaxies	OB in LMC, M31	10 - 18
Cool stars		
Stellar chromospheres	cool stars	4 - 10
Winds and mass loss	cool stars	4 - 8
Activity in pre-main sequence stars	A, F and G stars	4 - 10
Interacting binaries		
Flows of warm gas in binaries	OB X-binaries	6 - 15
Accretion of low-mass binaries	Sco X-1 stars	8 - 12
Mass transfer in cataclysmic binaries	AM Her stars	14 - 16

nearby galaxies (1 to 20 Mpc), because they provide a new independent estimate of the Hubble constant.

Quasat will also provide, with unprecedented detail, observations of the nuclei of radio galaxies and quasars on resolution scales approaching those expected for accretion discs around massive black holes.

The observing frequencies selected for Quasat are 22, 5 and 1.6 GHz with an additional low frequency, 327 MHz, chosen for the investigation of refractive scattering as a possible cause of low-frequency variable radio sources as well as observation of pulsars. All the Quasat receivers were to be dual circular-polarisation systems.

The telescope size is 10 m and the mission will be launched on a shared Ariane launcher in a Geostationary Transfer Orbit (GTO). At apogee, a solid motor boosts the inclination to 30° and raises the perigee altitude to 5 000 km. In the present concept, the apogee altitude can be lowered from 36 000 km down to 22 000 km by the use of hydrazine thrusters.

3.3. GRASP

The GRASP (Gamma Ray Astronomy with Spectroscopy and Positioning) telescope will be the first high-resolution spectral imager to operate in the gamma-ray region. It will have the following features :
- A wide bandwidth (15 keV to 100 MeV) which, for the first time, links X-ray and gamma-ray astronomy ;
- High-resolution spectroscopy from 15 keV to 10 MeV (E/Δ E around 1000 at 1 MeV) ;
- Accurate source positioning (typically ± 1 arc min) within a field of view of $\simeq$ 50 square degrees ;
- High sensitivity for both extended and point sources (typically 10mCrab at 3 sigma in 10^5s).

The telescope employs a coded aperture mask (∅ - 130 cm) at about 4 m distance from the detector, which consists of an array of position-sensitive CsI detectors (3300 cm^2).

The following mission requirements were identified :
- A three-axis stabilised spacecraft with moderate (1 arc min) pointing accuracy ;
- Accommodation of a 1000 kg payload ;
- Telemetry capability of 70 kbps.

Among many scientific objectives GRASP will address several major astrophysical subjects such as :
- Explosive nucleosynthesis in local (less than 10 Mpc) supernovae ;
- Identification and understanding of existing unidentified gamma-ray objects as a class ;
- Discovery and identification of new gamma-ray sources ;
- Physical environment close to compact objects ;
- Galactic mapping of recent (less than 10^6 yr) nucleosynthesis;
- Physical environment of the centre of the Galaxy ;
- Explosive nucleosynthesis in novae ;
- Search for recent (less than 100 yr) undetected Super Novae Remnants ;
- Localised and extended Inter Stellar Medium/Cosmic Ray interaction regions.

5. THE CORNERSTONE MISSIONS

ESA's Long Term Plan in Space Science includes four major missions in the cost class of a half billion dollars called the cornerstones of the plan, and which are programmed in the time frame 1994 - 2004. They ensure the stability of the plan and are evenly distributed between astronomy and solar system exploration. Two are major X-ray and submillimetre observatories. One, the Solar Terrestrial Science

Programme is studying the sun itself, the variations of its luminosity and spectral radiance together with the reaction of the magnetospheric plasma to these variations. The last one is a Comet Nucleus Sample Return mission. We will dwell here on the first three because of their relevance to the topic of this NATO Institute.

5.1. The Solar Terrestrial Science Programme (STSP)

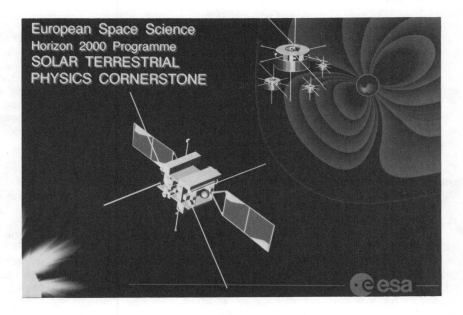

Figure 4

The STSP as shown on Figure 4 is made in fact of two projects: Soho, the Solar Heliospheric Observatory and Cluster, a set of four identical spacecraft orbiting inside the magnetospheric plasma. It is the first 'Cornerstone' element of ESA's Horizon 2000 -Space Science Long-Term Plan to be implemented. The prime STSP objective is to attack outstanding scientific problems in solar, heliospheric and space-plasma physics in a unified and co-ordinated approach.

STSP involves major contributions to the Programme by NASA, for example provision of the Soho launch and flight operations.

The joint ESA/NASA Announcement of Opportunity for scientific investigations on Cluster and Soho was released in March 1987, and the selection of the investigations is planned to take place by the end of 1987. The launches of the Cluster and Soho missions are foreseen for the 1994-1995 time frame. The following describes Soho in more detail, a mission dedicated to the study of a particular star: the sun.

The main objectives of the Solar and Heliospheric Observatory are:
a)   to investigate the solar interior structure by methods of
     helioseismology ;
b)   to investigate variations of the sun's luminosity through
     high-precision solar-irradiance measurements ;
c)   to investigate and understand the physical processes that form and
     heat the solar corona, maintain it and give rise to the expanding
     solar wind, by means of plasma diagnostics of the chromosphere,
     transition region and corona ;
d)   to investigate solar-wind streams in situ and their relationship
     to observed solar phenomena.

Helioseismology investigations from Soho should allow derivation,
by inversion techniques, of structural and dynamical parameters
throughout most of the sun's interior by covering a wide range of solar
oscillation modes. Soho helioseismology investigations emphasise those
observations that are not feasible from ground observatories owing to
the earth's atmospheric and rotation effects. For example, the core of
the sun can be investigated by studying the low-order p- and g- modes
whose periods and very small amplitudes can be derived with extreme
precision from the analysis of very long series of uninterrupted
measurements of large-scale oscillations of solar-surface velocity
fields or variations in luminosity. Similarly, the outermost convection
zone can be analysed in detail through observations of the
intermediate- and high-degree p-modes of oscillation, whose resolution
from the ground is severely compromised by the effects of seeing
through the Earth's atmosphere.

The Soho orbit should provide continuous visibility of the sun
with no attenuation, it will be ideal for a continuation and
improvement of the investigation of solar-irradiance (solar-constant)
variations. Such variations are a major source of information on which
to base an understanding of the sun's interior ; they may help to
explain the blocking of energy by sunspots in the convection zone, and
may have an important bearing on terrestrial climatic research.

Soho will be launched in October 1994 and will be located in a
halo orbit around the L1 Lagrangian point at about 1.5 million km
sunward from the earth. The spacecraft will therefore be constantly in
the solar wind well outside the earth's magnetosphere. The design
lifetime is two years, with on-board consumables sufficient for mission
extension by a further four years.

Besides helioseismology, Soho will provide the necessary plasma
diagnostics to permit modelling of plasma heating, solar wind
acceleration, the transport of mass, momentum and energy from the
photosphere up through the corona, and the mechanisms that produce
differences in chemical composition in the source region of the solar
wind. Soho will investigate such phenomena by means of spectroscopic
measurements in the EUV region of the electromagnetic spectrum with
adequate spectroscopic, spatial and temporal resolution.

Because the solar wind is essentially collisionless over most of
its journey from the sun, it retains important information about its
source regions in the corona, despite the evolution of the plasma flow,
even out to 1 AU. Moreover, because of its expected capabilities in the

field of improved time resolution in measuring plasma parameters, Soho is intended to be an important step forward in the understanding of the small-scale plasma physics phenomena in the solar wind.

The model payload used for the definition of the Soho spacecraft is an optimised combination of instruments dedicated to solar-corona observations, solar-wind in-situ measurements and solar-oscillation observations. Table 4 gives a brief description of the envisaged instruments.

Soho will be a three-axis stabilised spacecraft that will make it possible for instruments to be continuously pointed at the Sun. The spacecraft's is 3.7 m diameter and 3.6 m high. Its dry mass is about 1350 kg, and the available power about 750 W. The resources allocated to the payload are about 650 kg, 350 W and 40 kbps. The high-accuracy and high-stability sun-pointing design goals are 10 arc sec absolute pointing and 1 arc sec/15 minutes relative stability.

Science data will be either stored on tape on board and played back during station passes, or real-time data which will be transmitted from the spacecraft at the same time as playback data. High-rate data (160 kbps) is necessary for the Solar Oscillation Imager (SOI).

5.2. The Submillimetre Spectroscopy Mission (SMM)

This 'cornerstone' of ESA's 'Horizon 2000' long-term space-science programme will explore the 50 μm - 1 mm region of the electromagnetic spectrum. This region, apart from providing unique information on the continuum, hosts a large number of very important atomic and molecular spectral lines.

The objectives of the mission are the following :
(1) Study the physics of the interstellar medium and its fragmentation into protostellar clouds :
- determination of the physical conditions and chemical composition of dense clumps through high excitation, rotation-vibration transitions ;
- the energy balance of the various phases of the interstellar gas, including shocks (role of H_2, hydrides, fine-structure lines etc.) ;
- mapping of star-forming regions in external galaxies with much better angular resolution to investigate triggering mechanisms, search for low-mass stars and protostars in molecular clouds.
(2) Study the physics of star formation :
- the dynamics and physical conditions (through studies of high-excitation transitions) in accretion disks and bipolar flows as clues to the main problem of star formation ; the mechanism which allows the gas of a protostellar cloud to get rid of its angular momentum and to be accreted to the protostar.
(3) Cosmological studies like the detection of star-burst galaxies at large redshifts and the search for small-angular-scale ($\simeq$ 10 arc min) fluctuations of the cosmological background.
(4) Study the properties of primitive solar-system material through

Table 4 : List of potential instruments on SOHO

Experiment	Description
Solar atmosphere remote sensing	
Two EUV imaging telescopes	30.4 nm He II and 17.0 nm Fe IX lines. Spectral resolution: 2 nm. Spatial resolution: 5 arc sec.
Grazing incidence spactrometer	Wavelength range: 10 - 130 nm. Spectral resolution: 10^{-2} nm. Spatial resolution: 1.5 arc sec.
Normal incidence spectrometer	Wavelength range: 58 - 130 nm. Spectral resolution: 2×10^{-3} nm. Spatial resolution: 1.5 arc sec.
UV coronal spectrometer	102.5 nm (H I Ly-beta), 121.6 nm (H I Ly-alpha), 103.2 nm (O VI), 103.7 nm (O VI) lines. Spectral resolution: 8×10^{-3} nm. Spatial resolution: 10 -60 arc sec.
White-light coronagraph	Wavelength range: 400 - 700 nm. Spectral resolution: 20 nm. Spatial resolution: 10 - 60 arc sec.
Solar-wind in-situ measurements	
Solar wind plasma analyser	Electrons (1 eV $< E_e <$ 1.5 keV), ions (0.1 - 20 keV/q). 3D velocity distribution.
Solar wind composition analyser	Ion composition and charge state (10 - 100 keV/q).
Suprathermal particle analyser	Ion and electron fluxes (20 keV -1 MeV), directional distribution and energy spectrum.
Energetic particle analyser	Electrons and ion (1 - 50 MeV) charge state and composition, energy spectrum.
Magnetometer	DC magnetic field. Sensitivity: 0.1 nT.
Helioseismology	
High-resolution spectrometer	Doppler measurement of velocity oscillations. Integrated sunlight. Sensitivity: 1 cm s^{-1}. Time resolution: ~ 60 s.
Solar oscillation imager	Doppler measurement of velocity oscillations. Spatial resolution: 2 arc sec. Sensitivity: 3 m s^{-1} per pixel. Time resolution: ~ 60 s.
Solar-irradiance monitor	Solar irradiance, integrated sunlight. Relative precision of photometers: 10^{-6}. Time resolution: ~ 60 s. Absolute accuracy of radiometer: 10^{-3}.

molecular (especially H_2O) studies of comets.

The telescope diameter will be 8-m. The applicability of the spectral and spatial resolution of an 8-m diameter telescope equipped with heterodyne receivers to different classes of astrophysical objects is demonstrated in Figures 5 and 6.

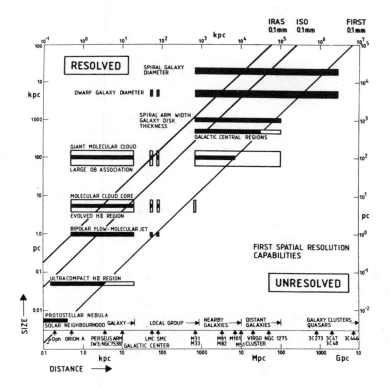

Figure 5. *Typical velocities associated with astrophysical phenomena are plotted versus spectral resolution. The diagonal line separates resolved and unresolved spectral lines.*

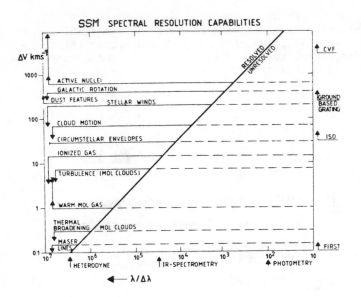

Figure 6. Astronomical objects that can be studied by SSM, as a function of distance and size. The diagonal lines indicate typical SSM resolutions in the 0.1 to 1.0 mm wavelength range. At the bottom of the diagram, typical representatives of each class of object are given.

The performance characteristics required of the mission are summarised in Table 5.

Unlike IRAS and ISO, the SSM will be based on a non-cryogenic, large, passively cooled telescope equipped with high-spectral-resolution ($R = 10^6$) heterodyne receivers. Not only will it fill the wavelength gap, but it will also provide increased spatial and spectral resolution and an improved sensitivity to point sources.

The payload consists of coherent and non-coherent high-resolution spectrometers. It also includes a photometer.

The baseline coherent spectrometer consists of heterodyne receivers (operating between 0.1 mm and 1 mm), using superconducting mixers to as high a frequency as possible.

The non-coherent spectrometer operates between 50 and 300 µm. The baseline spectrometer is designed for a resolution of 5×10^3 and consists of a Fabry-Perot étalon, a grating order sorter and an array of detectors.

**Table 5 : Submillimetre spectroscopy mission
Summary of performance characteristics**

Antenna

Size	8 m, *f*/10
Surface quality	8 - 10 μm
Temperature quality	$\dfrac{d^2T}{dt^2} \leq 10^{-6}$ K s^{-2}
Chopper throw	10 arc min
Field of view	3 arc min (goal: 10 arc min)
Diffraction limit	down to 150 μm
Pointing accuracy	1 arc sec
Pointing stability	0.5 arc sec

Coherent spectroscopy

Wavelength	down to 157 μm (C II line)
Resolution	0.3 km s^{-1} ($\nu/\Delta\nu = 10^6$)
Instantaneous frequency coverage	1%
Sensitivity	0.5 $\cdot \nu_{GHz}$ K

Noncoherent spectroscopy and photometry

Line spectroscopy and imaging	50 < λ <350 μm
Sensitivity	1 mJy (wideband)
Line imaging	30 $\times$ 30 pixels, resolution up to 10^4
Photometry	> 350 μm
Sensitivity	0.2 MJy sr^{-1} at 1 mm
Beam throw	10 arc min

The photometer is designed to provide high-resolution (3.5 arc sec at 100 µm) imaging at short wavelengths (50 - 350 µm) and high-throughput photometry at wavelengths longward of 350 µm. The detector consists of a combination of photoconductors and bolometers. The Ge:Ga bolometers are designed to operate at a temperature of 300 mK, which is provided by small closed-cycle liquid-^{3}He refrigerators.

The mission has been studied in two configurations : one in an elliptical 12-hour orbit and one in a 500 km altitude 28° inclination orbit. In that option, it might be possible to use the future space station for servicing the mission. Figure 7 shows the telescope in orbit. The launch is foreseen at the turn of the century.

Figure 7. *The Submillimetre Spectroscopy Mission (SSM) in orbit. Note the 8 m diameter Telescope, the large solar arrays and the inflatable baffle, deployed and polymerised in orbit under solar UV radiation.*

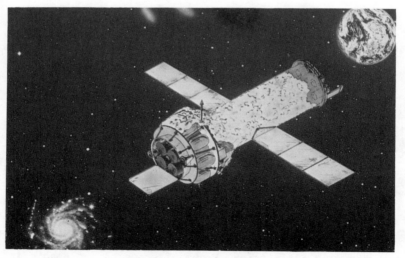

Figure 8. *The High-Throughput X-ray spectroscopy mission (XMM) in orbit. Note the four nested telescopes, each made of 58 nested individual shells.*

5.3. The High-Throughput X-ray spectroscopy mission

The High-Throughput X-ray spectroscopy 'cornerstone' is the second one in the implementation of Horizon 2000 (Figure 8). The scientific objectives require a powerful imaging instrument with the largest possible collecting area for high-quality spectral measurements on faint sources, i.e. down to a level of 10^{-15}ergs cm^{-2}s^{-1}, and fast, low- and medium-resolution spectroscopy on the brighter objects.

The mission will have a major impact on several topics like :
- The determination of the large- and medium-scale structure of (hot) matter in the universe ;
- The diffuse X-ray background and active galactic nuclei ;
- The study of inner accretion disk and the surface of compact objects.

However, one major part of the observing programme will be devoted to the study of stellar coronae. X-rays from stellar coronae originate from optically thin thermal plasmas in collisional equilibrium. Since the relevant temperatures are in the range from 10^6 to several times 10^7K, the spectrum is rich in emission lines produced by the most abundant elements. If suitably resolved, these lines are powerful diagnostics of temperature, density, ionisation equilibrium, mass motions and element abundances. The instrument flown on EINSTEIN and EXOSAT had in general insufficient spectral resolution to take full advantage of the wealth of information contained in X-ray spectral lines. Even in the best cases, only a crude estimate of the relevant parameters could be obtained.

With the improved resolution of the instruments planned for the mission, it will be possible to fully resolve the temperature structure of stellar coronae, and to determine the amount of coronal plasma present at each temperature.

The early observations from the Einstein observatory showed that there was a dichotomy between early type stars and late type stars with regards to the dependence of coronal emission on basic stellar parameters. In order to test the predictions of different models, it is now necessary to acquire spectra with a resolution and a signal over noise ratio substantially better than those obtained in all previous missions. Both the reflecting grating spectrometer and the CCD detectors on the mission will be able to perform these observations for a large number of sources.

Another important diagnostic of stellar coronal emission is time variability. Observations of time variations are crucial for understanding the instabilities occurring in the radiatively-driven winds of early-type stars, as well as for investigating the stochastic nature of magnetic activity in heating, especially if, as recently suggested, 'quiescent' coronal emission results from the superposition of a large number of small-amplitude flare-like events. Observations of time variations are also crucial to resolve the spatial structure of eclipsing binary systems or to infer the inhomogeneous distributions of surface activity on single, rapidly rotating stars. EXOSAT has already demonstrated the importance of time variability as a diagnostic tool of stellar coronal emission. However, the small collecting area of EXOSAT greatly limited the number of coronal sources that could be studied with sufficient counting statistics. The X-ray cornerstone with a grasp more than two orders of magnitude higher, will allow the observation of objects located at larger distances, thus increasing appreciably the number of interesting objects that it will be possible to study in this way.

As shown by EXOSAT an essential requirement for this technique to be successful is continuous monitoring for periods of at least one orbital cycle and, preferably longer. The highly eccentric orbit of the X-ray cornerstone will make this type of observation possible. This is an area in which complementarity with the US AXAF mission is particularly evident.

The prime design drivers for the satellite are the following :
- an energy range of 0.2 - 10 keV ;
- an effective area of 10 000 and 5 000 cm^2 at 8 and 2 keV respectively ;
- an angular resolution of around 30" ;
- broad-band spectrophotometry ($\lambda /\Delta\lambda \simeq 10 - 50, \sim 2A$) ;
- medium-resolution spectroscopy ($\lambda /\Delta\lambda \simeq 250$) ;
- high-resolution spectroscopy ($\lambda /\Delta\lambda \simeq 1000$) limited to specific line complexes, e.g. oxygen and iron.

The major element of the mission is a set of grazing-incidence X-ray optics which will provide the required grasp (cm^2) and angular resolution. The principal characteristics of these optics are summarised in Table 6.

Table 6 : The X-ray optics

Telescope design	Nested Wolter I
Number of telescopes	4
Number of shells/telescope	58
Maximum shell thickness	1.40 mm
Minimum shell thickenss	0.64 mm
Focal length	8 m
Mirror outer diameter	70 cm
Mirror inner diameter	32 cm
Mirror length	60 cm
Telescope material	'Carbon fibre'
X-ray mirror coating	Gold
Mirror surface area/telescope module	52.3 m^2
Field of view	30 arc minutes
Spatial resolution	30 arc seconds (HEW)

Figure 9 shows a schematic of a single-mirror module. The effective area of the cluster of four mirror modules is shown in Figure 10 as a function of X-ray photon energy. For comparison, the effective areas of other missions past and future are also indicated.

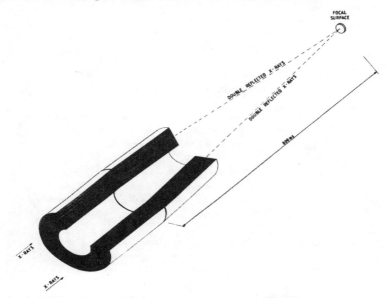

Figure 9. *A schematic of a single XMM telescope module.*

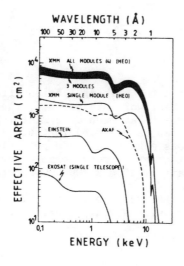

Figure 10. *The effective area of the XMM telescope array as a function of photon energy. The three modules correspond to the case where 50 % of two telescopes' beams are diverted by two reflection gratings.*

Table 7 : The Model Payload*

Objective	Instruments
Broad-band imaging spectrophotometry	Prime instrument: a CCD camera on each telescope with a field of view of 20′ × 15′. The camera has resolving powers of 7 and 40 at 1 and 6 keV respectively.
	Secondary instrument: an imaging GSPC with a field of view of 30′ × 30′. The camera has resolving powers of 5 and 13 at 1 and 6 keV respectively.
Medium-resolution spectroscopy	Reflection grating module fitted to 2 telescopes. The readout element of this grating is a strip CCD array. The gratings cover the waveband 4 - 25 Å with a resolving power of 200 - 400.
High-resolution spectroscopy	Bragg crystal spectrometers fitted to 2 telescopes provide a high-resolution spectroscopic capability in two limited wavebands (O VII, 21 Å and Fe XXV, 1.8 Å) with a resolving power of over 1000.

*In addition to the basic X-ray instrumentation an optical monitor (2000 - 6000 Å) provides the dual rôle of a bore-sight tracker and a high-sensitivity spectrophotometric instrument. The limiting magnitude is ~ 24. Alternative methods of providing medium-/high-resolution spectroscopy are also under investigation.

A model payload which will achieve the principal scientific objectives has been defined. On the basis of this payload, the initial outline design of the spacecraft has been determined. A summary of the instruments in the model payload is provided in Table 7.

As shown by EXOSAT, the continuous monitoring made possible by the use of a high excentricity orbit offers a considerable advantage. This is the reason why a 60° inclination, 1000 x 100 000 km orbit has been selected for the mission.

It is intended to launch XMM in 1997/98. The mission will operate as a general astronomical observatory with a goal lifetime of 10 years.

6. CONCLUSION

We have shown the great potentiality of ESA's science programme for the study of stellar atmospheres in nearly all stages of stellar evolution. The missions we have described above, cover indeed many more astrophysical objectives, mainly in the area of cosmology and extragalactic astronomy. Most of these missions are unique and represent major steps in the development of space astronomy. Others are complementary to other missions planned or under development in the American, Soviet or Japanese programmes. This is the case for Lyman and the Hubble Space Telescope, Quasat, the Soviet Radioastron and the Japanese VSOP missions, the submillimetre spectroscopy mission and the US Large-Deployable Reflector or the smaller AELITA Soviet telescope, the X-ray cornerstone and the Soviet spectrum-X, US AXAF and Japanese ASTRO-D missions.

This set of future observatories will provide extremely powerful tools for the study of stellar atmospheres and of their evolution, opening a new era in the development of this discipline.

7. REFERENCES

For more information, the reader may consult :
- Report on the Scientific satellites of the European Space Agency, ESA SP-1090, May 1987 ;
- R.M. Bonnet, 1987, 'Future Prospects of Stellar and Solar Physics from space', proceedings of the EPS Solar physics section, symposium on 'Solar and Stellar Activity', Titisee, FRG, 27th-30th April 1987, E.H. Schröter Editor ;
- 'Future missions in Solar, Heliospheric and Space plasma physics', ESA SP-235 ;
- 'An ESA Workshop on a Cosmic X-ray Spectroscopy mission', ESA SP-239 ;
- ISO Info, Newsletter on the Infrared Space Observatory n° 1 and 2 March 1986, June 1987, M. Kessler, ESTEC, Ed.

NASA MISSIONS TO STUDY STAR FORMATION AND EVOLUTION

David. R. Soderblom
Space Telescope Science Institute
3700 San Martin Drive
Baltimore, MD 21218 U.S.A.

ABSTRACT. This paper briefly describes several planned NASA missions with important capabilities for studying star formation problems. These missions have the ability to observe in the infrared or near-infrared, the far ultraviolet, and the X-Ray portions of the spectrum. A number of other NASA missions, (for example, *IHST* and *EUVE*) will surely influence future research in this area, but cannot be included in this short review.

1. THE ADVANCED X-RAY ASTROPHYSICS FACILITY (AXAF)

AXAF and SIRTF will continue the series of NASA's Great Observatories begun by the Hubble Space Telescope. *AXAF* will be a space-based observatory designed for high sensitivity to X-rays with a large area, high resolution grazing incidence telescope, accompanied by a complement of sensitive imaging and spectroscopic instruments. The majority of observing time will be set aside for general observers to be selected by peer review.

Our knowledge of the low luminosity galactic X-ray sources has changed drastically over the past few years, particularly so since the flight of the *Einstein* Observatory. *AXAF* will be capable of measuring the X-ray emission of stars throughout the HR diagram, including the pre-main sequence stars. It is believed that stellar X-ray emission very likely reflects the evolutionary state of the star, so that highly sensitive X-ray observations join the more traditional tools of stellar astronomy in constraining theories of stellar structure and evolution. Because of its 100-fold increased sensitivity to X-rays over the *Einstein* mission, *AXAF* should be able to detect vast numbers of X-ray sources - since the available volume for sampling increases roughly by a factor of 1000. In addition, the quality of low-resolution spectral information should be substantially improved. *AXAF* therefore provides the only means for generating an X-ray HR diagram.

Certain subclasses of stars (such as T Tauri stars, RS CVn, and W UMa binaries) have for some time been recognized as classes of low luminosity X-ray sources. In several of these cases, substantial studies covering a variety of spectral regimes and involving detailed modeling have been carried out in the past; the principal limitation upon the X-ray diagnostic observations have been the relatively low sensitivity of spectrally resolved observations with resolution better than that obtainable with gas proportional counters. The sensitivity of *AXAF* for high resolution spectroscopy will be at

433

A. K. Dupree and M. T. V. T. Lago (eds.), Formation and Evolution of Low Mass Stars, 433–445.
© 1988 by Kluwer Academic Publishers.

Fig. 1: Image of the *Advanced X-ray Astrophysics Facility* (AXAF) in orbit.

least 1000 times that of *Einstein*, thus substantially enlarging the number of stars accessible for detailed studies.

Several attributes of stars, in addition to the extent of surface convective activity must be important in determining the level and character of stellar X-ray emission. The principal additional processes are: stellar rotation, stellar convection, and dynamo processes leading to stellar surface magnetic fields. X-ray data may provide a unique tool for studying these stellar attributes and their variation as stars evolve. In addition, another major scientific goal includes the determination of the X-ray luminosity function for stars of distinct spectral types and luminosity classes. This will allow study of the variation of coronal emission with changing convective zone structure, and the extent of solar-like activity cycles. Correlations can be constructed of the level of coronal emission with other stellar parameters such as stellar age, rotation period, and surface magnetic fields. The contribution of stars (and here late-type dwarfs are important) to diffuse emission in star clusters, our galaxy, and other galaxies and galaxy clusters can be evaluated.

The detailed high sensitivity spectral data can be used, in conjunction with optical and ultraviolet spectroscopy, to elucidate the detailed structure of extended stellar atmospheres. Variability of cool stars in X-rays on both short and long time scales will allow the placement of limits upon the spatial inhomogeneity of coronal emission as a function of stellar type.

2. THE SPACE INFRARED TELESCOPE FACILITY (SIRTF)

SIRTF is to be a one-meter class telescope with a cryogen lifetime of five years or more. At these low temperatures, SIRTF's background is limited by astrophysical effects such as zodiacal light, so that its sensitivity can be 1000 or more times that of a large ground-based telescope.

SIRTF will address a variety of astrophysical questions, including many problems in the study of young stars, protoplanetary disks, and related phenomena. The attributes of SIRTF and its three focal plane instruments are now being defined by university-based teams. Those qualities that make it particularly well-suited to the study of problems in this area include:

> Ultra-high sensitivity to both point sources and extended emission over the wavelength band from 2 to 700 microns.
> The use of large format arrays - up to 128 x 128 pixels covering a 5 x 5 arcmin field of view - which will provide efficient imaging of star-forming regions.
> High angular resolution (1 arcsec for wavelengths below 5 microns) and a very stable modulation transfer function which will permit maximum use of image processing techniques to distinguish faint companions and debris clouds near stars.
> Polarimetric capability at selected wavelengths.
> Spectroscopic capability from 2 to 200 microns, in both low (R=100) and moderate (R=2000) resolving power modes, using two-dimensional arrays to provide spectroscopic imaging.
> Long on-orbit lifetime (5 years or more) and a vigorous general investigator program which will permit systematic study of young stellar objects.

Fig. 2: Artist's rendition of the *Space Infrared Telescope Facility* (SIRTF) in orbit.

2.1. The Earliest Stages of Star Formation.

SIRTF's sensitivity to diffuse emission at sub-millimeter wavelengths between 100 and 700 micron will allow a study of density fluctuations in dense clouds, which are the earliest manifestations of the collapse processes which culminate in the formation of stars. For a cloud at a typical temperature of 20 K, SIRTF can easily detect dust column density fluctuations corresponding to a visual absorption of less than 0.1 magnitude or a 1% fluctuation in a typical cloud with ten magnitudes of extinction. SIRTF can provide such data with a resolution of a few arcminutes over a region degrees in extent. The assembled data for an entire cloud would reveal the distribution of fragment size, which depends on the mechanism of fragmentation in the cloud and determines the mass distribution of the forming stellar populations. These studies of fragmentation can be combined with polarimetry of both the submillimeter emission and of the infrared light of background stars (down to 15th magnitude at 5microns, for example). The polarimetry will delineate the magnetic field direction in even the densest fragments and allow the study of the influence of the magnetic field on the collapse process.

2.2. The Protostellar Phases.

SIRTF can detect objects with luminosities as low as one-tenth that of the Sun throughout the Galaxy and correspondingly fainter objects in the nearest star-forming complexes. These limits are well below the luminosities of even the lowest mass stars. Thus SIRTF's sensitivity and broad wavelength coverage will permit determination of the initial mass function of forming stars through a series of "snapshots" of such star-forming complexes. These observations will identify the principal luminosity sources in a cloud and determine the total luminosity of the entire forming cluster. They will provide a basis for detailed modeling of the cluster which should allow the luminosities of the individual objects to be determined, even in the presence of complicating radiation diffusion effects. Different star formation models predict different forms for the luminosity function, so that determination of the luminosity functions of populations of protostars and pre-main sequence stars in molecular clouds could lead to an understanding of the specific mechanisms which produce stars from clouds.

These photometric observations will point the way for spectroscopic studies which will probe in detail the stellar birth phenomena from the earliest inflow and collapse stages to the mass-loss, outflow, and shock phenomena which are now known to characterize the later stages of protostellar evolution. The best probes of density and temperature in these regions lie in the mid- and far-infrared, where the extinction is small and the atomic and molecular excitation energies are a good match to the thermal energies. Particularly useful diagnostic lines which SIRTF can study are: 1. The far-infrared rotational lines of CO, the relative strengths of which reveal density and temperature; 2. The pure rotational lines of molecular hydrogen in the mid-infrared, which will be preferentially excited in shocks; 3. The far-infrared rotational lines of water vapor, inaccessible even from aircraft altitudes, which will be important in the cooling of post-shock gas; 4. The 63 micron line of forbidden neutral oxygen, which probes the stellar wind and yields information on the mass flux in a protostellar outflow. As an example, SIRTF's spectrograph can readily measure objects in nearby dark clouds in the [OI] 63 micron and molecular hydrogen 17 micron lines to a flux limit sufficient to detect mass-loss rates in the range of $10-7$ to $10-8$ solar masses per year, well below the current limits. These observations will provide an understanding of the role of outflows in mediating or inhibiting the collapse of clouds and allow a study of the chemical transformations thought to occur in shocked interstellar gas.

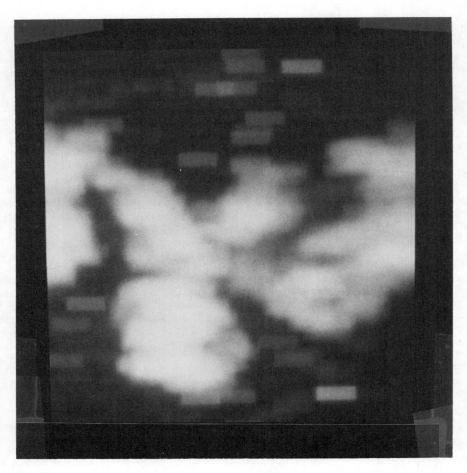

Fig. 3: IRAS image of infrared galaxies and the infrared "cirrus." The field of view is 42.7 minutes on a side.

2.3. Protoplanetary and Circumstellar Material

Recent work has highlighted the prevalence of circumstellar disks in young stellar objects, and SIRTF's broad spectral coverage will permit searches for the spectral signatures of such disks. Detailed spectroscopic and imaging studies of selected objects in the near infrared will reveal the presence of specific chemical constituents through such features as the ice band at 3.1micron and the 3.4 micron "organic" emission bands recently identified in cometary dust.

Of particular importance for SIRTF will be the study of clouds of planetary debris such as those discovered by IRAS around Vega and some nearby solar-type stars. The shells around Vega, Fomalhaut, and Beta Pictoris have diameters around 20

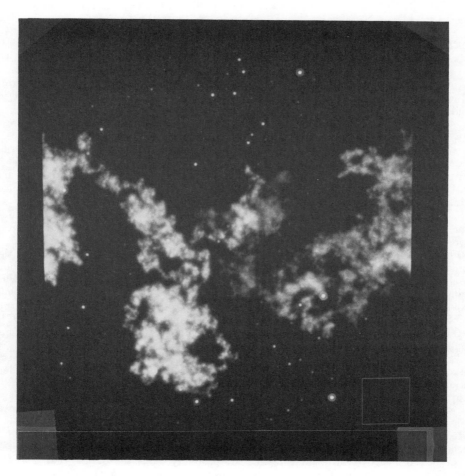

Fig. 4: Calculated SIRTF image at 60 μ diffraction limit of the same field as in Figure 3. (Courtesy of E. Wright - UCLA).

arcsec at 60 microns. For these and other bright, nearby objects, SIRTF can image the shells directly and search for the presence or absence of inner disks of warmer material and for rings or voids which would suggest resonance effects attributable to planetary companions. Spectroscopic studies of these systems may provide invaluable information related to the chemical conditions in the early solar nebula. On a broader scale, SIRTF can detect Vega-like shells around stars as distant as several kpc and test directly the exciting indication, based on the results from IRAS, that planetary debris disks are associated with more than 15% of the low-mass stars in the solar neighborhood.

3. THE STRATOSPHERIC OBSERVATORY FOR INFRARED ASTRONOMY (SOFIA)

SOFIA is a proposed 3-meter class, airborne observatory, anticipated as a joint project of NASA and the West German Science Ministry. SOFIA would fly in a modified 747 and replace the Kuiper Airborne Observatory, which is a 91 cm telescope in a C-141.

SOFIA would operate above 41,000 feet and provide access to wavelengths from 0.3 microns to 1.6 mm, as well as convenient changes of focal-plane instruments. It would be diffraction-limited beyond about 30 microns, and seeing-limited (2 to 3 arcsec) at visible and near-infrared wavelengths. Pointing stability should be 0.2 arcsec rms, as has been achieved with the KAO.

SOFIA'S lifetime should be twenty years or more, and it can start operating in the early 1990's. SOFIA uniquely complements SIRTF and ISO because of its high spatial and spectral resolution, and its wavelength range complements HST, GRO, and AXAF. Thus SOFIA is proposed as supporting the orbiting observatories, and to provide an important step leading to FIRST and LDR.

SOFIA can also take advantage of advances in infrared technology over its lifetime. Focal plane instruments for SOFIA will be provided largely by individual investigators, and will include photometers, array cameras, and polarimeters for the 0.3 to 350 micron range, as well as spectrometers with resolving powers up to 300,000 for the 2 micron to 1.6 mm range. At a rate of 120 flights per year, SOFIA can support about 15 instrument teams and 25 guest investigator groups, selected by peer review, as with the KAO. This vigorous observing program with state-of-the-art instrumentation will provide images and spectra of an enormous variety of objects. For example, all of the point sources that IRAS detected at 60 and 100 microns will be detectable with SOFIA. Because of its far-infrared sensitivity, SOFIA will be particularly useful for star formation studies.

Star formation, as now understood, begins with the contraction of a cool, dusty, gaseous region of interstellar space. As this contraction accelerates, a compact condensation forms that is surrounded by an orbiting disk. Because of the large optical depths of the dust, most of these objects' luminosity is detectable only in the far infrared. Some isolated examples of these sources were identified from KAO observations of globules and other clouds, prior to the flight of IRAS, which apparently detected a cluster of such young stellar sources in the Taurus Cloud. Ground-based observations at radio and near-infrared wavelengths indicate that individual protostellar objects in nearby star forming regions are probably separated by less than 2 to 10 arcsec. Such separations cannot be resolved in the far infrared by the KAO, or IRAS, or ISO, or SIRTF. But SOFIA's large aperture enables just such resolution.

As an example, SOFIA should be able to detect for the first time the infall of material onto a protoplanetary disk that surrounds a central condensed star. Such disks are of order 1 to 100 AU in diameter. As gases strike the disk, a shock is produced that heats and collisionally excites the atoms and molecules of the gas, causing them to radiate. These shocked regions are so deep within the disk that the only radiation that can escape unhindered is at wavelengths longer than about 100 microns. Fortunately, there are important diagnostic spectral features at just such wavelengths (e.g., carbon monoxide, atomic oxygen, and water vapor). A shocked protoplanetary disk in the nearby Taurus Cloud is expected to produce a weak signal from individual emission lines, but one that could be detected by a spectrometer on SOFIA. Velocity resolution of these lines would provide unique information on the dynamics of the disk and the characteristics of the shock.

SOFIA's excellent spatial resolution may allow far-infrared photometry that can discriminate between competing models of disk formation in young stellar objects.

Some current theories predict continuum flux distributions which differ substantially in the 100 to 300 micron range, depending on the details of the disk physics. Since confusion between nearby objects can be a serious problem for this type of observation, SOFIA can play a crucial role in establishing the character of this emission and its underlying significance.

High spatial resolution mapping with SOFIA can help determine the more global characteristics of star formation. Observations of the far-infrared continuum and lines emitted by the nuclei of obscured external galaxies can provide unique evidence on the distribution of stellar types, the morphology of the ionized gas, and its relation to the distribution of the dust and luminous stars. For example, velocity structure in [OIII] lines observed with the KAO suggest strong variations in the emissions from different components of the source. SOFIA in this case could readily isolate what KAO can't. Many interesting questions--such as whether a massive burst of star formation or an obscured central object is the ultimate luminosity source in an infrared galaxy--will be addressed by SOFIA and its observers.

4. ADVANCED INSTRUMENTS FOR THE HUBBLE SPACE TELESCOPE

The Hubble Space Telescope (HST) is NASA's first satellite to incorporate in-orbit repair and upgrades as part of the overall mission plan. For scientists, the key part of this program is the provision for Orbital Replacement Instruments (ORIs). These new instruments will be able to use new technology and design to keep HST functioning well for its full 15 years. Three ORIs have been selected for the design phase.

4.1 Space Telescope Imaging Spectrograph (*STIS*)

The Principal Investigator of STIS is Bruce Woodgate of the Goddard Space Flight Center. STIS improves substantially on the launch instruments because of its capabilities for ultraviolet, visible, and near-infrared (to 11,000 Å) spectroscopy, and for long-slit imaging. Thus the STIS instrument is well-matched for studying a wide variety of stellar environments.

Jets and winds in pre-main sequence objects will be particularly attractive targets for STIS. Systematic, often bipolar, highly collimated outflows of material appear to be common in pre-main sequence objects, and flows may dominate the dynamics of the surrounding cloud, thereby regulating further star formation. Detailed studies of the flows offer unique insights into the processes that shape the formation of stars and, quite probably the formation of proto-planetary disks. The jets associated with young stellar objects are nearby; they can be spatially resolved with the STIS instrument to achieve spatially resolved spectroscopy that will take advantage of the limiting HST angular resolution of about 0.06 arc sec (FWHM). Spectral data obtained with the resolution of HST may even allow measurements of the temperature stratification of interstellar shocks to test rigorously shock models that are non-planar and that are time truncated. This latter case will be important in distinguishing between excitation by a continuous stellar wind and by eruptive events that are known to be associated with T Tauri stars.

STIS can critically study the collimating mechanism itself. A fundamental question is whether beaming occurs at the stellar surface or is imposed on an isotropic flow at some distance from the star. Ground-based ($\sim$2 arcsec) observations indicate that the jets near young stellar objects have typical opening angles of five to ten degrees. At typical distances, this angular size corresponds to collimation at 0.001 pc, and is maintained over a distance of 0.01 pc. The STIS should offer a factor of ten higher spatial resolution combined with the capability of measuring the velocity field within the jet. In some cases speckle interferometry has succeeded in measuring disk-like structures several hundred AU in extent. The dimensions and masses are plausible for nebulae which may form planets. With the HST resolution, it should be possible to examine directly and on comparable spatial scales the relationship between mass outflows and these disks.

A common theme addressed by the STIS instrument is the study of winds and the shocked interstellar medium around pre-main sequence stars. These scientific goals impose the instrumental requirement for simultaneous spectral and spatial resolution at the HST limit in either the long slit or nebular spectrography (wide slit) mode. High excitation spectra expected from these objects are only accessible in the ultraviolet spectral region. Because the first generation instruments on HST do not have wide slit capability, STIS will be a welcome and significant addition to the suite of focal plane instruments.

4.2 Near-Infrared ORIs

Two near-IR ORIs were chosen for the design phase: HIMS (the Hubble Imaging Michelson Spectrometer), whose P.I. is D.N.B. Hall of the University of Hawaii, and NICMOS (the Near-Infrared Camera and Multi-Object Spectrometer), whose P.I. is R.I. Thompson of Steward Observatory.

Only one of these two instruments will be built and deployed. The choice will be made during 1988, and the instrumental designs are not yet final. It is therefore appropriate to avoid discussing the particular characteristics of these instruments, and instead concentrate on their science goals.

Both near-IR ORIs intend to address major problems in the study of star formation and star-forming regions. It is important to realize that these instruments will provide the first-ever access to this wavelength regime that is free of the earth's atmosphere. Their sensitivity starts at about 8000 Angstroms, where the visible instruments drop off, and continues to about 2.5 microns. For some kinds of observations, even longer wavelengths may be profitably observed, but, in general, above about 2.5 microns the thermal background of the telescope begins to dominate, and there is little advantage over ground-based telescopes unless the whole telescope is cooled.

In the study of star formation, these new instruments can use good spectral resolution, together with HST's high spatial resolution to address these areas:

> The dynamics of disks around low-mass stars.
> Infall and outflow in star-forming regions.
> The mineralogy of primeval "solar nebulae" around low-mass stars.
> Star formation in external galaxies of different types.
> Formation and evolution of double-star systems.

For many of these studies, the combination of HST's relatively large aperture with the relatively short wavelengths of the near-IR mean higher absolute spatial resolution than is possible with the far-IR instruments discussed above. Moreover, these ORIs are likely to become available to users before SIRTF or SOFIA are operational.

As an example, when used at the highest (diffraction-limited) resolution, the near-IR ORIs can resolve an angle corresponding to about 15 AU at the distance of the Taurus-Auriga or Ophiuchus Clouds. Polarimetric and/or coronagraphic modes can be used to help reject scattered light from the central star. Disks appear to have typical dimensions of 100 to 200 AU, and so will be well resolved. The variation of disk brightness with radius, for instance, will help determine the density distribution in the disk.

Both ORIs should have long lifetimes: five years or more. This too provides an important capability when coupled to HST's high resolution. At outflow velocities of 50 to 100 km/s (15 AU/yr), features can move by a pixel per year. Thus after about five years the detailed appearance of the disks should change markedly. This will help us develop a three-dimensional picture of how star-forming regions change with time.

Disk dynamics can be probed through studies of spectral line emission. Paschen-alpha is ten times brighter than Brackett-gamma, but is blocked by the earth's atmosphere. With a near-IR ORI, there is the potential of a high spatial/spectral resolution probe of ionized material that is almost 100 times more sensitive than the VLA. This can enable the study of ionized material in outflows to get crucial velocity information, and can help solve the critical problem of angular momentum loss in very young stars. The interaction of the disk or outflow with the surrounding molecular cloud can be studied by observing molecular hydrogen directly, instead of a proxy.

The mineralogy of the primitive solar nebula is a potentially fascinating field of study. In the temperature range of 100 to 200 K, large-scale chemical differentiation and freeze-out of ices occurs. A solar-type protostar passes through a maximum luminosity of about 50 times the Sun, so that the radius of freeze-out is 15 to 60 AU for large grains, and even more for freeze-out onto small grains. This means that variations in mineral properties can be resolved by the ORIs in Taurus, Ophiuchus, etc.

Going beyond the Galaxy, the unparalleled spatial resolution of HST will enable the study of star formation in other galaxies as well, from nearby normal galaxies to more distant Seyferts and starburst galaxies. For example, the relations between starbursts and activity in the nuclei of galaxies is poorly understood; it's not even clear if Seyferts and starbursts galaxies are different stages of the same series of events or whether they are unrelated. Galactic nuclei and their surroundings can be probed with the near-IR ORIs to address these questions. These instruments can also study the gentler forms of star formation seen in the spiral arms of most galaxies. Several mechanisms have been suggested as triggers for star formation in our own Galaxy. These can be distinguished by looking at surface brightness distributions relative to star formation sites in nearby galaxies.

Two other problems related to star formation can be addressed. One is the question of the formation and evolution of multiple star systems. Although we observe most field stars to be part of multiple systems, few such systems have been found in star forming regions. This may be largely because of the distance of those regions, a difficulty that HST's high spatial resolution can overcome. The other problem relates to the stellar mass function for very low masses. By surveying random portions of the sky to faint levels, a near-IR ORI should find very-low-mass stars and brown dwarfs, if significant numbers of them exist. Having a firm grasp on the mass function will be crucial for understanding star formation.

5. *LYMAN-FUSE* - THE FAR ULTRAVIOLET SPECTROSCOPIC EXPLORER

This mission concept was accepted by NASA in 1988 for a Phase A study. Professor H. Warren Moos of The Johns Hopkins University is the Principal Investigator of the *LYMAN-FUSE* study. This mission, if implemented will be unique in its ability to address fundamental problems of low mass stars. It will be operated in the Guest Observer mode, much like the successful *International Ultraviolet Explorer*.

As proposed, the prime wavelength interval to be covered by *LYMAN-FUSE* will be the 912 to 1200Å region; the option of coverage between 100 to 900Å will be studied with particular emphasis on the 100 to 350Å interval. Pre-Phase A studies by NASA and ESA scientists have led to an innovative baseline design exploiting proven technology. The scientific goals of *LYMAN-FUSE* require a spectral resolving power of 30,000, wide simultaneous wavelength coverage, and high throughput across the 900-1200Å region. High throughput is best achieved with a glancing incidence telescope (with $\sim 10°$ graze angles) of 80-cm aperture and a Rowland circle spectrograph with an aspheric toroidal grating.

$\quad$ *LYMAN-FUSE* can contribute in many areas of study relating to low mass stars and star formation. The ultraviolet bands of the outer regions of star-forming molecular clouds will provide detailed knowledge of H_2, HD, and CO abundances, temperatures, and velocities, thus complementing the optical and microwave studies of rare molecules. Shocks in the interstellar medium, as evidenced by the presence of Herbig-Haro objects, can reveal their dynamics through ultraviolet spectroscopy. HH objects are faint (15-18 mag, visual), they emit 90 percent of their flux in the ultraviolet, and some have anomalously low ultraviolet extinction. The heretofore unobserved spectrum below 1200Å contains key diagnostic lines (C III, O VI, and H_2, and HD lines). Furthermore, the far-UV continuum in these objects can be used to measure properties of shocked H (2-photon continuum) and shocked processed grain extinction in these star forming regions.

$\quad$ Accretion disks are central to our current ideas of star formation, and indeed they are pervasive throughout astrophysics. They characteristically have power-law spectra rising to the shortest presently observed ultraviolet wavelengths (1200 Å) but must reach a maximum flux at shorter wavelengths, determined principally by the temperature in their innermost regions. *LYMAN-FUSE* will be able to measure the velocity structure through direct measurement of the Doppler shifts of emission and absorption lines giving further clues into the location of the emitting gas. The essence of the theoretical problem is the transfer of angular momentum. While the effects of transfer of angular momentum through a disk are parameterized in the disk models well enough for general predictions of the emission spectra, the physics of the turbulent or magnetic viscosity which drives the angular momentum transfer is not understood. *LYMAN-FUSE* will observe the time variability of emission from the inner regions of the accretion disk enabling the time scales of the variations to be related to the physical characteristics of the disk.

$\quad$ The high throughput and spectral resolution of *LYMAN-FUSE* allows one to address the structure and energy balance of the hot plasma in the outer atmospheres of cool stars such as T Tauri objects and the low mass part of the main sequence. The ions with strong resonance transitions in the *LYMAN-FUSE* spectral range are formed in gas with temperatures ranging from 10^4 K to over 10^7 K. Since each atom or ion in a collisionally dominated plasma is formed over a narrow range of temperature, the detection of a particular species signals the presence of plasma at a specific temperature. The broad spectral coverage of *LYMAN-FUSE* gives a continuous probe of emission measure as a function of temperature in stellar atmospheres. Thus detailed physical models can be constructed of the heterogeneous structures that appear ubiquitous in stellar atmospheres. The spectral resolution of the instrument will allow direct measurement of the outflow of gas from low mass objects. The presence and character of mass loss can be defined as a function of stellar temperature and gravity. We now have no direct evidence for mass loss from any solar-type star other than the Sun, but studies of solar coronal holes show that detectable mass outflow begins near the formation level of the O VI lines. Thus *LYMAN-FUSE* measurements of Doppler shifts in the O VI and higher temperature lines in the EUV will permit direct measurement of the onset of coronal winds for the first time. Vigorous stellar winds found in young stars may be accelerated by quite different mechanisms than the winds of solar-like stars and red giants. Observations of nearby galactic cluster stars can constrain the evolution of stellar winds and their impact on stellar spindown as well as the evolution of stellar magnetic activity.

I am grateful to C. Beichman, Andrea Dupree, Ed Erickson, Don Hall, Warren Moos, Harvey Tananbaum, Rodger Thompson, Mike Werner and Bruce Woodgate for supplying material essential to this brief report.

The Voyagers: A Cosmic Saga

One day, they sat by the Lago...
They said: "this time, we should go".
They put on their Shus, and also their Bonnets,
But the stars were cool, and in spite of blankets,
The temperature was not warm enough,
And someone had a Gough.

Then came the crew, some on foot,
Like the Walkers, with Constance.
All brave men and women, not for loot,
With high spirits, not counting on chance.
Adventurous, Arcoragious, all would know fame.
Arrived from all horizons, many wore a strange name:
Collier, Santo, Thierry, Jan, René, Gösta, Bjorn,
Between family and exploration they were torn.
There was also Jérome,
Who had been in Rome;
From Athens came Caillault,
At first to say hello.
Bernard brought along his fiddle,
So that no one would stay idle;
Pat took his Herbig-Haronica,
I mean, his harmonica.
There were people from the Middle East,
In case there would be a moonshine feast.
From the infrared emerged Emerson,
From the CO appeared Hjalmarson.
So that turbulence could be checked,
They of course called upon Wojcek.
To find directions, and follow the arrow,
Many volunteered, but they chose little Bo.
With the songs to be sung by Singh,
The Voyagers were sure to have everything.

The starships were low mass
(They had been designed by Baliunas!),
When leaving the planet, molecular clouds
Were threatening, but did not Dieter the crowds.
Up went the sails, taking wind from the stars...
They feared a maëlStrom, but it never came;
Encoutered gravity holes, but the Black one was afar.
It was hard work, and for a Hartmann not a game,
Bodenheimer on the verge of collapse!
As many stars went ultra-violent,
No one saw the time elapse,
In magnetic fields and strong currents,

As the starships rotated in violation
Of energy and momentum conservation...

And then, on their approach towards the Main Sequence,
They sighted the goal of their saga,
Which for all would be a lifetime experience.
Yes, it was it, the green planet of Viana!!

What they first saw was rated "X":
Surfing on radio waves, naked T Tauris,
Infrared sources playing with accretion disks
(Guinan immediately looked for close binaries!).
The Voyagers were greeted with H-alpha jets,
Within minutes, they were completely wet!
Zinnecker went ashore to shake Hans,
And to please the people of the new lands,
Everyone, including Pringle
Gave them presents, including jingles.

The landscape was strange: outflows of green liquor,
Which were first spotted by Gibor,
Were pouring out of a Grand Kenyon;
Grapes everywhere, and tomatoes, and onions.

The planet, thay had been told,
Was ruled by Queens, and a Lady Mayor;
Their "castelo" was in a "parque" by the river.
Was the legend true? As they were bold,
Regained activity, under an oscillating sun,
Left their ships, and began to run!
They knew they were close, when they saw
The Princesses, some dressed in extravaganza:
Sara and Martha, in a vineyard below,
Nuria and Joanna, from nearby planets,
Jessica and Steffa, caught in fishermen's nets,
And Brigitta, Rikka, Maria Gabriela, Monica,
Maria Adriana, Antonella, Ligia, Catarina...et coetera!

The tension rose as the doors flung open;
They were in search of the Sky, they found Heaven!
There stood the Queens,
Fulfilling their dreams:
The fair Andrea,
The dark Teresa.

The Voyagers had arrived.

Thierry Montmerle, Viana do Castelo, Portugal, October, 1987

AUTHOR INDEX

OBJECT INDEX

SUBJECT INDEX

Absorption efficiency, 23
Accretion
  episodic, 178
Accretion disks, 13, 14, 153ff, 169ff
  Boundary layer, 14, 170
  Dust disks, 153, 164
  Dwarf novae, 156, 159
  Evolution, 163
  Flaring, 15, 166, 168, 204
  FU Ori phenomenon, 18, 104
  Isothermal, 166
  Line doubling, 174
  Line profiles, 174
  Luminosity, 166, 169
  Observational constraints, 163ff
  Opaque disks, 165, 203
  Origin of solar system, 153
  Pre-main sequence, 59
  Remnant disk, 159, 175
  Reprocessing disks, 164, 165, 204
  Scale height, 166
  Self-gravitating, 158
  Spectroscopic diagnostics, 163
  Spectrum, 165, 169
  Steady-state disks, 154, 184
  Structure, 14, 185
  Temperature distribution, 164, 165, 204
  Thin disks, 153
  Time dependent, 155, 170
  T Tauri stars, 19, 103, 183, 193, 247
  Turbulent, 184ff
  Viscosity, 156, 170, 175
  Young stellar objects, 163
Accretion luminosity, 140, 151, 169
  Mass accretion rates, 107, 170
Accretion processes, 93
Accretion shock, 157
Acoustic waves, 4, 5
AELITA, 431
Age
  Determination, 379
  Lithium abundance, 379, 382
  Chromospheric emission, 382
  W UMa stars, 346, 354, 366, 367
Alfvén point, 132
Alfvén radius, 177, 361, 383, 390, 407
Alfvén waves, 5, 9, 18, 243
Alfvénic turbulence, 124

Ambipolar diffusion, 12, 100, 123
  Evolution of cloud core, 124
Ammonia core, 125, 199
Amorphous material, 30
Angular momentum, 13
  Accretion disks and, 153, 157, 163
  Loss during star formation, 331
  Loss from coronal clouds, 407
  Loss from spin-orbit coupling, 360
  Loss in solar-type stars, 390, 408
  Low mass stars, 377
  "Problem", 333, 390
  Transfer, 139, 163, 385
ASTRO-D, 431
Astroseismology, 343
AXAF, 428, 433ff, 440

Balmer
  Continuum jump, 251
  Decrement, 253
  Lines, 249
Bidet, 284
Binary systems, 112
  Chromospheric emissions, 397
  Contact, 345ff
  Eclipsing, 309
  Formation of, 306
  Main sequence, 307
  Pre-main sequence stars in, 306ff, 315
  Rotational velocity, 341
  Separations, 308
  Spectroscopic, 309, 316
  Visual, 313-315
  W UMa type, 345ff
Bipolar outflows, 16, 47, 48
  Collimation, 51
  Elias 21 Nebula, 115
  Giant molecular clouds and, 79
  Molecular, 257
  Neutral, 126
Birthline, 131, 298
  Deuterium birthline, 131
Boundary layer, 14, 60, 62, 170ff
  Chromospheres, 171
  Emission models, 172
  Energy generation, 171
  Luminosity, 173
  Temperature, 172